T0254722

Reviews of Environmental Contamination and Toxicology

VOLUME 231

For further volumes:
http://www.springer.com/series/398

Ecological Risk Assessment for Chlorpyrifos in Terrestrial and Aquatic Systems in the United States

Editors

John P. Giesy and Keith R. Solomon

VOLUME 231

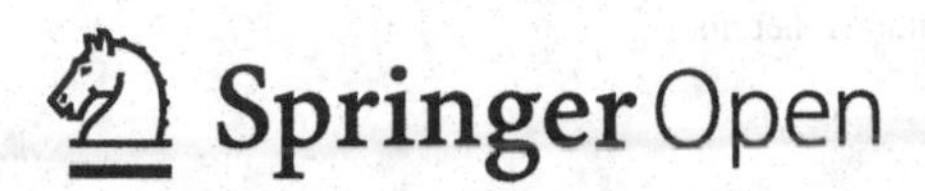

Editors
John P. Giesy
Department of Veterinary Biomedical
Sciences and Toxicology Centre
University of Saskatchewan
Saskatoon, SK, Canada

Keith R. Solomon
Centre for Toxicology
School of Environmental Sciences
University of Guelph
Guelph, ON, Canada

ISSN 0179-5953
ISSN 2197-6554 (electronic)
ISBN 978-3-319-35531-3
ISBN 978-3-319-03865-0 (eBook)
DOI 10.1007/978-3-319-03865-0
Springer Cham Heidelberg New York Dordrecht London

Softcover reprint of the hardcover 1st edition 2014

Printed on acid-free paper

Springer is part of Springer Science+Business Media (www.springer.com)

Special Foreword

This volume of Reviews in Environmental Contamination and Toxicology (RECT) is devoted to an ecological risk assessment of the insecticide, chlorpyrifos (CPY). Chlorpyrifos (*O,O*-diethyl *O*-(3,5,6-trichloro-2-pyridinyl) phosphorothioate; CAS No. 2921-88-2) is a widely used organophosphorus insecticide that is important for the control of a large number of insect pests of crops. Applications are made to soil or foliage and can occur pre-plant, at-plant, post-plant, or during the dormant season.

Under the enabling legislation of the Federal Insecticide, Fungicide, and Rodenticide Act (FIFRA), the US Environmental Protection Agency (USEPA) is charged with assessing the potential for agricultural chemicals to cause undue harm to nontarget plants and animals while considering beneficial uses in protection of crops. Registrations for labeled insecticides and other pesticides are reevaluated approximately every 15 years. Since CPY was last reevaluated, there have been many changes that affect its environmental profile:

- The EPA has changed how risk assessments of chemicals used to protect agricultural crops are conducted.
- The data available on concentrations and fate in the environment and toxicity of CPY to animals have increased.
- Methods and models used to estimate concentrations in the environment and exposure doses to biota have improved.
- Use patterns have changed in response to changes in cropping patterns, pest pressures, and new agricultural technologies.
- The crops for which CPY can be used and the amounts and frequencies of application permitted on the label have also changed, all of which are the primary determinants of the entry of CPY into the environment and its subsequent fate in the regions of use and beyond.

In anticipation of the next reregistration review, the registrant for most of the formulations containing CPY, Dow AgroSciences, requested that we convene an expert panel to review the ecological risks of CPY with a specific focus on current patterns of use of CPY in the US agriculture.

An independent panel of experts was assembled to reassess the risks posed by CPY to aquatic life and wildlife and consider newly available information, assessment techniques, label requirements and restrictions, changes in market-driven use patterns, and issues and uncertainties that have arisen since similar assessments were conducted for aquatic and terrestrial organisms in 1999 and 2001. As is typical of risk assessments of pesticides, this exercise required a range of expertise. The panel consisted of seven scientists with experience in environmental fate of organic molecules and associated simulation modeling, toxicity of pesticides to aquatic and terrestrial organisms, environmental chemodynamics, and ecological risk assessment. The panel developed conceptual models and an analysis plan that addressed the issues identified by the registrant, the panel, and in the problem formulation process developed by the USEPA. All panel members participated in all aspects of the problem formulation and goal-setting for the assessment. Individual experts took the lead in conducting various portions of the assessment. The assessment built upon the findings of the previous assessments, updated the data sets available, and applied the most current methods and models for assessing risks. The panel met several times over the period 2011–2013, discussed the findings of the various teams, and then prepared a series of peer-reviewed manuscripts, which are published together as a series in this volume of RECT. All of the papers were reviewed by anonymous reviewers. In particular, the panel addressed the following specific issues raised by the Environmental Fate and Effects Division of the USEPA and others:

- The potential for long-range atmospheric transport (LRT)
- Occurrence and environmental risks of the oxon metabolite of CPY
- The potential for missing significant exposures during monitoring programs
- Risk of CPY to pollinators

The assessment built upon the conclusions of the previous published assessments and did not repeat the lower tiers of the risk assessment process. In the refined assessments, probabilistic techniques were used. For birds, aquatic organisms, and bees, the results of the risk assessments were placed into the context by comparison to incident reports and results of field studies.

All of this information is presented in seven papers in this volume of RECT. The first of these is a general overview of the main conclusions and some of the sources of uncertainty that were identified. The other six papers address the following topics:

- Uses and key properties
- Fate and transport in the atmosphere
- Fate and transport in water
- Risks to aquatic organisms
- Exposures and risks to birds
- Risks to pollinators

These papers are augmented by extensive supplemental information (SI) in PDF format. The objective of providing this SI was to make the environmental risk assessment of this agrochemical as transparent as possible.

As volume editors and co-chairs of the panel, we enjoyed the process and are very pleased with the high level of scientific rigor used by the panel and the breadth of the data used in the process. We thank the panel for their contributions to these papers and also to the anonymous reviewers for ensuring the quality and clarity of the papers. We trust that readers will find this compilation useful and will stimulate further scientific endeavors.

Saskatoon, SK, Canada — John P. Giesy
Guelph, ON, Canada — Keith R. Solomon

Foreword

International concern in scientific, industrial, and governmental communities over traces of xenobiotics in foods and in both abiotic and biotic environments has justified the present triumvirate of specialized publications in this field: comprehensive reviews, rapidly published research papers and progress reports, and archival documentations. These three international publications are integrated and scheduled to provide the coherency essential for nonduplicative and current progress in a field as dynamic and complex as environmental contamination and toxicology. This series is reserved exclusively for the diversified literature on "toxic" chemicals in our food, our feeds, our homes, recreational and working surroundings, our domestic animals, our wildlife, and ourselves. Tremendous efforts worldwide have been mobilized to evaluate the nature, presence, magnitude, fate, and toxicology of the chemicals loosed upon the Earth. Among the sequelae of this broad new emphasis is an undeniable need for an articulated set of authoritative publications, where one can find the latest important world literature produced by these emerging areas of science together with documentation of pertinent ancillary legislation.

Research directors and legislative or administrative advisers do not have the time to scan the escalating number of technical publications that may contain articles important to current responsibility. Rather, these individuals need the background provided by detailed reviews and the assurance that the latest information is made available to them, all with minimal literature searching. Similarly, the scientist assigned or attracted to a new problem is required to glean all literature pertinent to the task, to publish new developments or important new experimental details quickly, to inform others of findings that might alter their own efforts, and eventually to publish all his/her supporting data and conclusions for archival purposes.

In the fields of environmental contamination and toxicology, the sum of these concerns and responsibilities is decisively addressed by the uniform, encompassing, and timely publication format of the Springer triumvirate:

Reviews of Environmental Contamination and Toxicology [Vol. 1 through 97 (1962–1986) as Residue Reviews] for detailed review articles concerned with any aspects of chemical contaminants, including pesticides, in the total environment with toxicological considerations and consequences.

Bulletin of Environmental Contamination and Toxicology (Vol. 1 in 1966) for rapid publication of short reports of significant advances and discoveries in the fields of air, soil, water, and food contamination and pollution as well as methodology and other disciplines concerned with the introduction, presence, and effects of toxicants in the total environment.

Archives of Environmental Contamination and Toxicology (Vol. 1 in 1973) for important complete articles emphasizing and describing original experimental or theoretical research work pertaining to the scientific aspects of chemical contaminants in the environment.

Manuscripts for Reviews and the Archives are in identical formats and are peer reviewed by scientists in the field for adequacy and value; manuscripts for the Bulletin are also reviewed, but are published by photo-offset from camera-ready copy to provide the latest results with minimum delay. The individual editors of these three publications comprise the joint Coordinating Board of Editors with referral within the board of manuscripts submitted to one publication but deemed by major emphasis or length more suitable for one of the others.

Coordinating Board of Editors

Preface

The role of Reviews is to publish detailed scientific review articles on all aspects of environmental contamination and associated toxicological consequences. Such articles facilitate the often complex task of accessing and interpreting cogent scientific data within the confines of one or more closely related research fields.

In the nearly 50 years since *Reviews of Environmental Contamination and Toxicology* (formerly *Residue Reviews*) was first published, the number, scope, and complexity of environmental pollution incidents have grown unabated. During this entire period, the emphasis has been on publishing articles that address the presence and toxicity of environmental contaminants. New research is published each year on a myriad of environmental pollution issues facing people worldwide. This fact, and the routine discovery and reporting of new environmental contamination cases, creates an increasingly important function for *Reviews*.

The staggering volume of scientific literature demands remedy by which data can be synthesized and made available to readers in an abridged form. *Reviews* addresses this need and provides detailed reviews worldwide to key scientists and science or policy administrators, whether employed by government, universities, or the private sector.

There is a panoply of environmental issues and concerns on which many scientists have focused their research in past years. The scope of this list is quite broad, encompassing environmental events globally that affect marine and terrestrial ecosystems; biotic and abiotic environments; impacts on plants, humans, and wildlife; and pollutants, both chemical and radioactive; as well as the ravages of environmental disease in virtually all environmental media (soil, water, air). New or enhanced safety and environmental concerns have emerged in the last decade to be added to incidents covered by the media, studied by scientists, and addressed by governmental and private institutions. Among these are events so striking that they are creating a paradigm shift. Two in particular are at the center of everincreasing media as well as scientific attention: bioterrorism and global warming. Unfortunately, these very worrisome issues are now superimposed on the already extensive list of ongoing environmental challenges.

The ultimate role of publishing scientific research is to enhance understanding of the environment in ways that allow the public to be better informed. The term "informed public" as used by Thomas Jefferson in the age of enlightenment conveyed the thought of soundness and good judgment. In the modern sense, being "well informed" has the narrower meaning of having access to sufficient information. Because the public still gets most of its information on science and technology from TV news and reports, the role for scientists as interpreters and brokers of scientific information to the public will grow rather than diminish. Environmentalism is the newest global political force, resulting in the emergence of multinational consortia to control pollution and the evolution of the environmental ethic. Will the new politics of the twenty-first century involve a consortium of technologists and environmentalists, or a progressive confrontation? These matters are of genuine concern to governmental agencies and legislative bodies around the world.

For those who make the decisions about how our planet is managed, there is an ongoing need for continual surveillance and intelligent controls to avoid endangering the environment, public health, and wildlife. Ensuring safety-in-use of the many chemicals involved in our highly industrialized culture is a dynamic challenge, for the old, established materials are continually being displaced by newly developed molecules more acceptable to federal and state regulatory agencies, public health officials, and environmentalists.

Reviews publishes synoptic articles designed to treat the presence, fate, and, if possible, the safety of xenobiotics in any segment of the environment. These reviews can be either general or specific, but properly lie in the domains of analytical chemistry and its methodology, biochemistry, human and animal medicine, legislation, pharmacology, physiology, toxicology, and regulation. Certain affairs in food technology concerned specifically with pesticide and other food-additive problems may also be appropriate.

Because manuscripts are published in the order in which they are received in final form, it may seem that some important aspects have been neglected at times. However, these apparent omissions are recognized, and pertinent manuscripts are likely in preparation or planned. The field is so very large and the interests in it are so varied that the editor and the editorial board earnestly solicit authors and suggestions of underrepresented topics to make this international book series yet more useful and worthwhile.

Justification for the preparation of any review for this book series is that it deals with some aspect of the many real problems arising from the presence of foreign chemicals in our surroundings. Thus, manuscripts may encompass case studies from any country. Food additives, including pesticides, or their metabolites that may persist into human food and animal feeds are within this scope. Additionally, chemical contamination in any manner of air, water, soil, or plant or animal life is within these objectives and their purview.

Manuscripts are often contributed by invitation. However, nominations for new topics or topics in areas that are rapidly advancing are welcome. Preliminary communication with the editor is recommended before volunteered review manuscripts are submitted.

Summerfield, NC, USA David M. Whitacre

Contents

Ecological Risk Assessment of the Uses of the Organophosphorus Insecticide Chlorpyrifos, in the United States

John P. Giesy, Keith R. Solomon, G. Christopher Cutler, Jeffrey M. Giddings, Don Mackay, Dwayne R.J. Moore, John Purdy, and W. Martin Williams

1 Introduction

As explained in the foreword, this volume of Reviews of Environmental Contamination and Toxicology is devoted to an assessment of the ecological risks posed by chlorpyrifos (*O,O*-diethyl *O*-(3,5,6-trichloro-2-pyridinyl) phosphorothioate; CAS No. 2921-88-2; CPY) as used in the United States (U.S.). CPY is a widely used organophosphorus insecticide that is available in a granular formulation for treatment in soil, or several flowable formulations that can be applied to foliage,

J.P. Giesy (✉)
Department of Veterinary Biomedical Sciences and Toxicology Centre,
University of Saskatchewan, 44 Campus Dr., Saskatoon, SK S7N 5B3, Canada
e-mail: john.giesy@usask.ca

K.R. Solomon
Centre for Toxicology, School of Environmental Sciences, University of Guelph,
Guelph, ON, Canada

G.C. Cutler
Department of Environmental Sciences, Faculty of Agriculture, Dalhousie University,
Truro, NS, Canada

J.M. Giddings
Compliance Services International, Rochester, MA, USA

D. Mackay
Trent University, Peterborough, ON, Canada

D.R.J. Moore
Intrinsik Environmental Sciences (US), Inc., New Gloucester, ME, USA

J. Purdy
Abacus Consulting, Campbellville, ON, Canada

W.M. Williams
Waterborne Environmental Inc., Leesburg, VA, USA

J.P. Giesy and K.R. Solomon (eds.), *Ecological Risk Assessment for Chlorpyrifos in Terrestrial and Aquatic Systems in the United States*, Reviews of Environmental Contamination and Toxicology 231, DOI 10.1007/978-3-319-03865-0_1, © The Author(s) 2014

soil, or dormant trees (Solomon et al. 2014). CPY can be applied by use of aerial spraying, chemigation, ground boom or air-blast sprayers, tractor-drawn spreaders, or hand-held equipment.

Since the registration of CPY was last re-evaluated (USEPA 2004, 2008), there have been changes in how assessments of risks of chemicals used to protect agricultural crops are conducted. The amount of data available on mobility, persistence, and concentrations in the environment and toxicity of CPY to animals has increased. Most importantly, many methods and models for estimating concentrations in the environment and exposures to wildlife have improved significantly since the results of the last assessments were published (Giesy et al. 1999; Solomon et al. 2001). Also, patterns of use have changed in response to changes in cropping, pest pressure, introduction of genetically modified crop (GMO) technology, and competing pesticides. Uses of CPY are the primary determinants of the entry of CPY into the environment and its subsequent fate in the regions of use and beyond. The purpose of this paper is to provide a synopsis of uses and properties of CPY and the results risk assessments conducted for aquatic life and terrestrial biota. Mammals were not addressed in any of these risk assessments because they are less sensitive to CPY and do not have as large a potential for exposure as do birds. It was previously concluded that, if birds are not affected by a particular pattern of use, then mammals occurring in the same environment would also not be adversely affected (Solomon et al. 2001).

2 Uses and Properties of Chlorpyrifos

The second paper in the series reviews the current uses permitted under the current label and patterns of use in various crops (Solomon et al. 2014). The data on physical and chemical properties were reviewed and a set of consensus values were selected for use in the environmental fate assessments, which included modeling of long-range transport and assessment of bioaccumulation (Mackay et al. 2014), characterizing routes of exposure to CPY through soil, foliage, and food items in terrestrial systems (Cutler et al. 2014; Moore et al. 2014), and in surface-water aquatic systems (Giddings et al. 2014; Williams et al. 2014). Currently-registered formulations of CPY and their uses in the U.S. were the basis for the development of the exposure scenarios and the conceptual models used in assessing risks to aquatic organisms (Giddings et al. 2014; Williams et al. 2014), birds (Moore et al. 2014), and pollinators (Cutler et al. 2014). These data on use were based on the current labels and reflect changes in labels and use-patterns that have occurred since 2000. Important changes included removal of all residential and termiticide uses and changes in buffers. CPY is now registered only for use in agriculture in the U.S. but is an important tool in management of a large number of pests, mainly insects and mites. CPY is used on a wide range of crops, although applications to corn and soybeans account for 46–50% of annual use in the U.S. Estimates of total annual use in the U.S. from 2008 to 2012 range from 3.2 to 4.1 M kg y^{-1}, which is about 50% less than the annual use prior to 2000.

Large amounts of data are available on the environmental properties of CPY. These data were summarized and key values were selected for modeling fate in the environment. The vapor pressure of CPY is 1.73×10^{-5} torr, solubility in water is <1 mg L^{-1}, and its log K_{OW} is 5. The mean water-soil adsorption coefficient normalized to fraction of organic carbon in the soil (K_{OC}) of CPY is 8.2×10^3 mL g^{-1}. Negligible amounts enter plants via the roots, and CYP is not translocated in plants. Chlorpyrifos has short to moderate persistence in the environment as a result of several dissipation pathways that may proceed concurrently. Primary mechanisms of dissipation include volatilization, photolysis, abiotic hydrolysis, and microbial degradation. Under laboratory conditions, estimates of half-lives of CPY in soils range from 2 to 1,575 d (N=126), depending on properties of the soil and rate of application. As with other pesticides in soil, dissipation of CPY is often biphasic with an initial rapid dissipation followed by slower breakdown. Laboratory and field dissipation half-lives are often calculated by assuming 1st order kinetics, which might over-estimate persistence and potential for runoff into surface waters. At rates of application that were used historically for control of termites, the degradation rate is slower than at rates used in agriculture. In agricultural soils under field conditions, half-lives are shorter (2–120 d, N=58) than those measured in the laboratory. Half-lives for hydrolysis in water are inversely related to pH, and range from 16 to 73 d.

CPY is an inhibitor of acetylcholinesterase (AChE) and is potentially toxic to most animals. *In vivo* and in the environment, CPY is converted (activated) to chlorpyrifos oxon (CPYO), which is more reactive with AChE. Similar activation reactions occur with other phosphorothioate insecticides. Co-exposure to other chemicals can induce mixed-function oxidase enzymes responsible for activation. However, concentrations required to induce this synergism are large and co-occur rarely. Thus, this phenomenon is not an issue at environmentally relevant concentrations.

Timing of the use of CPY depends on occurrence of the pests it is used to control. There is no predominant seasonal use of CPY, although there is a somewhat greater usage in the winter for tree crops in California and greater use in summer for certain field crops (e.g., corn).

3 Fate of Chlorpyrifos and Its Oxon in the Atmosphere and Long-Range Transport

The third paper in the series characterized the fate of CPY and CPYO in a number of environmental compartments with a focus on transport through the atmosphere (Mackay et al. 2014). Detectable concentrations of CPY in air, rain, snow and other environmental media have been measured in North America and other locations at considerable distances from likely agricultural sources. Thus, there is a potential for long-range transport (LRT) in the atmosphere. A simple mass balance model was developed to quantify likely concentrations of CPY and CPYO at locations ranging from local sites of application to more remote locations up to hundreds of km distant. The characteristic travel distance (CTD) is defined as the distance at which

63% of the original mass of volatilized substance is degraded or deposited. Based on a conservative concentration of •OH radicals of 0.7×10^6 molecules cm^{-3} in the atmosphere which would result in a half-life of 3 h, the CTD for CPY was estimated to be 62 km. At lesser concentrations of •OH radical, such as occur at night and at lesser temperatures or in less urbanized regions, the CTD would be proportionally longer. The calculated fugacities of CPY in air and other media decrease proportionally with increasing distance from sources. This information provided an approximate prediction of downwind concentrations that are generally consistent with concentrations measured in nearby and semi-remote sites. This analysis was an improvement over previous estimates of LRT of CPY and CPYO, but a need for improved estimates of the chemical-physical properties of CPYO was identified.

The properties of CPY were assessed against criteria for classification as a persistent organic pollutant (POP) under the Stockholm convention or as persistent, bioaccumulative and toxic (PBT) under the European Community regulation EC 1107/2009 (Mackay et al. 2014). CPY and CPYO do not trigger criteria for classification as a POP or LRT under the Stockholm convention or a PB chemical under EC 1107/2009. Although CPY is toxic at concentrations less than the trigger for classification as "T" under EC1107/2009, this simple trigger needs to be placed in the context of low risks to non-target organisms close to the areas of use. Overall, neither CPY nor CPYO trigger the criteria for PBT under EC 1107/2009.

CPYO is not predicted to persist in the environment, and indeed is not found in surface waters. Because CPYO is the metabolically activated toxic form of CPY, the toxicity of CPYO is implicitly measured when testing CPY. For these reasons, we concluded that additional fate studies for CPYO in the environment by either modeling or monitoring or additional studies of toxicity are not warranted. There is sufficient monitoring and toxicity testing to determine that the uncertainties in conclusions about the oxon are not large.

4 Chlorpyrifos in Surface Water

The fourth paper in the series characterizes the measured and modeled concentrations of CPY in surface waters of the U.S. (Williams et al. 2014). The frequencies of detection and 95th centile concentrations of CPY in surface waters in the U.S. have decreased more than five-fold between 1992 and 2010. Detections of CPY in 1992–2001 ranged from 10 to 53% of samples. In the period 2002–2010, detections were 7 to 11%. The 95th centile concentrations ranged from 0.007 to 0.056 μg L^{-1} in 1992–2001 and 0.006 to 0.008 μg L^{-1} in 2002–2010. The greatest frequency of detections and 95th centile concentrations occurred in undeveloped and agricultural land-use classes.

The two classes of land-use with the most urban land (urban and mixed) had the smallest frequency of detections and 95th centile concentrations. This result is consistent with cessation of sale of CPY-based products for residential uses in December 2001. Overall, the U.S. Geological Survey (USGS) database with the greatest number of samples (more than 10,000) and broadest geographical representation showed that CPY was detected in 9% of samples between 2002 and 2010 and that 95% of samples contained less than 0.007 μg L^{-1}, and the maximum was 0.33 μg L^{-1}. Regional databases maintained by the California Department of Pesticide Regulation (CDPR) and Washington State Department of Ecology (WDOE), which were more focused on areas of pesticide use than the USGS database, had more frequent detections (13–17%) and greater concentrations (95th centiles 0.010–0.3 μg L^{-1}). No detections were reported in the 42 samples of saltwater and few data (total = 123) were available for analyses of CPY in sediments. Only three sediments had concentrations above the limit of detection (LOD = 2 μg kg^{-1} dwt) and the largest concentration detected was 59 μg kg^{-1}. Overall, the results indicate decreasing trends in concentrations of CPY that are explained largely by corresponding decreases in annual use and removal of residential uses from the labels.

Detections of CPYO were infrequent and all were less than the level of quantitation (LOQ = 0.011–0.054 μg L^{-1}). Only 25 detections were reported in a total of 10,375 samples analyzed between 1999 and 2012. The low frequency of detection and the small concentrations found are consistent with the reactivity of CPYO and its shorter hydrolysis half-life (Mackay et al. 2014). These findings suggest that concerns for the presence of CPYO in drinking-water (USEPA 2011a), where CPYO may be formed during chlorination, are not transferable to surface waters.

Collectively, the monitoring data on CPY provide relevant insight for quantifying the range of concentrations in surface waters. However, relatively few monitoring programs have sampled at a frequency sufficient to quantify the temporal pattern of exposure. Therefore, numerical models were used to characterize concentrations of CPY in water and sediment for three representative high exposure environments in the U.S. (Williams et al. 2014). The environments were selected by parallel examination of patterns and intensity of use across the U.S. Simulations were conducted to understand relative vulnerabilities of CPY to runoff with respect to soil, and weather variability across the U.S. From the analyses, three geographical regions, one each in central California, southwestern Georgia, and the Leelanau peninsula of Michigan, were identified as having greater potential exposure to CPY and were used as focal scenarios for detailed modeling.

A small watershed, defined as a 3rd order stream, was selected from each region based on high density of cropland eligible for receiving applications of CPY according registered uses. The modeling used two versions of PRZM, one (V-3.12.2) for modeling applications to field-crops and the other (WinPRZM), which was modified for use of CPY in fields irrigated by flood and furrow. Additional models used were EXAMS, AgDRIFT®, and SWAT. Models were configured for each watershed and simulated for up to 30 yr of consecutive use of CPY using historical weather records for those geographical areas of the country. Daily mean concentrations of CPY in water and sediment from runoff, erosion, and drift sources were predicted at the outlet of the watersheds. Conservative assumptions were used in the configuration of the Georgia and Michigan watersheds. For example, all eligible crop acreage in each watershed was assumed to be treated, and the soil properties and number and frequency of applications of CPY were those of the use pattern that produced the greatest exposure estimates. Model simulations for Orestimba Creek in California used actual reported applications of CPY, but field-specific management practices were not represented in the simulations. Two half-lives for aerobic soil metabolism of CPY in soil, 28 and 96 d, were selected for the purposes of modeling. These half-lives conservatively represent the first and second phases of bi-phasic degradation in aerobic soil metabolism studies.

Estimated concentrations of CPY in water were in general agreement with ambient monitoring data from 2002 to 2010. Maximum daily concentrations predicted for the watersheds a in California, Georgia, and Michigan were 3.2, 0.041, and 0.073 μg L^{-1}, respectively, with the 28-d aerobic soil metabolism half-life and 4.5, 0.042, and 0.122 μg L^{-1}, respectively, with the 96-d soil half-life. These estimated values compared favorably with maximum concentrations measured in surface water, which ranged from 0.33 to 3.96 μg L^{-1}. For sediments, the maximum daily concentrations predicted for the watersheds in California, Georgia, and Michigan were 11.2, 0.077, and 0.058 μg kg^{-1} dwt, respectively, with the 28-d half-life, and 22.8, 0.080, and 0.087 μg kg^{-1}, respectively, with the 96-d soil half-life. Twelve detections out of 123 analyses (10%) were contained in the USGS, CDPR, and WDOE databases with concentrations reported from <2.0 to 19 μg kg^{-1}, with the exception of one value reported at 58.6 μg kg^{-1}. Again, the modeled values compared favorably with measured values.

Duration and recovery intervals between peak concentrations of CPY influence the potential for recovery from sublethal exposures in aquatic organisms. Recovery intervals were characterized by using threshold values derived from toxicity data. Based on modeling with the 28-d half-life value, no toxicologically significant exposure-recovery events were identified in the focal watersheds in Georgia and Michigan. Using the 96-d half-life value, three exposure-recovery events of 1 d duration only were identified in the Michigan focal watershed. Frequency of significant events was greater in the focus watershed from California and the probability of shorter recovery events was greater. However, even in the worst-case focus-watershed in California the median duration was 1 d.

5 Risks of Chlorpyrifos to Aquatic Organisms

The fifth paper in this series addressed the risks of CPY to aquatic organisms. In contrast to the previous lower-tier assessments that indicated potential adverse effects in aquatic organisms (Giesy et al. 1999), this paper relied on higher and more refined tiers of risk assessment. Effects of CPY on aquatic organisms were evaluated by comparing measured or modeled concentrations of CPY in aquatic environments to species sensitivity distributions (SSDs), cosm no observed ecologically adverse effect concentrations ($NOAEC_{eco}$), or individual toxicity values where sufficient data to derive a SSD were not available (Giddings et al. 2014). Toxicity data included in the SSDs were all of high quality. The ranges for acute toxicity endpoints for 23 species of crustaceans ranged from 0.04 to 457 μg L^{-1}; for 18 species of aquatic insects, from 0.05 to >300 μg L^{-1}; and for 25 species of fish, from 0.53 to >806 μg L^{-1}. The concentrations affecting 5% of species (HC5) derived from the SSDs were 0.034, 0.091, and 0.820 μg L^{-1} for crustaceans, insects, and fish, respectively. Limited toxicity data for amphibians suggested that they were less sensitive to CPY than fish. The $NOAEC_{eco}$ in 16 micro- and meso-cosm studies conducted in a variety of climatic zones was consistently close to 0.1 μg L^{-1}. These results indicated that measured concentrations of CPY in surface waters (see Williams et al. 2014) are rarely greater than the thresholds for acute toxicity to even the most sensitive aquatic species. Comparison of limited toxicity data for benthic organisms to measured concentrations of CPY in sediments suggested *de minimis* risks. These conclusions are consistent with the small number (4) of kills of fish and/or invertebrates reported for use of CPY in U.S. agriculture between 2002 and 2012. The four incidents over that period of time were the result of misuse.

Analysis of risks from measured exposures showed that the decline in CYP concentrations in surface waters after labeled use-patterns changed in 2001 resulted in decreased risks for crustaceans, aquatic stages of insects, and fish. A probabilistic analysis of 96-h time-weighted mean concentrations, predicted by use of model simulation for three focus-scenarios selected for regions of more intense use of CPY and vulnerability to runoff, showed that risks from individual and repeated exposures to CPY in the Georgia and Michigan watersheds were *de minimis*. Risks from individual exposures in the intense-use scenario from California were *de minimis* for fish and insects. Risk was small for crustaceans, which are the most sensitive class of organisms.

Risks from repeated exposures in the California intense-use scenario were judged not to be ecologically relevant for insects and crustaceans, but there were some risks to fish. Limited data show that CPYO is of similar toxicity to the parent compound. Concentrations of CPYO in surface waters are smaller than those of CPY and less frequently detected (Williams et al. 2014). Risks for CPYO in aquatic organisms were found to be *de minimis*.

Limited data on recovery of AChE activity after inhibition with CPY suggested that conservative intervals between sublethal exposures of 2 weeks for arthropods and 4–8 weeks for fish would be sufficient to mitigate against cumulative toxicity. In the focus scenarios in Michigan and Georgia, the likelihood of cumulative toxicity was very small, although some cumulative toxicity might occur in the high-use focus scenario in California. Lack of good information on recovery of AChE in relevant species of fish and arthropods was identified as a source of uncertainty.

6 Risks of Chlorpyrifos to Birds

The sixth paper in this series (Moore et al. 2014) evaluated the risks of CPY to birds and built upon past assessments of CPY, including the most recent EPA re-registration assessment (USEPA 1999), and a refined probabilistic assessment of risk to birds by Solomon et al. (2001). Since these assessments were completed, there have been a number of amendments to the label. These included reductions in single and seasonal application rates, reductions in number of applications per season, and increases in minimum re-treatment intervals (USEPA 2009). These changes and their effects on risk to birds were addressed in this paper (Moore et al. 2014).

Refined risk assessments for birds exposed to CPY were conducted for a range of current use patterns for each formulation in the U.S. The assessments relied on focal bird species that commonly occur in and around areas where CPY might be applied and for which adequate data were available to quantify their foraging behavior and diets.

A refined version of Liquid Pesticide Avian Risk Assessment Model (LiquidPARAM) was developed for the assessment of the risks of repeated uses of flowable formulations of CPY. Flowable formulations of CPY are registered for a variety of field and tree crops in the U.S. Focal species of birds associated with these crops were selected for inclusion in the model. The major routes of exposure for birds to flowable CPY were consumption of treated dietary items and drinking water. For acute exposure, LiquidPARAM was used to estimate the maximum retained dose in each of 20 birds on each of 1,000 treated fields over the 60-d period following initial application to account for multiple applications. For each bird, the standard normal Z score for the maximum retained dose was determined from the appropriate probit dose–response curve and was then compared to a randomly drawn value from a uniform distribution with a range of 0 to 1 to determine whether the bird survived or died. For species lacking acceptable acute oral toxicity data (all focal species except northern bobwhite and red-winged blackbird), a SSD approach was used to generate hypothetical dose–response curves assuming high, median and low sensitivity to CPY. For acute risk, risk curves were generated for each use pattern and exposure scenario. The risk curves show the relationship between exceedance probability and percent mortality.

The results of the LiquidPARAM modeling indicated that flowable CPY poses an acute risk to some bird species, particularly those species that are highly sensitive and forage extensively in crops with large maximum application rates (e.g., grapefruit and orange; 6.3 kg a.i. ha^{-1}). Overall, most species of birds would not experience significant mortality as a result of exposure to flowable CPY. The results of a number of field studies conducted in the U.S. and EU at application rates similar to those on the Lorsban® Advanced label indicated that flowable CPY rarely causes avian mortality and suggest that LiquidPARAM is likely over-estimating acute risk to birds for flowable CPY. A lack of well-documented bird-kill incidents associated with normal use since 2002 support the conclusions of the field studies. Of the two bird-kill incidents reported between 2002 and 2009, one was from a misuse and the other lacked sufficient information to make a determination of causality.

For estimating chronic exposure risks, the maximum average total daily intake was compared to the chronic no-observed-effect-level (NOEL) and lowest-observed-effect-level (LOEL) from the Mallard. The probabilities of exceeding the LOEL were very small, thus indicating that CPY is not a chronic risk concern for birds.

Risks resulting from the use of granular CPY were estimated using the Granular Pesticide Avian Risk Model (GranPARAM) model. Granular CPY is registered for a wide variety of field and row crops under the trade name Lorsban® 15G. Consumption of granular pesticides is a route of exposure that is specific to birds. Grit is dietary requirement of many birds to aid in digestion of hard dietary items such as seeds and insects. Because granules of CPY are in the same size range as natural grit particles that are consumed by birds, there is a potential for birds to mistakenly ingest granular CPY instead of natural grit. The GranPARAM model accounts for the proportion of time that birds forage for grit in treated fields, relative proportions of natural grit versus pesticide granules on the surface of treated fields, rates of ingestion of grit, attractiveness of pesticide granules relative to natural grit, variability in rates of ingestion of grit, foraging behavior between birds within a focal species, and variability in soil composition between fields for the selected use pattern. Analysis of a wide variety of use patterns of the granular formulation found that CPY posed little risk to bird species that frequent treated fields immediately after application. The predictions of the model were consistent with the results of several avian field studies conducted with Lorsban 15G at application rates similar to or exceeding maximum application rates on the label.

7 Risks to Pollinators

The seventh and last paper in this series used a tiered approach to assess risks posed by CPY to insects that serve as pollinators (Cutler et al. 2014). The assessment focused on bees, although other groups of insects were also considered. Because there have been recent reports of adverse effects of some pesticides on pollinators, assessing risks of pesticides to pollinators is an important topic. A recent SETAC workshop (Fischer and Moriarty 2011) proposed changes to the assessment process (USEPA 2011b), and these served as guidance for this assessment.

Pollinators are important for both natural and agricultural ecosystems (Cutler et al. 2014). In the U.S., production of crops that require or benefit from pollination by the European honey bee, *Apis mellifera* L. (Apidae) has been estimated to have a monetary value greater than $15 billion annually, while the value of non-*Apis* pollinators to crop production is estimated to be more than $11 billion.

CPY is considered to be highly toxic to honey bees by direct contact exposure. However, label precautions and good agricultural practices prohibit application of CPY when bees are flying and/or when flowering crops or weeds are present in the treatment area. Therefore, the risk of CPY to pollinators through direct contact exposure should be small. The primary routes of exposure for honey bees are dietary and contact with flowers that were sprayed during application and remain available to bees after application.

The main pathways for secondary exposure to CPY are through pollen and nectar brought to the hive by forager bees and the sublethal body burden of CPY carried on forager bees. Foraging for other materials, including water or propolis, are not important routes of exposure. Because adult forager honey bees are most exposed, they are expected to be most at risk compared to other life stages and castes of honey bees in the hive. Although there were data on the acute oral toxicity of CPY to honeybees, this was not the case for non-*Apis* pollinators, where no data on toxicity or exposures were found and risks could not evaluated.

An assessment of concentrations reported in pollen and honey from monitoring in North America indicated that there was little risk of acute toxicity from CPY through consumption of these food sources. Several models were also used to estimate upper-bound exposure of honey bees to CPY through consumption of water from puddles or dew. All models suggest that the risk of CPY is minimal for this pathway. Laboratory experiments with field-treated foliage, and semi-field and field tests with honey bees, bumble bees, and alfalfa leaf-cutting bees indicate that exposure to foliage, pollen, and/or nectar is hazardous to bees up to 3 d after application of CPY to a crop. Pollinators exposed to foliage, pollen, or nectar after this time should be minimally affected.

Overall, the rarity of reported bee kill incidents involving CPY indicates that there is compliance with the label precautions and good agricultural practice with the product is the norm in North American agriculture. We concluded that the use of CPY in North American agriculture does not present an unacceptable risk to honeybees, provided label directions and good agricultural practices are followed. The lack of data on toxicity of and exposures to CPY in non-*Apis* pollinators was identified as an uncertainty. However, this issue is not specific to CPY and applies to all foliar-applied insecticides.

Acknowledgments The authors of the papers thank Jeff Wirtz of Compliance Services International, Julie Anderson of the University of Saskatchewan; J. Mark Cheplick, Dean A. Desmarteau, Gerco Hoogeweg, William J. Northcott, and Kendall S. Price, all of Waterborne Environmental for their contributions. We also acknowledge Yuzhou Luo of the California Department of Pesticide Regulation for providing the SWAT dataset for the Orestimba Creek watershed that was developed when he was associated with the University of California-Davis. We also thank Lou Best, Larry Brewer, Don Carlson of FMC Corporation, Dylan Fuge from Latham and Watkins LLP, and Nick Poletika and Mark Douglas from Dow AgroSciences for their contributions to the assessment of risks to birds. We thank the anonymous reviewers of the papers for their suggestions and constructive criticism. This independent evaluation was funded by Dow AgroSciences. Prof. Giesy was supported by the Canada Research Chair program, a Visiting Distinguished Professorship in the Department of Biology and Chemistry and State Key Laboratory in Marine Pollution, City University of Hong Kong, the 2012 "High Level Foreign Experts" (#GDW20123200120) program, funded by the State Administration of Foreign Experts Affairs, the P.R. China to Nanjing University and the Einstein Professor Program of the Chinese Academy of Sciences.

References

Cutler GC, Purdy J, Giesy JP, Solomon KR (2014) Risk to pollinators from the use of chlorpyrifos in the United States. Rev Environ Contam Toxicol 231:219–265

Fischer D, Moriarty T (2011) Pesticide risk assessment for pollinators: summary of a SETAC Pellston Workshop. Society of Environmental Toxicology and Chemistry, Pensacola, FL

Giddings JM, Williams WM, Solomon KR, Giesy JP (2014) Risks to aquatic organisms from the use of chlorpyrifos in the United States. Rev Environ Contam Toxicol 231:119–162
Giesy JP, Solomon KR, Coates JR, Dixon KR, Giddings JM, Kenaga EE (1999) Chlorpyrifos: ecological risk assessment in North American aquatic environments. Rev Environ Contam Toxicol 160:1–129
Mackay D, Giesy JP, Solomon KR (2014) Fate in the environment and long-range atmospheric transport of the organophosphorus insecticide, chlorpyrifos and its oxon. Rev Environ Contam Toxicol 231:35–76
Moore DRJ, Teed RS, Greer C, Solomon KR, Giesy JP (2014) Refined avian risk assessment for chlorpyrifos in the United States. Rev Environ Contam Toxicol 231:163–217
Solomon KR, Giesy JP, Kendall RJ, Best LB, Coats JR, Dixon KR, Hooper MJ, Kenaga EE, McMurry ST (2001) Chlorpyrifos: ecotoxicological risk assessment for birds and mammals in corn agroecosystems. Human Ecol Risk Assess 7:497–632
Solomon KR, Williams WM, Mackay D, Purdy J, Giddings JM, Giesy JP (2014) Properties and uses of chlorpyrifos in the United States. Rev Environ Contam Toxicol 231:13–34
USEPA (1999) Reregistration eligibility science chapter for chlorpyrifos: fate and environmental risk assessment chapter. United States Environmental Protection Agency, Washington, DC
USEPA (2004) Overview of the ecological risk assessment process in the office of pesticide programs: endangered and threatened species effects determinations. United States Environmental Protection Agency, Office of Prevention, Pesticides, and Toxic Substances, Office of Pesticide Programs, Washington, DC
USEPA (2008) Registration review—preliminary problem formulation for ecological risk and environmental fate, endangered species and drinking water assessments for chlorpyrifos. United States Environmental Protection Agency, Office of Pesticide Programs, Washington, DC
USEPA (2009) Chlorpyrifos final work plan. Registration review. United States Environmental Protection Agency, Office of Pesticide Programs, Washington, DC
USEPA (2011a) Revised chlorpyrifos preliminary registration review drinking water assessment. United States Environmental Protection Agency, Office of Chemical Safety and Pollution Prevention, Washington, DC. PC Code 059101 http://www.epa.gov/oppsrrd1/registration_review/chlorpyrifos/EPA-HQ-OPP-2008-0850-DRAFT-0025%5B1%5D.pdf
USEPA (2011b) Interim guidance on honey bee data requirements. United States Environmental Protection Agency, Environmental Fate and Effects Division, Office of Pesticide Programs
Williams WM, Giddings JM, Purdy J, Solomon KR, Giesy JP (2014) Exposures of aquatic organisms to the organophosphorus insecticide, chlorpyrifos resulting from use in the United States. Rev Environ Contam Toxicol 231:77–118

Properties and Uses of Chlorpyrifos in the United States

Keith R. Solomon, W. Martin Williams, Donald Mackay, John Purdy, Jeffrey M. Giddings, and John P. Giesy

1 Introduction

The physical and chemical properties of chlorpyrifos (*O, O*-diethyl *O*-3,5, 6-trichloro-2-pyridinyl phosphorothioate, CPY; CAS No. 2921-88-2) are the primary determinants that govern fate (movement, adsorption, degradation, and catabolism) in the environment and in biota. The uses of chlorpyrifos in locations of interest, such as the United States in the case of this paper, are the primary determinants of the entry of chlorpyrifos into the environment and its subsequent fate in the regions of use and beyond. The uses and manner of use are addressed in this paper.

The data on physical and chemical properties provided here were the basis for modeling long range transport and assessing bioaccumulation (Mackay et al. 2014), characterizing routes of exposure to chlorpyrifos in terrestrial systems such as soil,

The online version of this chapter (doi:10.1007/978-3-319-03865-0_2) contains supplementary material, which is available to authorized users.

K.R. Solomon (✉)
Centre for Toxicology, School of Environmental Sciences, University of Guelph, Guelph, ON, Canada
e-mail: ksolomon@uoguelph.ca

W.M. Williams
Waterborne Environmental Inc., Leesburg, VA, USA

D. Mackay
Trent University, Peterborough, ON, Canada

J. Purdy
Abacus Consulting, Campbellville, ON, Canada

J.M. Giddings
Compliance Services International, Rochester, MA, USA

J.P. Giesy
Department of Veterinary Biomedical Sciences and Toxicology Centre, University of Saskatchewan, 44 Campus Dr., Saskatoon, SK S7N 5B3, Canada

J.P. Giesy and K.R. Solomon (eds.), *Ecological Risk Assessment for Chlorpyrifos in Terrestrial and Aquatic Systems in the United States*, Reviews of Environmental Contamination and Toxicology 231, DOI 10.1007/978-3-319-03865-0_2, © The Author(s) 2014

foliage, and food items (Cutler et al. 2014; Moore et al. 2014), and in surface-water aquatic systems (Giddings et al. 2014; Williams et al. 2014). The currently-registered formulations of chlorpyrifos and their uses in the United States were the basis for the development of the scenarios of exposure and the conceptual models used in assessing risks to birds (Moore et al. 2014), pollinators (Cutler et al. 2014), and aquatic organisms (Williams et al. 2014). These data on use are based on the current labels and reflect changes in labels and use-patterns since the earlier assessments of risks to aquatic (Giesy et al. 1999) and terrestrial organisms (Solomon et al. 2001). Physical and chemical properties of chlorpyrifos were extensively reviewed by Racke (1993) and, rather than repeat all of this information, relevant values from Racke 1993 are included in this paper and supplemental material (SI) with updates as appropriate.

2 Physical and Chemical Properties of Chlorpyrifos

Fundamental to assessing and predicting the general fate of chlorpyrifos in the environment are having reliable data on physical chemical and reactivity properties that determine partitioning and persistence in the environment. In the following sections, some of the key properties are discussed in more detail.

2.1 Properties Affecting Fate in Air and Long-Range Transport

The fate of CPY and chlorpyrifos-oxon (CPYO; CAS No. 5598-15-2; CPY's biologically active metabolite, degradate, and minor technical product component) in air, with respect to short- and long-distance transport are discussed in detail in a companion paper (Mackay et al. 2014). The physical and chemical properties specific to fate in air are presented in Tables 5–8 in Mackay et al. (2014) and are not repeated here except in the context of biological relevance and fate and movement in other matrices.

2.2 Properties Affecting Fate in Soil, Water, and Sediment

An extensive review of the data on half-lives of CPY in soils and has shown the high variability attributed to soil organic carbon content, moisture, application rate and microbial activity (Racke 1993). Fewer data are available for water and sediments, but processes related to soils and sediments have been summarized in a recent review (Gebremariam et al. 2012). The key physical and chemical properties of CPY are listed in Tables 1 and 2.

Chlorpyrifos has short to moderate persistence in the environment as a result of several dissipation pathways that might occur concurrently (Fig. 1). Primary mechanisms of dissipation include volatilization, photolysis, abiotic hydrolysis, and

Table 1 Physicochemical properties of chlorpyrifos

Parameter	Values for Chlorpyrifos	Source
Chemical Name	O,O-diethyl o-(3,5,6-trichloro-2-pyridyl phosphorothioate	USEPA (2011b)
Chemical Abstracts Service (CAS) Registry Number	2921-88-2	
Empirical formula	$C_9H_{11}Cl_3NO_3PS$	
USEPA Pesticide Code (PC #)	59101	
Smiles notation	S=P(OC1=NC(=C(C=C1Cl)Cl)Cl)(OCC)OCC	
Molecular mass	350.6 g mol^{-1}	Mackay et al. (2014)
Vapor pressure (25 °C)	1.73×10^{-5} torr	
Water solubility (20 °C)	0.73 mg L^{-1}	
Henry's Law Constant	1.10×10^{-5} atm m^{-3} mol^{-1}	
Log K_{OW}	5.0	

Table 2 Environmental fate properties of chlorpyrifos

Parameter	Values	Source
Hydrolysis (t½)	pH 5: 73 d pH 7: 72 d pH 9: 16 d	USEPA (2011b)
Aqueous photolysis (t½)	29.6 d	
Aerobic soil metabolism (t½)	2–1,576 d, N=68 (next highest value is 335 d)	See SI Table A-1
Aerobic aquatic metabolism (t½)	22–51 d, N=3	See SI Table A-5
Anaerobic soil metabolism (t½)	15 and 58 d	USEPA (2011b)
Anaerobic aquatic metabolism (t½)	39 and 51 d	
Soil adsorption coefficient K_{OC}	973–31,000 mL g^{-1}, N=33	See SI Table A-4
Terrestrial field dissipation (t½)	2–120 d, N=58	See SI Table A-3

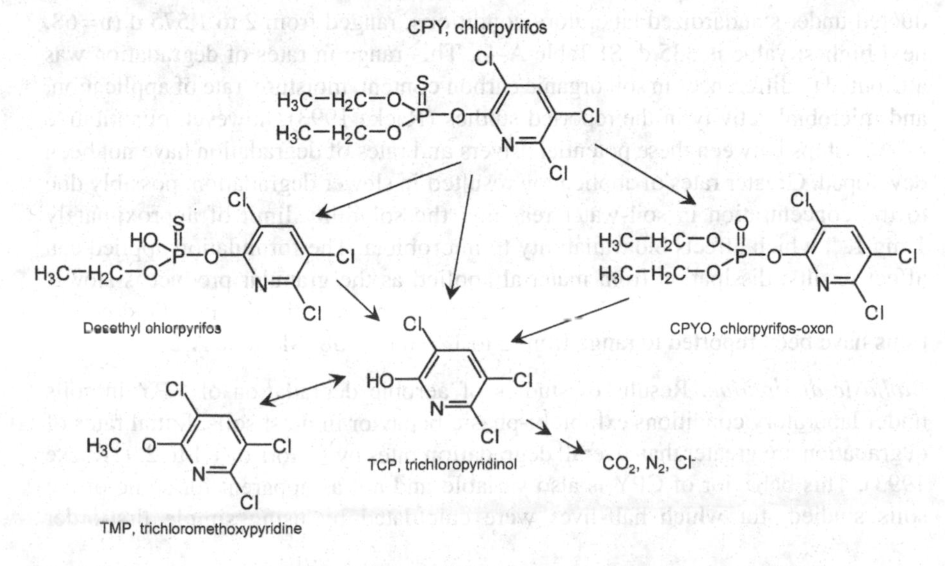

Fig. 1 Pathways for degradation of chlropyrifos in the environment (after Racke 1993)

microbial degradation. Volatilization dominates dissipation from foliage in the initial 12 h after application, but decreases as the formulation adsorbs to foliage or soil (Mackay et al. 2014). In the days after application, CPY adsorbs more strongly to soil, and penetrates more deeply into the soil matrix, and becomes less available for volatilization; other degradation processes become important.

Dissipation from soil. Factors affecting degradation of CPY in soil have been reviewed by Racke (1993). The key values that affect soil dissipation have been updated and are presented in SI Table A-1. Photolysis and oxidation are known to form CPYO in air (Mackay et al. 2014) and on foliar surfaces. These routes are either insignificant in soil or CPYO degrades as quickly as it is formed, since CPYO has only been formed in undetectable or small amounts in studies that have used radiotracers to investigate degradation in soils in the laboratory (de Vette and Schoonmade 2001; Racke et al. 1988) or field (Chapman and Harris 1980; Rouchaud et al. 1989). The primary degradation pathway in soil involves hydrolysis to yield 3,5,6-trichloropyridinol (TCP, Fig. 1) from either CPY or CPYO. Results of several studies have shown that this step can be either abiotic or biotic, and the rate is 1.7- to 2-fold faster in biologically active soils. Both modes of hydrolysis can occur in aerobic and anaerobic soil. The rate of abiotic hydrolysis is faster at higher pH. Hydrolysis is also faster in the presence of catalysts such as certain types of clay (Racke 1993). Degradation of the intermediate, TCP, is dependent on biological activity in soil, and leads to formation of bound residues and reversible formation of 3,5,6-trichloro-2-methoxypyridinol (TMP; Fig. 1). Under aerobic conditions, the primary, terminal degradation product of CPY is CO_2. Since TCP and TMP are not considered to be residues of concern (USEPA 2011b), they were not included in characterizations of exposures presented here or the assessment of risk in the companion papers. Because of rapid degradation in soil (see above), CPYO (Fig. 1) was not included in the characterization of exposures via soil.

The half-life for degradation of CPY in soils, based on results of studies conducted under standardized laboratory conditions, ranged from 2 to 1,575 d (n=68, next highest value is 335 d; SI Table A-1). This range in rates of degradation was attributed to differences in soil organic carbon content, moisture, rate of application, and microbial activity in the reported studies (Racke 1993); however, quantitative relationships between these potential drivers and rates of degradation have not been developed. Greater rates of application resulted in slower degradation, possibly due to the concentration in soil-water reaching the solubility limit of approximately 1 mg L^{-1}, which affects bioavailability to microbiota. The formulation applied can affect results; dissipation from material applied as the granular product is slower (Racke 1993). Half-lives for dissipation from soils determined under field conditions have been reported to range from 2 to 120 d (N=58; SI Table A-2).

Biphasic dissipation. Results of studies of aerobic degradation of CPY in soils under laboratory conditions exhibit bi-phasic behavior in most soils. Initial rates of degradation are greater than overall degradation rates by factors of 1.1 to 2.9 (Racke 1993). This behavior of CPY is also variable and not as apparent for some of the soils studied, for which half-lives were calculated by using simple first-order

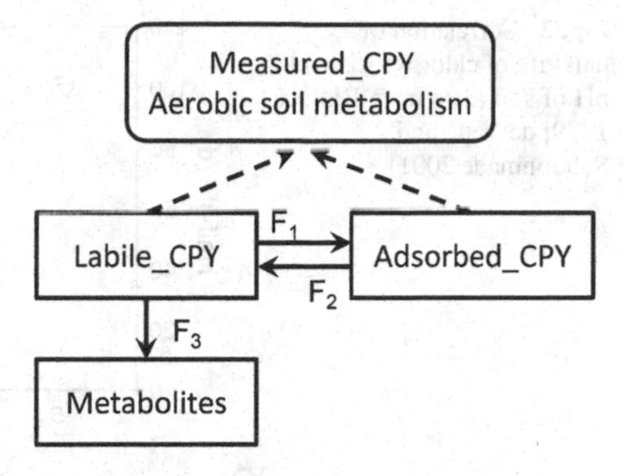

Fig. 2 Schematic diagram of a two–compartment kinetic model for chlorpyrifos (CPY) degradation

kinetics (de Vette and Schoonmade 2001). Nonetheless, some of the half-lives reported in SI Table A-2 that have been derived from 1st-order degradation kinetics might overestimate the persistence of CPY in the environment.

There have been several approaches to calculate rate constants of degradation for this biphasic degradation of CPY. The DT_{50} values reported by Bidlack were calculated using the Hamaker two-compartment kinetic model (Nash 1988), but details of the goodness of fit were not provided and the DT_{50} values do not correspond to degradation rate constants (Bidlack 1979). Also, bi-phasic degradation, described by use of the double first-order parallel (DFOP) model, best characterized the data from three dissipation studies performed in terrestrial environments (Yon 2011).

To obtain the biphasic rate constants for the available aerobic soil degradation results, a dissipation model was structured with two compartments for the parent compound; one adsorbed in such a manner that was not available for biological degradation or abiotic hydrolysis, and the other in which these processes can occur (Fig. 2). The initial thought was to consider these as adsorbed and dissolved compartments, respectively. However, it is known that partitioning of CPY between soil and soil pore water reaches equilibrium within hours (Racke 1993), whereas the biphasic degradation process observed for CPY occurs over a period of several days. The two compartments were identified as *Labile CPY* and *Adsorbed CPY*. Reversible movement of parent CPY between these compartments was represented as two simple first-order processes shown by arrows F_1 and F_2 in Fig. 2, with rate constants k_{ads} and k_{des}. This model has advantages over older two-compartment models in that simple first-order equations are used and the rate constants are not concentration-dependent as they are in the Hamaker kinetic equations (Nash 1988). Since the reported concentrations of CPY include both compartments, the model was configured so that measured values are entered as the sum of the amounts in these two compartments at each time point (Fig. 2). The sum of processes that degrade CPY was also described as a first-order kinetic process F_3, but was non-reversible. The rate constant for this process was designated k_m. The resulting set of three first-order equations was integrated numerically using Model-Maker Version 4.0 software from Cherwell Scientific Software Ltd. UK. Metabolism data from 11 soils reported in two studies (Bidlack 1979; de Vette and Schoonmade 2001) were fit to this model. It was assumed that the CPY was entirely in the labile compartment at time-zero, and the rate of degradation was determined by k_m and the concentration

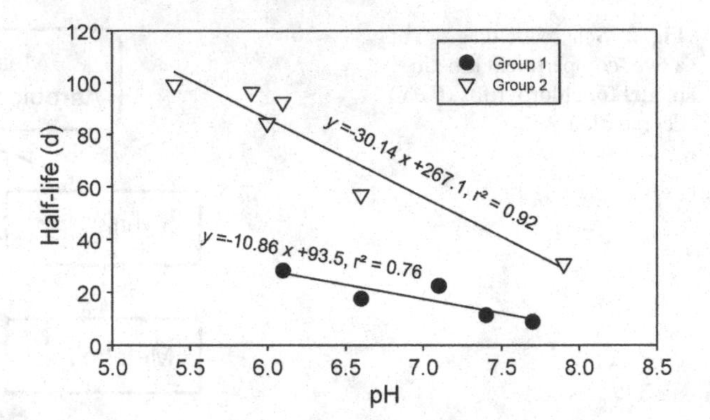

Fig. 3 Correlation of half-life of chlorpyrifos with pH of soil (data from Bidlack 1979; de Vette and Schoonmade 2001)

in this compartment. As CPY partitions into the adsorbed compartment, less is available for degradation, and the rate of desorption, described by the rate constant k_{des} becomes the rate limiting step. This transition from k_m to k_{des} creates the biphasic behavior in the model. Further details on the equations used the model set-up and typical results are given in SI Appendix C.

The model results fit the data well (SI Appendix C; SI Table C-2) (Bidlack 1979). The resulting rate constant represents the entire data set for each soil, optimized simultaneously and represents a consistent model across all the soils considered. This provides a better representation of the half-life than the values in the original reports. As noted above, it is expected that the rate constants might be correlated with the physical and chemical properties of the soils such as % organic matter, etc. No significant correlation could be found among rate constants or half-lives with the K_{OC}, or water-holding capacity. It has been suggested that there might be a correlation between the rate constant k_m for degradation of CPY, and pH (Bidlack 1979). This is expected, given the dependence of the abiotic hydrolysis on pH, which contributes to this process, but the correlation is not simple. A graph of half-life vs. pH is shown (Fig. 3). It is possible to consider the data in two groups; one group of soils has half-lives >30 d, which were pH dependent; the other group had shorter half-lives with a much weaker correlation to pH.

The correlations for the two groups in the range from pH 5 to 8 are given in (1) and (2).

$$Group\ 1\ half\text{-}life = 93.5 - 10.86 \times pH\left(r^2 = 0.76\right) \quad (1)$$

$$Group\ 2\ half\text{-}life = 267 - 30.14 \times pH\left(r^2 = 0.92\right) \quad (2)$$

The mean half-life in the Group-1 was 17.6 d with a 90th centile of 25.9 d and for Group-2 was 77.7 d with a 90th centile of 97.7 d. The greatest half-life among the U.S. soils in each group was selected as a conservative value to represent the group in simulations with the PRZM/EXAMS model runs used to characterize concentrations in surface waters (Williams et al. 2014). These values were 96 d from the Stockton soil and 28 d from the Catlin soil (Table 3).

Table 3 Half-lives of chlorpyrifos in selected soils recalculated using a two-compartment model

Soil	Reported T½[a]	Calculated T½[1] from two-compartment model[b]	Group	Reference
Commerce, MI	11	11	2	Bidlack (1979)
Barnes, ND	22	22	2	Bidlack (1979)
Miami, IN	24	18	2	Bidlack (1979)
Caitlin, IL	34	28	2	Bidlack (1979)
Marcham, UK	43	9	2	de Vette and Schoonmade (2001)
Thessaloniki, GR	46	31	1	de Vette and Schoonmade (2001)
Charentilly, FR	95	93	1	de Vette and Schoonmade (2001)
Norfolk, VA	102	57	1	Bidlack (1979)
Stockton, CA	107	96	1	Bidlack (1979)
Cuckney, UK	111	84	1	de Vette and Schoonmade (2001)
German 2:3	141	99	1	Bidlack (1979)

[a]Rounded to nearest day
[b]For detailed derivation of the data, see SI Appendix B in (Williams et al. 2014)

Adsorption to soil. Based on reported water-soil adsorption coefficients (K_{OC}) of 973 to 31,000 mL g^{-1}; mean 8,216 mL g^{-1} (SI Table A-3), CPY has a large potential to adsorb to soil and would not likely be biologically available for uptake by roots of plants. Possible uptake by roots, translocation, and metabolism of CPY in plants also has been investigated (summarized in Racke 1993). In general, negligible amounts enter the plant via the roots. Thus, CPY is not systemic and this pathway of exposure need not be considered in exposure assessments for CPY.

Dissipation from plants. CPY rapidly dissipates from foliar surfaces of plants, primarily due to volatility and secondarily due to photolysis, with most reported dissipation half-lives on the order of several days (Racke 1993). In a field study performed in California that examined mass loss of CPY to air, maximum volatility fluxes occurred in the first 8 h after application to recently cut alfalfa (Rotondaro and Havens 2012). Total mass loss of CPY, based on the calculated fluxes, ranged between 15.8 and 16.5% of applied mass, as determined by the Aerodynamic (AD) and Integrated Horizontal Flux (IHF) methodologies, respectively. Data on dissipation of CPY from various crops are provided in SI Table A-4.

Dissipation in aquatic systems. In aquatic systems, abiotic degradation of CPY due to aqueous hydrolysis has been reported to occur with half-lives at 25 °C of 73, 72, and 16 d at pH 5, 7, and 9, respectively (summarized in Racke 1993). The U.S. EPA (2011a) used an aqueous hydrolysis half-life of 81 d at pH 7 in modeling to estimate concentrations of CPY in drinking water. Half-lives of 22–51 d have been reported from metabolism studies conducted in aerobic aquatic systems (Kennard 1996; Reeves and Mackie 1993). A half-life of 30 d was reported in an aqueous photolysis study of CPY that was conducted under natural sunlight in sterile pH 7 phosphate buffered solution (Batzer et al. 1990). Data on the dissipation of CPY from aquatic systems are summarized in SI Table A-5.

Field-scale analyses of runoff have demonstrated little potential for CPY to be transported with runoff water (Racke 1993). Chlorpyrifos has been extensively examined in field studies under varying conditions, including greater and lesser antecedent soil moisture, incomplete and full canopy development stages, 2 h to 7 d intervals between application and rainfall, maximum soil erosion conditions, different soils properties, and a range of rainfall events up to a 1-in-833 year return frequency (Cryer and Dixon-White 1995; McCall et al. 1984; Poletika and Robb 1994; Racke 1993). Resulting concentrations of CPY in runoff ranged from 0.003 to 4.4% of the amount applied (McCall et al. 1984; Poletika and Robb 1994). A field runoff study conducted in Mississippi indicated that the majority of chemical mass was transported in the dissolved chemical phase (Poletika and Robb 1994), while a study conducted in Iowa under record high rainfall conditions concluded that the majority of compound was transported attached to eroded sediment (Cryer and Dixon-White 1995).

3 Toxicity of CPY

The primary mode of action of organophosphorus insecticides, such as CPY, is well known and has been characterized in mammals (Testai et al. 2010) and in aquatic organisms, particularly fish (Giesy et al. 1999). Chlorpyrifos inhibits the enzyme acetylcholinesterase (AChE) in synaptic junctions of the nervous system. As a result of this inhibition, acetylcholine accumulated in the synapse causes repeated and uncontrolled stimulation of the post-synaptic axon. Disruption of the nervous system that results is the secondary effect that causes the death of the animal. The amino acid sequence of acetylcholinesterase is highly conserved in animals, with the result that CPY is toxic to most groups of animals, although differences in toxicokinetics (adsorption, distribution, metabolism, and excretion—ADME) account for differences in susceptibility among taxa (Timchalk 2010).

3.1 Mechanism of Action

The mechanism of action (toxicodynamics) of CPY involves activation by biotic transformation to CPYO, followed by covalent binding to the serine-hydroxyl in the active site of the acetylcholinesterase molecule (Testai et al. 2010) (Fig. 4). While this can occur in the environment (Mackay et al. 2014), in animals this reaction is

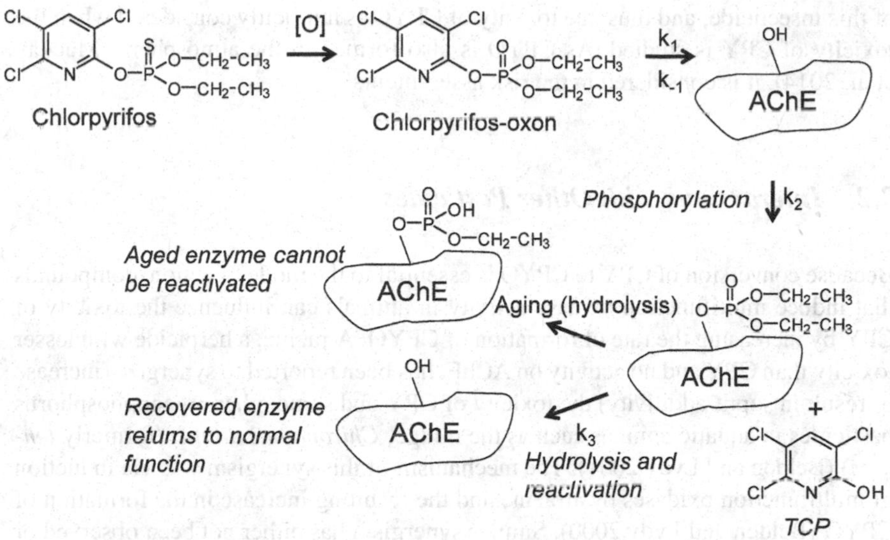

Fig. 4 Diagrammatic representation of the mechanism of action of chlorpyrifos in the nerve synapse

catalyzed by multifunction oxidase enzymes (MFO) and is important in the mode of action of CPY. For example, inhibition of MFOs by the synergist piperonyl butoxide resulted in a decreased toxicity of CPY by up to sixfold in aquatic organisms (El-Merhibi et al. 2004). Chlorpyrifos itself is not a strong inhibitor of AChE, but when transformed to CPYO, the phosphorus atom in the molecule becomes more susceptible to nucleophilic attack by the serine hydroxyl in the active site of AChE. The initial association of CPYO with AChE is reversible (k_1, k_{-1}; Fig. 4) and is modified by the tertiary structure of the enzyme and the inhibitor. During phosphorylation of the serine–OH (k_2; Fig. 4), CPYO is hydrolyzed to release the leaving group TCP (Fig. 4), the reaction is no longer reversible, and AChE is inhibited for as long as it remains phosphorylated. The phosphonic acid moiety is covalently bound to the serine in AChE but the bond can be cleaved by hydrolysis, unless the phosphorylated enzyme ages. If the serine-O-P bond is hydrolyzed by water, AChE is reactivated and normal function returns. If aged via hydrolysis of one of ethyl-ester bonds (Fig. 4), the reactivity of the serine-O-P bond is greatly reduced, AChE cannot be reactivated, and recovery essentially requires the synthesis of new AChE.

The leaving group, TCP, is several orders of magnitude less toxic than CPY or CPYO (Giesy et al. 1999) and is not of toxicological significance (USEPA 2011a). The phosphonic acid released by reactivation of AChE is of low toxicity and is easily excreted from animals (Timchalk 2010). For this reason, the focus of the risk assessments in this series of papers (Cutler et al. 2014; Giddings et al. 2014; Moore et al. 2014) is only on CPY and CPYO. It should be noted that CPYO is the activated form of CPY and its formation in the animal is integral to the mode of action of this insecticide, and thus, the toxicity of CPYO is implicitly considered when the toxicity of CPY is studied. As CPYO is also formed in the atmosphere (Mackay et al. 2014), it is considered in the risk assessments.

3.2 Interactions with Other Pesticides

Because conversion of CPY to CPYO is essential to the mode of action, compounds that induce multifunction oxidase activity in animals can influence the toxicity of CPY by increasing the rate of formation of CPYO. Atrazine, a herbicide with lesser toxicity than CPY and no activity on AChE, has been reported to synergize (increase or result in supra-additivity) the toxicity of CPY and some other organophosphorus pesticides in aquatic animals such as the midge, *Chironomus dilutus (*formerly *tentans*) (Belden and Lydy 2001). The mechanism of this synergism was via induction of multifunction oxidases by atrazine and the resulting increase in the formation of CPYO (Belden and Lydy 2000). Similar synergism has either not been observed or was observed only at small synergistic ratios (<2) in other invertebrates (Trimble and Lydy 2006) and vertebrates (Tyler Mehler et al. 2008; Wacksman et al. 2006). In addition, synergism was only observed at greater concentrations of atrazine and CPY, which rarely co-occur (Rodney et al. 2013). For this reason, synergistic interactions between CPY and other chemicals were not included in the assessment of the risks of CPY to aquatic organisms (Giddings et al. 2014).

Synergism of CPY by the sterol-inhibiting fungicide prochloraz was reported to occur in the red-legged partridge (Johnston et al. 1994), but this was only observed in birds pretreated at a large dose of 180 mg prochloraz kg^{-1} (bwt), an extremely unlikely exposure in birds. The synergism was attributed to induction of multifunction oxidases and an increase in the formation of CPYO. As for aquatic organisms, interactions of this type were judged to be very unlikely to occur in terrestrial organisms and were not included in the risk assessment.

4 Use of Chlorpyrifos and Its Formulations

CPY is a widely used organophosphate pesticide with broad spectrum insecticidal activity. It is used against a broad array of insects and mites, primarily as a contact insecticide, although it does have some efficacy through ingestion. It provides control for many adult and larval forms of insects. Foliar pests for which CPY provides control include: aphids, beetles, caterpillars, leafhoppers, mites, and scale. CPY is also effective against many soil insects, including rootworms, cutworms, wireworms, and other grubs. Although it does not translocate readily, CPY can effectively control boring insects in corn, fruit, and other crops through contact exposure. It can also provide contact control of such insects as case-bearers, orange-worms, and other flies that damage fruits and nuts. The diversity of arthropod pests subject to control with CPY has made it one of most widely used insecticides.

4.1 Formulations of Chlorpyrifos

CPY is currently available as a granular formulation and as several spray formulations. CPY is widely effective against many different insects in various habitats that may attack crop throughout the year. Therefore, it has a wide variety of applications and may be applied to foliage, soil, or dormant trees. Application might occur pre-plant, at-plant, post-plant or during the dormant season using aerial equipment, chemigation systems, ground-boom sprayers, air-blast sprayers, tractor-drawn spreaders, and hand-held equipment. Dow AgroSciences (and its predecessors) originally developed CPY, but it is now also produced and/or marketed by other registrants of pesticides. The analysis of uses covered in this paper addresses only those CPY products that are registered by Dow AgroSciences, including Special Local Needs labels (SLNs, FIFRA section 24c) for specific States in the U.S. that are based upon these products.

Lorsban 15G® is a granular formulation that contains 15% (wt/wt) CPY (a.i.) in a solid matrix (Dow AgroSciences 2008). It is used primarily as a soil insecticide, although it can be applied into the whorls of corn to control European corn borer. Applications are in-furrow, banded, and broadcast. One "special local needs" label (FIFRA section 24c State label) was found for use on ginseng in Michigan.

Lorsban 4E® is an emulsifiable concentrate that contains 44.9% (wt/wt) a.i. (479 g L^{-1} = 4 pounds of per gallon) (Dow AgroSciences 2004). It is used both directly on plants and as a soil treatment. Foliage and woody parts of plants can be treated. Treatments of soil are by broadcast, banded, side-dress, or, for onions and radishes, applied in-furrow. Chemigation is specified for some treatments. There are a few special local needs (24c) labels for the Lorsban 4E, but many old ones have expired and appear to have been replaced by similar labels for Lorsban Advanced®.

Lorsban Advanced® is a newer, low odor, water-based version of Lorsban 4E that contains 40.18% a.i. (wtwt) (450 g L^{-1} = 3.755 lb. a.i. per gallon) (Dow AgroSciences 2010). It is used in the same ways as the 4E formulation but contains smaller quantities of volatile solvents, thus reducing air pollution by VOCs. There are a number of special local needs (24c) labels for Lorsban Advanced that both modify application methods and rates and for several additional crops.

Lorsban 75WG® was registered by EPA late in 2011 (Gowan 2011), but is not yet listed among Dow AgroSciences products. It contains 75% a.i. (wt/wt) as water dispersible granules for use in many of the same crops as the Lorsban 4E and Lorsban Advanced formulations. One special local needs (24c) label for peppers in Florida was found that referenced Dow AgroSciences as the registrant, although Gowan Company was the distributor.

Lorsban 50 W® is a water soluble formulation that contains 50% a.i. (wt/wt) and is used for treating seeds in commercial establishments (Dow AgroSciences 2007). It is not permitted for such use on farms and other agricultural sites. It does, however, have a supplemental label for use on unspecified trees in the eastern U.S. The treatment is to trunks of trees at a rate of 3 lb a.i./100 gallons of spray, but no amount or limit per acre is specified. A similar use for Lorsban Advanced is only for apple trees in the eastern U.S., but the Lorsban 50 W label is not limited to any species of tree.

Rates and methods of application for Lorsban 15G are summarized in SI Table B-1. Flowable formulations of Lorsban Advanced, Lorsban 4E, and Lorsban 75WG are summarized (SI Tables B-2, B-3, and B-4). The crops, pests, methods, and rates are very similar for these three two flowable formulations. Because Lorsban 50 W does not have a federal label for application in agricultural settings, it was not included in the tabular information.

4.2 Environmental Precautions

All Lorsban products have the standard precautionary labeling involving risks to aquatic organisms, birds, small mammals, and bees. It is not to be applied to water or below the mean high tide level or when bees are visiting the area; dusk to dawn applications are allowed for many uses when bees are active during the day. Labels advise that drift and runoff might be hazardous in water adjacent to treated areas.

Lorsban 15G has a limitation on aerial application; rates >1.121 kg a.i. ha^{-1} (=1 lb. a.i. A^{-1}) are not permitted. Lorsban 4E, Lorsban Advanced, and Lorsban 75WG have mandatory buffers in their sections on drift-management: Setback buffers from aquatic habitats ("permanent bodies of water such as rivers natural ponds lakes, streams, reservoirs, marshes, estuaries and commercial fish ponds") "must" be utilized: 7.6 m (25 ft) for ground application and chemigation, 15 m (50 ft) for orchard air blast, and 45 m (150 ft) for aerial applications. Aerial applications must follow nozzle and boom width requirements, and applications must neither be made more than 3 m (10 ft) above the height of the plants (unless required for aircraft safety), nor when wind speed exceeds 16 km h^{-1} (10 mph). The above buffers are mandatory. In addition, there are numerous additional recommendations on the label(s) meant to reduce drift. Lorsban Advanced, Lorsban 4E, and Lorsban 75WG may only be applied by ground spray equipment in Mississippi.

4.3 Use of Chlorpyrifos in U.S. Field Crops

Chlorpyrifos is one of the most widely used insecticides in the world. Estimates of annual use in the U.S. since 2008 range from 3.2 to 4.1 M kg y^{-1} (7 to 9 M lb a.i. per annum) (Gomez 2009; Grube et al. 2011). Because of withdrawal of domestic uses, changes in agricultural production, and the introduction of new insecticides, current use is less than 50% of estimated amounts used in the early 2000s (USEPA 2001). Although there are selected survey data from some states on certain crops, and quantitative usage data from California, there were no other recent applicable data on national usage. Estimates of use vary with the amounts of crops planted or harvested, with climate and pest pressure, and sometimes with recent or local occurrences of new or resistant pests.

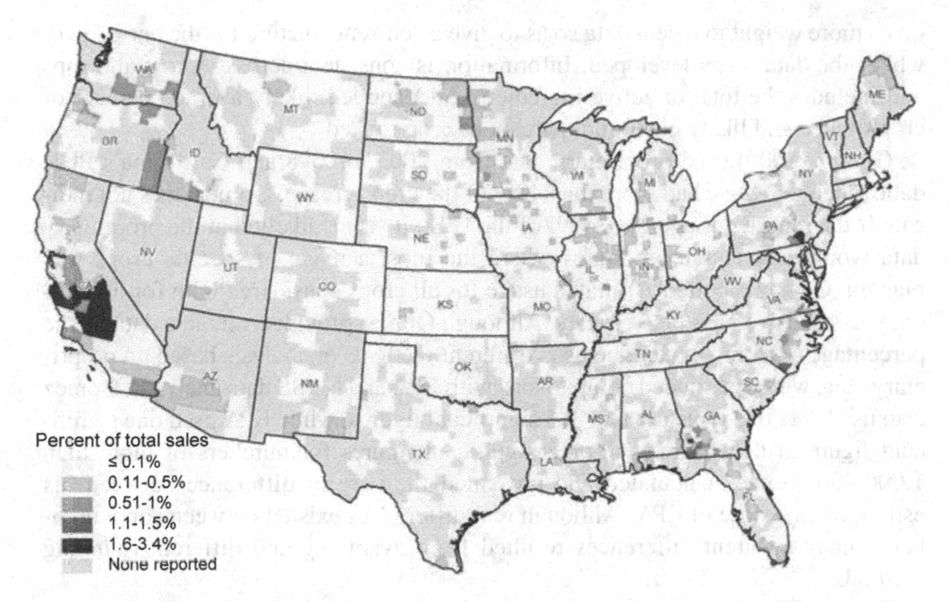

Fig. 5 Geographical distribution of use of granular and liquid formulations of chlorpyrifos in the United States from 2010 to 2011 as % of total. Derived from unpublished sales data from Dow AgroSciences, Indianapolis, IN

Data on sales of granular and flowable CPY, presented as percent of total use across the U.S. from 2010 to 2011, are provided in Fig. 5 (developed from unpublished sales data from Dow AgroSciences). Regions with the highest percentage of total sales (depicted in blue), include Kern, Tulare, Santa Cruz, Fresno counties in central California; Lancaster County in southeastern Pennsylvania; and Calhoun, Decatur, and Mitchell counties in southeast Georgia.

Since purchases of CPY might not be made close to areas of use, data on sales might not accurately reflect use. Several agencies estimate pesticide use on crops but these estimates are derived from a variety of imprecise sources. Although California's Pesticide Use Reporting (PUR) is based upon the actual amounts reported by pesticide users, all others are derived from sampling and statistical analyses. For specific crops, analysis of CPY use was undertaken by EPA (2008), Gomez (2009), and USDA's National Agricultural Statistics Service (USDA 2012). Usage data for CPY are inclusive of all products from any manufacturer or registrant.

Data on amounts of pesticides used were collected differently by Gomez, USEPA, and USDA. EPA acquired their data from USDA/NASS from 2001 to 2006, proprietary market research data from 2001 to 2006, data from the CropLife Foundation's National Pesticide Use Database, only when other data were not available, and California Pesticide Use Report (PUR) data from 2000 to 2005 when 95% of the crop was grown in California (USEPA 2008). EPA noted that their estimates included only data from states that were surveyed, rather than for the entire U.S. The reported figures were derived from an algorithm that covers many years but gives more weight to recent data so as to give a "current" picture for the period over which the data were developed. Information is consistent across almost all crops and includes the total of active ingredient (a.i.) applied, likely average percent of crop treated and likely maximum percent of crop treated.

Gomez (2009) used proprietary data from 2003 to 2008 and California's PUR data, when that State had more than 40% of the crop acreage. Gomez does not indicate if the reported usage covers all of the U.S.; it seems likely that the proprietary data would concentrate on states where the most acreage of specific crops was planted. Gomez reported estimated usage for all crops considered both for individual years and averages of 4–5 years. Although Gomez provided valuable data on the percentage of a crop treated, he used different methods of analysis based on proprietary data, which precluded comparisons with EPA and NASS data analyses. Gomez also used data that typically had 3–5 significant figures, while EPA used one significant figure in their estimates, or 2 significant figures for numbers of more than 1,000,000. Gomez calculated and presented the percent difference between his estimates and those of EPA. Although real differences existed between some numbers, many apparent differences resulted from averaging and different rounding methods.

The NASS performed usage surveys of individual pesticides on certain crops in selected states ("program states") where those crops were most important (http://www.nass.usda.gov/). These surveys are not performed every year. The frequency is dependent upon the crop and typically varies from 3 to 5 y. Methods used by NASS are publically available; but, because they are required to protect individual privacy, data are aggregated in ways that sometimes hides useful information. NASS maintains databases of known growers that are stratified in several ways. They typically send out questionnaires to selected samples of growers. Depending upon the nature of the survey, they might follow up by letter, telephone or computer. They analyze these data using standard statistical aggregation. Therefore, the collected data are representative rather than actual, and only apply to the selected states. As a result, the amounts presented as total pesticide applied nationally are likely to be underestimates, the magnitude of which depends upon how much of a particular crop is grown in the states selected for analysis. However, the percentage of crops treated and the amount applied by acre are likely to be comparable in non-selected states. NASS data are reliable for specific states, at least for years that are sampled. Although annual data might be skewed, the comparisons are fairly close among sources when averaged over several years.

A summary of data from the three national sources on the amount of CPY used on various crops is given in Table 4. EPA estimates usage from existing stocks on some crops that are no longer labeled, but these are not included in Table 4. NASS usage estimates are only given for the latest year, although the amount of CPY used will vary considerably from year to year, depending upon pest pressure.

Table 4 Summary of amount of chlorpyrifos used and percentage of crop treated for selected crop sites

Crop	Ave. lbs.[a] a.i. applied (Gomez 2009) from Doane	Ave. lbs. a.i. applied (USEPA 2008)	Ave. % crop treated (USEPA 2008)	Ave. lbs. a.i. applied (NASS program states –latest year)[b] (USDA 2012)	Ave. % crop treated (NASS) (USDA 2012)
Alfalfa	374,750	400,000	5		
Almonds	341,991	500,000	30		
Asparagus	22,104	20,000	25	211,100	44
Apples	414,600	400,000	55		
Beans, green	4,119	3,000	<1		
Beans & peas, dry		4,000	<1		
Broccoli	60,385	90,000	45		
Brussels Sprouts		6,000	n/c		
Cabbage	7,055	10,000	10		
Carrots (SLN-WA[c])		1,000	<2.5		
Cauliflower	15,239	20,000	40		
Cherries, all	80,140	60,000	30	36,300	16–23
Christmas trees				26,600	16–20
Corn	2,617,432	3,000,000	5	478,000	1
Cotton	285,350	200,000	<1		
Cranberries		50,000	70		
Grapefruit	54,855	60,000	15	42,500	19
Grapes, wine	68,603			64,500	4
Grapes, table	60,428			40,000	12
Grapes, all		100,000	5		
Hazelnuts	7,286	7,000	15		
Lemons	47,033	90,000	35	22,800	12
Mint		50,000	25		
Nectarines		20,000	20	3,400	5
Onions, dry	68,805	60,000	35	51,100	30–32
Oranges	241,735	300,000	10	194,800	12
Peaches	69,853	70,000	30	8,900	7
Peanuts	119,213	200,000	5		
Pears	29,564	30,000	20	11,300	10
Peas, green		<500	<1		
Pecans	296,596	300,000	35		
Peppers (SLN-FL)		2,000	<1		
Plums & Prunes	18,674	40,000			
Plums			15	2,400	7
Prunes			10		
Sod/turf		2,000	n/c		
Sorghum		30,000	<1		

(continued)

Table 4 (continued)

Crop	Ave. lbs.[a] a.i. applied (Gomez 2009) from Doane	Ave. lbs. a.i. applied (USEPA 2008)	Ave. % crop treated (USEPA 2008)	Ave. lbs. a.i. applied (NASS program states –latest year)[b] (USDA 2012)	Ave. % crop treated (NASS) (USDA 2012)
Soybeans	1,017,953	700,000	<1		
Strawberries	10,043	9,000	15	7,700	15
Sugar Beets	138,020	100,000	10		
Sunflowers	34,857	20,000	<1		
Sweet Corn	120,881	100,000	10	36,500	13–23
Sweet potatoes		100,000	65		
Tangelos & tangerines		8,000	10	8,300	7–19
Tobacco	98,468	100,000	15		
Walnuts	195,505	400,000	45		
Wheat	288,751	300,000	<1	577,000	2–3
Total					

[a]To maintain consistency with uses and the labels of formulated products sold in the U.S., amounts of CPY applied are given in imperial units (pounds (lbs.))
[b]Generally this is 2011 for fruits and 2010 for other crop
[c]Special Label Needs

4.4 Timing of the Use of Chlorpyrifos

CPY is normally applied to coincide with infestations of pests, which vary from one location to another. Timing of application of CPY in relation to local climatic conditions, rainfall, and patterns of weather might have significant effects on the degradation, potential for movement, and exposures of non-target organisms. To properly characterize timing of the use of CPY, we relied on the USDA publication "Usual Planting and Harvesting Dates for U.S. Field Crops" (USDA 2010) and other sources (i.e., mainly state extension services and the internet). These data are summarized in Table 5 for crops that are in the field year round, and in Table 6 for crops that are seasonal.

From these data, it is apparent that there is no strong seasonal use of CPY, although there is a somewhat greater usage in winter months for tree crops in California and greater use in summer for certain field crops (e.g., corn). These use patterns and how they affect scenarios for exposures are discussed in more detail in the companion papers of this volume (Moore et al. 2014; Williams et al. 2014).

5 Summary

Physical properties and use data provide the basis for estimating environmental exposures to chlorpyrifos (CPY) and for assessing its risks. The vapor pressure of CPY is low, solubility in water is <1 mg L^{-1}, and its log K_{OW} is 5. Chlorpyrifos has

Table 5 Timing of chlorpyrifos use for crops in the U.S. that are in the field all year (Jan to Dec)

Crop and location	Months of the year in which CPY is applied											
	J	F	M	A	M	J	J	A	S	O	N	D
Alfalfa, warmer states (CA, AZ, etc.)		■	■	■	■	■	■	■	■	■	■	
Cooler states				■	■	■	■	■	■			
Southern MO			■	■								
Northern MO				■				■	■			
Apple tree trunks				■	■	■	■	■	■	■	■	
Asparagus CA only Southern desert	■	■	■	■								
Delta		■	■	■	■							
Central coast		■	■	■	■	■						
Other U.S.			■	■	■	■	■	■	■	■	■	
Brassica (cole) leafy vegetables			■	■	■				■	■	■	
Brussels sprouts			■	■	■	■	■	■	■	■		
Carrots for seed OR & WA						■	■	■	■			
Christmas tree				■	■	■	■	■	■			
Citrus orchard floors	■	■	■	■	■	■	■	■	■	■	■	■
Citrus fruits				■	■	■	■	■	■	■	■	
Cranberry				■	■	■	■	■	■			
Fig (CA only)		■	■									
Ginseng (MI, WI-SLN[a])			■	■	■	■	■	■	■			
Grape (E of Continental Divide only)			■	■	■	■	■	■	■			
Grapes (CA-SLN)	■	■	■							■	■	■
Grass and clover for seed (NV, ID, OR, WA-SLNs)			■	■	■			■	■	■		
Legume vegetables (except soybeans)	■			■	■	■	■	■	■	■	■	■
Onion (dry bulb)			■	■	■			■	■	■		
Pears (CA, OR, & WA only)							■	■	■	■		
Peppers (FL only – special local need)	■	■	■	■	■	■	■	■	■	■	■	■
Pineapple (HI only – special local need)	■	■	■	■	■	■	■	■	■	■	■	■
Pulpwood (cottonwood & poplar, OR, WA-SLNs)	■	■	■	■	■	■	■	■	■			■
Strawberries			■	■	■	■	■	■	■	■		
Tree fruits and nuts – all applications, almond	■	■	■	■	■	■	■	■	■	■	■	■
Apples (all U.S.)	■	■	■									■
Apples (eastern U.S.)				■	■	■	■	■	■	■	■	
Cherry	■	■	■	■	■	■	■	■	■	■		■
Filbert				■	■	■	■	■	■	■		
Nectarine	■	■	■	■	■	■	■	■	■	■		■
Peach	■	■	■	■	■	■	■	■	■	■		■
Pear	■	■	■				■	■	■	■		■
Pecan	■	■	■	■	■	■	■	■	■	■	■	■
Plum	■	■	■									■
Prune	■	■	■									■
Walnut	■	■	■	■	■	■	■	■	■	■	■	■
Turfgrass	■	■	■	■	■	■	■	■	■	■	■	■
Wheat (W of the Mississippi River)	■	■	■	■	■	■	■	■	■	■	■	■

(continued)

Table 5 (continued)

[a]Special Label Needs
Data from:
USDA (2010)
Missouri Extension Service: http://extension.missouri.edu/p/G4550
http://ohioline.osu.edu/b826/b826_14.html
http://www.pickyourown.org/FLcitrus.htm; http://www.pickyourown.org/CAharvest calendar. htm
Seasonal Patterns of Citrus Bloom, by William A. Simanton, Florida Agricultural Experiment Station Journal Series No. 3426. Florida State Horticultural Society, 1969, pp 96–98
http://en.wikipedia.org/wiki/Cranberry; http://www.wenatcheeworld.com/news/2011/oct/07/cranberry-harvest-under-way-on-wa-coastal-bogs/; http://www.capecodtravel.com/attractions/ nature/cranberries0900.shtml
Morton, J. 1987. Fig. p. 47–50. In: Fruits of warm climates. Julia F. Morton, Miami, FL. @ http://www.hort.purdue.edu/newcrop/morton/fig.html; also http://www.latimes.com/features/ la-fo-market16-2008jul16,0,4856462.story
http://en.wikipedia.org/wiki/Cranberry; http://www.wenatcheeworld.com/news/2011/oct/07/cranberry-harvest-under-way-on-wa-coastal-bogs/; http://www.capecodtravel.com/attractions/nature/cranberries0900.shtml\
Monitoring and Control Tactics for Grape Root Borer Vitacea polistiformis Harris (Lepidoptera: Sesiidae) in Florida Vineyards. By Scott Weihman. Master's Degree Thesis, University of Florida, 2005 @ http://etd.fcla.edu/UF/UFE0009182/weihman_s.pdf
The Grape Root Borer in Tennessee, by P. Parkman, D. Lockwood, and F. Hale, University of Tennessee Extension Service publication W171, 2007. @ https://utextension.tennessee.edu/publications/documents/W171.pdf www.nysipm.cornell.edu/factsheets/grapes/pests/gcb.pdf
http://www.calagquest.com/BloomTime.php
http://sacramentogardening.com/edible_gardening.html
http://wiki.answers.com/Q/What_time_of_year_do_you_grow_peas
Pest Management Strategic Plan for Dry Bulb Storage Onions in Colorado, Idaho, Oregon, Utah, and Washington. Summary of a workshop held on February 26–27, 2004. Boise, ID. @ http://www.ipmcenters.org/pmsp/pdf/WesternONION.pdf
http://aggie-horticulture.tamu.edu/archives/parsons/publications/onions/oniongro.html
http://edis.ifas.ufl.edu/mv112
http://www.ctahr.hawaii.edu/oc/freepubs/pdf/f_n-7.pdf
http://www.strawberry-recipes.com/plant-strawberries.html
http://strawberryplants.org/2010/05/strawberry-varieties/

short to moderate persistence in the environment as a result of several dissipation pathways that may proceed concurrently. Primary mechanisms of dissipation include volatilization, photolysis, abiotic hydrolysis, and microbial degradation. Volatilization dominates dissipation from foliage in the initial 12 h after application, but decreases as CPY adsorbs to foliage or soil. In the days after application, CPY adsorbs more strongly to soil, and penetrates more deeply into the soil matrix, becoming less available for volatilization. After the first 12 h, other processes of degradation, such as chemical hydrolysis and catabolism by microbiota become important. The half-life of CPY in soils tested in the laboratory ranged from 2 to 1,575 d (N = 126) and is dependent on properties of the soil and rate of application. At application rates used historically for control of termites, the degradation rate is much slower than for agricultural uses. In agricultural soils under field conditions, half-lives are shorter (2 to 120 d, N = 58). The mean water-soil adsorption coefficient (K_{OC}) of CPY is 8,216 mL g^{-1}; negligible amounts enter plants via the roots, and it is not translocated in plants.

Table 6 Timing of chlorpyrifos use for crops in the U.S. that are in the field part of the year

	Months of the year in which CPY is applied											
Crop in field, location, and use of CPY	**J**	**F**	**M**	**A**	**M**	**J**	**J**	**A**	**S**	**O**	**N**	**D**
Corn, Southern states in field												
Use of CPY												
Northern states in field												
Use of CPY												
Cotton, Southern areas in field												
Use of CPY												
Northern areas + CA in field												
Use of CPY												
Peanuts, in field												
Use of CPY												
Peppermint and Spearmint, in field												
Use of CPY												
Sorghum, in field												
Use of CPY												
Soybeans, in field												
Use of CPY												
Sugarbeets, in field Imperial Valley, CA												
Use of CPY												
Other locations in field												
Use of CPY												
Sunflowers, in field CA												
Use of CPY												
TX & OK in field												
Use of CPY												
Other states in field												
Use of CPY												
Sweet potato, in field												
Use of CPY												
Tobacco, in field New England & PA												
Use of CPY												
Southern states in field												
Use of CPY												

Data from: (Chen et al. 2011; USDA 2010; Zheljazkov et al. 2010)

Half-lives for hydrolysis in water are inversely dependent on pH, and range from 16 to 73 d. CPY is an inhibitor of acetylcholinesterase and is potentially toxic to most animals. Differences in susceptibility result from differences in rates of adsorption, distribution, metabolism, and excretion among species. CPY is an important tool in management of a large number of pests (mainly insects and mites) and is used on a wide range of crops in the U.S. Estimates of annual use in the U.S. from 2008 to 2012 range from 3.2 to 4.1 M kg y^{-1}, which is about 50% less than the amount used prior to 2000. Applications to corn and soybeans accounts for 46–50% of CYP's annual use in the U.S.

Acknowledgements The authors thank Kendall Price, Waterborne, and Larry Turner, Compliance Services International, for compiling information on the properties and use of chlorpyrifos. We thank the anonymous reviewers of this paper for their suggestions and constructive criticism. Prof. Giesy was supported by the Canada Research Chair program, a Visiting Distinguished Professorship in the Department of Biology and Chemistry and State Key Laboratory in Marine Pollution, City University of Hong Kong, the 2012 "High Level Foreign Experts" (#GDW20123200120) program, funded by the State Administration of Foreign Experts Affairs, the P.R. China to Nanjing University and the Einstein Professor Program of the Chinese Academy of Sciences. Funding for this study was provided by Dow AgroSciences.

References

Batzer FR, Fontaine DD, White FH (1990) aqueous photolysis of chlorpyrifos. DowElanco, Midland, MI, Unpublished Report

Belden JB, Lydy MJ (2000) Impact of atrazine on organophosphate insecticide toxicity. Environ Toxicol Chem 19:2266–2274

Belden JB, Lydy MJ (2001) Effects of atrazine on acetylcholinesterase activity in midges (*Chironomus tentans*) exposed to organophosphorus insecticides. Chemosphere 44:1685–1689

Bidlack HD (1979) Degradation of chlorpyrifos in soil under aerobic, aerobic/anaerobic and anaerobic conditions. Dow Chemical, Midland, MI, Unpublished Report

Chapman RA, Harris CR (1980) Persistence of chlorpyrifos in a mineral and an organic soil. J Environ Sci Health B 15:39–46

Chen MZ, Trinnaman L, Bardsley K, St Hilaire CJ, Da Costa NC (2011) Volatile compounds and sensory analysis of both harvests of double-cut Yakima peppermint (*Mentha piperita* L.). J Food Sci 76:C1032–C1038

Cryer S, Dixon-White H (1995) A field runoff study of chlorpyrifos in southeast iowa during the severe flooding of 1993. Indianapolis, IN, DowElanco, Unpublished Report

Cutler GC, Purdy J, Giesy JP, Solomon KR (2014) Risk to pollinators from the use of chlorpyrifos in the United States. Rev Environ Contam Toxicol 231:219–265

de Vette HQM, Schoonmade JA (2001) A study on the route and rate of aerobic degradation of ^{14}C-chlorpyrifos in four European soils. Dow AgroSciences, Indianapolis, IN, Unpublished Report

Dow AgroSciences D (2004) Lorsban 4E insecticide label. Dow AgroSciences LLC, Indianapolis, IN

Dow AgroSciences (2007) Lorsban 50W wettable powder insecticide label. Dow Agrosciences Canada, Calgary, AB, Canada. http://msdssearch.dow.com/PublishedLiteratureDAS/dh_00e9/0901b803800e9ceb.pdf?filepath=ca/pdfs/noreg/010-20642.pdf&fromPage=GetDoc

Dow AgroSciences (2008) Lorsban 15G granular insecticide specimen label. Dow AgroSciences LLC, Indianapolis, IN

Dow AgroSciences (2010) Lorsban advanced insecticide label. Dow AgroSciences LLC, Indianapolis, IN

El-Merhibi A, Kumar A, Smeaton T (2004) Role of piperonyl butoxide in the toxicity of chlorpyrifos to *Ceriodaphnia dubia* and *Xenopus laevis*. Ecotoxicol Environ Safety 57:202–212

Gebremariam SY, Beutel MW, Yonge DR, Flury M, Harsh JB (2012) Adsorption and desorption of chlorpyrifos to soils and sediments. Rev Environ Contam Toxicol 215:123–175

Giddings JM, Williams WM, Solomon KR, Giesy JP (2014) Risks to aquatic organisms from the use of chlorpyrifos in the United States. Rev Environ Contam Toxicol 231:119–162
Giesy JP, Solomon KR, Coates JR, Dixon KR, Giddings JM, Kenaga EE (1999) Chlorpyrifos: ecological risk assessment in North American aquatic environments. Rev Environ Contam Toxicol 160:1–129
Gomez LE (2009) Use and benefits of chlorpyrifos in U.S. agriculture. Dow AgroSciences, Indianapolis, IN, Unpublished Report
Gowan (2011) Lorsban 75WG Label. Gowan Company, Yuma, AZ. http://www.keystonepestsolutions.com/labels/Lorsban_75WG.pdf
Grube A, Donaldson D, Kiely T, Wu L (2011) Pesticides industry sales and usage: 2006 and 2007 market estimates. United States Environmental Protection Agency, Office of Pesticide Programs, Biological and Economic Analysis Division, Washington, DC, EPA 733-R-11-001
Johnston G, Walker CH, Dawson A (1994) Interactive effects between EBI fungicides (prochloraz, propiconazole and penconazole) and OP insecticides (dimethoate, chlorpyrifos, diazinon and malathion) in the hybrid red-legged partridge. Environ Toxicol Chem 13:615–620
Kennard LM (1996) Aerobic aquatic degradation of chlorpyrifos in a flow-through system. DowElanco, Indianapolis, IN, Unpublished Report
Mackay D, Giesy JP, Solomon KR (2014) Fate in the environment and long-range atmospheric transport of the organophosphorus insecticide, chlorpyrifos and its oxon. Rev Environ Contam Toxicol 231:35–76
McCall PJ, Oliver GR, McKellar RL (1984) Modeling the runoff potential and behavior of chlorpyrifos in a terrestrial-aquatic watershed. Midland, MI, Dow Chemical, Unpublished Report
Moore DRJ, Teed RS, Greer C, Solomon KR, Giesy JP (2014) Refined avian risk assessment for chlorpyrifos in the United States. Rev Environ Contam Toxicol 231:163–217
Nash RG (1988) Dissipation from soil. In: Grover R (ed) Environmental chemistry of the herbicides, vol 1. CRC Press, Boca Raton, FL, pp 131–170
Poletika NN, Robb CK (1994) A field runoff study of chlorpyrifos in Mississippi delta cotton. Indianapolis, IN, DowElanco, Unpublished Report
Racke KD, Coats JR, Titus KR (1988) Degradation of chlorpyrifos and its hydrolysis product, 3,5,6-trichloro-2-pyridinol in soil. J Environ Sci Health B 23:527–539
Racke KD (1993) Environmental fate of chlorpyrifos. Rev Environ Contam Toxicol 131:1–151
Reeves GL, Mackie JA (1993) The aerobic degradation of ^{14}C-chlorpyrifos in natural waters and associated sediments. Dow Agrosciences, Indianapolis, IN, Unpublished Report
Rodney SI, Teed RS, Moore DRJ (2013) Estimating the toxicity of pesticide mixtures to aquatic organisms: A review. Human Ecol Risk Assess 19:1557–1575
Rotondaro A, Havens PL (2012) Direct flux measurement of chlorpyrifos and chlorpyrifos-oxon emissions following applications of Lorsban Advanced insecticide to alfalfa. Dow AgroScience, Indianapolis, IN, Unpublished Report
Rouchaud J, Metsue M, Gustin F, van de Steene F, Pelerents C, Benoit F, Ceustermans N, Gillet J, Vanparys L (1989) Soil and plant biodegradation of chlorpyrifos in fields of cauliflower and Brussels sprouts crops. Toxicol Environ Chem 23:215–226
Solomon KR, Giesy JP, Kendall RJ, Best LB, Coats JR, Dixon KR, Hooper MJ, Kenaga EE, McMurry ST (2001) Chlorpyrifos: ecotoxicological risk assessment for birds and mammals in corn agroecosystems. Human Ecol Risk Assess 7:497–632
Testai E, Buratti FM, Consiglio ED (2010) Chlorpyrifos. In: Krieger RI, Doull J, van Hemmen JJ, Hodgson E, Maibach HI, Ritter L, Ross J, Slikker W (eds) Handbook of pesticide toxicology, vol 2. Elsevier, Burlington, MA, pp 1505–1526
Timchalk C (2010) Organophosphorus insecticide pharmacokinetics. In: Krieger RI, Doull J, van Hemmen JJ, Hodgson E, Maibach HI, Ritter L, Ross J, Slikker W (eds) Handbook of pesticide toxicology, vol 2. Elsevier, Burlington, MA, pp 1409–1433
Trimble AJ, Lydy MJ (2006) Effects of triazine herbicides on organophosphate insecticide toxicity in *Hyalella azteca*. Arch Environ Contam Toxicol 51:29–34
Tyler Mehler W, Schuler LJ, Lydy MJ (2008) Examining the joint toxicity of chlorpyrifos and atrazine in the aquatic species: *Lepomis macrochirus*, *Pimephales promelas* and *Chironomus tentans*. Environ Pollut 152:217–224

USDA (2010) Usual Planting and Harvesting Dates for U.S. Field Crops. United States Department of Agriculture, National Agricultural Statistics Service pp 51
USDA (2012) National Agricultural Statistics Service. United States Department of Agriculture. Accessed 2012 April, August and September. http://www.nass.usda.gov/Surveys/Guide_to_NASS_Surveys/Chemical_Use/index.asp
USEPA (2001) Chlorpyrifos Interim Reregistration Eligibility Decision Document (IRED). United States Environmental Protection Agency, Washington, DC
USEPA (2008) CHLORPYRIFOS 059101 Screening Level Usage Analysis (SLUA). United States Environmental Protection Agency, Docket ID #EPA-HQ-OPP-2008-0850-0005 http://www.regulations.gov/#!documentDetail;D=EPA-HQ-OPP-2008-0850-0005
USEPA (2011a) Interim guidance on honey bee data requirements. United States Environtmental Protection Agency, Environmental Fate and Effects Division, Office of Pesticide Programs, Washington, DC. http://www.epa.gov/oppefed1/ecorisk_ders/honeybee_data_interim_guidance.htm
USEPA (2011b) Revised chlorpyrifos preliminary registration review drinking water assessment. United States Environmental Protection Agency, Office of Chemical Safety and Pollution Prevention, Washington, DC, PC Code 059101 http://www.epa.gov/oppsrrd1/registration_review/chlorpyrifos/EPA-HQ-OPP-2008-0850-DRAFT-0025%5B1%5D.pdf
Wacksman MN, Maul JD, Lydy MJ (2006) Impact of atrazine on chlorpyrifos toxicity in four aquatic vertebrates. Arch Environ Contam Toxicol 51:681–689
Williams WM, Giddings JM, Purdy J, Solomon KR, Giesy JP (2014) Exposures of aquatic organisms to the organophosphorus insecticide, chlorpyrifos resulting from use in the United States. Rev Environ Contam Toxicol 231:77–118
Yon D (2011) Modelling the kinetics of the degradation of chlorpyrifos and its metabolites in field soil—rate constant normalisation method. Dow AgroSciences European Development Centre, Abingdon, Unpublished Report
Zheljazkov VD, Cantrell CL, Astatkie T, Hristov A (2010) Yield, content, and composition of peppermint and spearmints as a function of harvesting time and drying. J Agric Food Chem 58:11400–11407

Fate in the Environment and Long-Range Atmospheric Transport of the Organophosphorus Insecticide, Chlorpyrifos and Its Oxon

Don Mackay, John P. Giesy, and Keith R. Solomon

1 Introduction

The fate and movement of the organophosphorus insecticide chlorpyrifos (CPY; CAS No. 2921-88-2) and its principal transformation product of interest, chlorpyrifos-oxon (CPYO; CAS No. 5598-15-2), are primary determinants of exposures experienced by animals in terrestrial and aquatic environments. Dynamics of the movement of CPY and CPYO are determined by the interactions between chemical and physical properties (Solomon et al. 2013a) and environmental conditions. Together, these properties provide the basis for developing and refining models of exposure for assessing risks. An extensive review of the environmental fate of CPY was published in 1993 (Racke 1993). The following sections build on this review, with updates exploiting relevant data from new studies and other reviews in the literature as these pertain to the assessment of risks in the ecosystem. This report addresses processes that affect fates of CPY and CPYO in various compartments of the environment and how these affect exposures of ecological receptors (Fig. 1) as discussed in companion papers (Cutler et al. 2014; Moore et al. 2014; Williams et al. 2014). This paper serves as an update on the environmental dynamics and

The online version of this chapter (doi:10.1007/978-3-319-03865-0_3) contains supplementary material, which is available to authorized users.

D. Mackay (✉)
Trent University, Peterborough, ON, Canada
e-mail: dmackay@trentu.ca

J.P. Giesy
Department of Veterinary Biomedical Sciences and Toxicology Centre,
University of Saskatchewan, 44 Campus Dr., Saskatoon, SK S7N 5B3, Canada

K.R. Solomon
Centre for Toxicology, School of Environmental Sciences, University of Guelph,
Guelph, ON, Canada

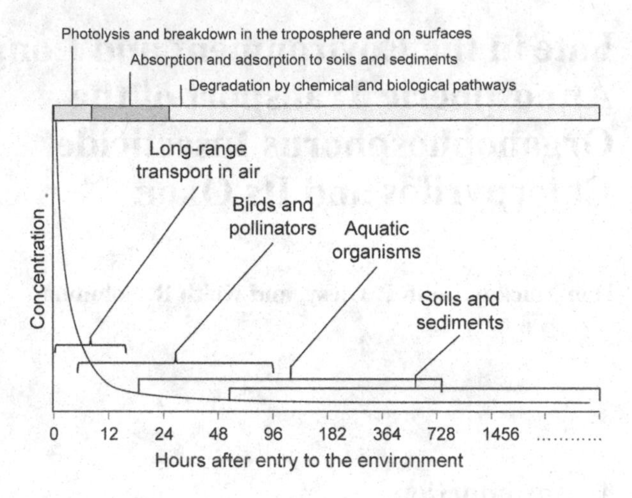

Fig. 1 Qualitative diagrammatic representation of the sequence of processes influencing the fate of CPY in various environmental compartments after release and their influence on exposures to biota

potential exposures to CPY that were presented previously (Giesy et al. 1999; Solomon et al. 2001) and includes additional information on environmental chemodynamics of CPY that have become available subsequent to those earlier publications. There have been and continue to be extensive studies on the presence of CPY and CPYO in environmental media near to and remote from sites of application. Many are prompted by concerns that these substances may have effects on distant sensitive organisms, such as amphibians and in remote food webs as have occurred with organo-chlorine pesticides.

2 Fate in the Atmosphere and Long-Range Transport

The potential for long-range transport (LRT) is a concern for synthetic chemicals of commerce, including pesticides. Concentrations of synthetic chemicals measured at locations distant from sources, in conjunction with mass-balance modeling, have combined to provide information on key contributing processes involved in LRT, especially for persistent organic pollutants (POPs) that have relatively long residence times in the atmosphere. A quantitative predictive capability has emerged in the form of simple mass-balance models such as TAPL3 and the OECD Tool (Beyer et al. 2000; Wegmann et al. 2009). These models have been used in regulatory contexts and characterize LRT as a Characteristic Travel Distance (CTD) over which some two-thirds of the mass of chemical transported from source regions is deposited or transformed to other chemicals, while the remaining third is transported greater distances through the atmosphere. The focus here is the organophosphate insecticide, CPY and its transformation product CPYO, in which the sulfur atom is replaced by oxygen (Giesy et al. 1999; Racke 1993).

Table 1 Reported concentrations of CPY in air in N America

Location and dates of sampling	Concentration (ng m^{-3})	Reference
CANCUP data for 8 sites in Canada 2004–2005, values range from areas of application to distant areas	0.08–22	Yao et al. (2008)
Passive air samples in Ontario 2003–2005	0.0003–0.06; median 0.007; 73% FOD[a]	Kurt-Karakus et al. (2011)
Iowa 2000–2002	Average 1.0, 19% FOD 1.4 at 1A—AM Site, 0.88 at Hills	Peck and Hornbuckle (2005)
Mississippi River Valley 1994	0.43	Majewski et al. (1998)
Chesapeake Bay 2000	0.015–0.670 Median 0.110 FOD 87%	Kuang et al. (2003)
Bratt's Lake (Saskatchewan) and Abbotsford (British Colombia)	Mostly 10–100 with 3 concentrations exceeding 100 and a maximum of 250 in Aug 2003 at Bratt's Lake near area of use but max only 1.38 in Jul 2005. Concentrations in Abbotsford, an area of lesser use all <0.26 in 2004 and 2005	Raina et al. (2010)
Mississippi River Valley	FOD 93, MS 35%, Iowa City 90%, Cedar Rapids 50%, Minneapolis 10%, Princeton MN 3%	Foreman et al. (2000)
Central Valley (CA) and Sierra Nevada	CPY up to 180 and CPYO up to 54, lesser concentrations in Sequoia NP	Aston and Seiber (1997)
CPY and CPYO in air in Sequoia NP	0.16–17.5	LeNoir et al. (1999)

[a]*FOD* = frequency of detection

2.1 Chlorpyrifos and Chlorpyrifos-Oxon

Evidence that CPY is subject to LRT is provided in reports of concentrations in air and other media at locations remote from sites where CYP is applied in agriculture (Tables 1, 2, 3 and 4). Notable are studies conducted in the intensely agricultural, Central Valley of California and adjacent National Parks. CTDs of several pesticides, including CPY have been estimated (Muir et al. 2004). Results of these modeling exercises have suggested a CTD of 280–300 km for CPY, the narrow range being the direct result of close similarities between the model equations. Monitoring observations of concentrations of CPY in air close to and carried downwind from application areas are in general accord with these distances. Also in accord are monitoring data reflecting deposition in foothill and mountainous terrain, especially in the Sierra Nevada of California.

Table 2 Reported concentrations of CPY in rain and snow and fluxes

Location and dates of sampling	Concentration in ng L^{-1}	Reference
Chesapeake Bay 2000	Rain; 0.97–29 average 4.8, FOD[a] 14% Wet flux, 190 ng m^{-2} event total 6,100 ng m^{-2}, 1.1 kg/yr	Kuang et al. (2003)
Svalbard Ice Cap 1979–1986	Ice, peak concentration 16.2 at ~15 m	Hermanson et al. (2005)
Delaware/Maryland April to Sept 2000–2003	Rain; 1.0–29 average 1.0 39% FOD Fluxes ~610–1750 average 1.0–4.5 ng m^{-2}	Goel et al. (2005)
7 US National Parks (NP) March–April 2003	Snow in Sequoia NP; 2.8, 1.3 Other NP ; 0.033 ~0.05 ~0.5 ~0.02 ~0.02 and 0.03 Deposition in Sequoia NP; 2,600 ng m^{-2} Other parks; 25, 65, 35, 30, 14, and 4 ng m^{-2} Correlations with altitude and distance, 75, 150, 300 km from sources	Hageman et al. (2006)

[a]*FOD* = frequency of detection

A comprehensive ecotoxicological risk assessment of CYP was developed for birds and mammals (Solomon et al. 2001) and aquatic environments (Giesy et al. 1999) that were near areas of application. The analysis of LRT of CPY and CPYO presented here extends those assessments to regions downwind of points of application. The approach taken in this study was to compile and evaluate data on concentrations of CPY and CPYO at locations both near to applications and remote from sources. This assessment of LRT thus goes beyond determination of CTD to include estimates of concentrations of CPY and CPYO in other environmental media such as rain, snow, and terrestrial phases as well as in the atmosphere at more remote locations, including high altitudes. This was accomplished by developing a relatively simple mass-balance model, predictions from which could be compared to available measured concentrations of CYP in air and other media. This can provide an order-of-magnitude test of the accuracy of the predictions of the model, and, in this way, make an indirect assessment of the relative importance of the included processes and parameters. The model can then serve as a semi-quantitative predictive framework that is consistent with observations. The equations included in the model enable examination of the effect of changes in parameters such as application rate, temperature, meteorology, distance from source and precipitation. Estimated concentrations in terrestrial and aquatic environments remote from areas of application can be used, in combination with toxicological data, to assess risk to organisms in those media and locations.

Monitoring data. Reports of concentrations of CPY in air at a variety of locations are presented in Table 1, with comments on other influencing factors such as altitude. Also included are reports of concentrations of toxicologically-relevant transformation products, such as CPYO, if and when such information was available. Reports of concentrations of CPY in precipitation (rain and snow) are given in Table 2, while Table 3 provides data for water bodies and other terrestrial media.

Table 3 Reported concentrations of CPY in aquatic and terrestrial media

Location and dates of sampling	Concentration in ng g^{-1}dwt unless indicated otherwise	Reference
Sierra Nevada CA 2,785–3,375 m elevation 2004/2005 and Yosemite National Park	Water: all concentrations in water were <0.07 ng L^{-1}	Bradford et al. (2010a)
	Sediment: 0.043–3.478 median 0.107 0.285–12.44 median 1.73 0.011–2.276 median 0.101 0.499–10.72 median 1.372 Tadpoles: 2.224–156, median 22.2 ng g^{-1} lipid 2.741–68.4, median 19.9 ng g^{-1} lipid	Bradford et al. (2013)
Lassen National Park, CA, 25 km from the San Joaquin Valley edge	Tadpoles of Pacific tree frog (*Hyla regilla*): 10–17 ng g^{-1}wwt	Datta et al. (1998)
Kaweah, Kings, and Kern Watersheds 43–85 km from the San Joaquin Valley edge	Tadpoles of Pacific tree frog (*Hyla regilla*): <0.6 ng g^{-1}wwt	Bradford et al. (2010b)
Thirty Canadian lakes 1999–2001	Lake water in application region: 0.28–1.0 mean 0.65 ng L^{-1} Lake water in remote regions: <0.017–2.9 mean 0.82 ng L^{-1} Lake water, subarctic: <0.017–<0.017 ng/L^{-1} in all samples Lake water, arctic: <0.017–1.6 mean 0.27 ng L^{-1}	Muir et al. (2004)
Ontario 2003/2005	Lake water: <0.002 to 0.5 ng L^{-1} median 0.02, 77% FOD Rain: <0.004 to 43 ng L^{-1} median 0.76, 80% FOD Zooplankton: <0.003–0.08, 0.004, 0.005 ng g^{-1}wwt (geometric means = GM) BAF in Zooplankton (GM) 70 (wwt), 3,300 (lipid wt)	Kurt-Karakus et al. (2011)
Chesapeake Bay, 2000	0.51–4.6	Kuang et al. (2003)
Lichen in Yosemite NP 2003–2005	0.92 ng g^{-1}, 1.7 ng g^{-1}	Mast et al. (2012)
Pine needles in Sequoia NP, 1994	10–125 ng g^{-1}dwt	Aston and Seiber (1997)
Sequoia NP, 1996–1997	Dry deposition: 0.2–24 ng m^{-2} d^{-1} Surface water: 0.2–122 ng L^{-1}	LeNoir et al. (1999)

A significant portion of these data are from the Sierra Nevada, including National Parks that are some 30–200 km and primarily downwind and up-gradient from the productive agricultural Central Valley of California, in which there is significant usage of CPY (Solomon et al. 2014).

The data in (Tables 1, 2, 3 and 4) confirm that measurable concentrations of CPY are found in air and other media remote from sources with a significant

Table 4 Numbers and approximate percentage distributions of reported concentration levels expressed as percentages of reported data

Phase	Air ng m^{-3}	Rain ng L^{-1}	Snow ng L^{-1}	Water ng L^{-1}	Sediment ng g^{-1}	Soil ng g^{-1}	Biota ng g^{-1}
Number and conc.	~100	15	18	30	8	2	10
% >10	15	20	11	0	0	0	0
% 1–10	14	40	17	46	20	0	10
% 0.1–1.0	36	20	11	27	40	100	10
% 0.01–0.1	27	20	61	10	40	0	80
% 0.001–0.01	8	0	0	17	0	0	0

Note that concentrations in air and rain appear similar numerically because the air-water partition coefficient of 0.00034 is similar to the 0.001 factor conversion from m^3 to L

frequency of detection. The key issue in this context is not one of presence/absence, because CPY and CPYO can be monitored in air at concentrations as little as 0.001 ng m^{-3}, which are much less than thresholds for adverse effects. Risk depends on the magnitude of concentrations, especially in media where organisms might be exposed and thus are potentially at risk. It can be difficult to assimilate ranges in concentrations in the atmosphere and the variety of concentration units of differing magnitudes in sampled media. Accordingly, here, the feasibility was assessed of compiling a more readily comprehendible depiction of multi-media environmental concentrations by expressing the concentrations as ranges and converting concentrations in various media to fugacities. Fugacity is essentially partial pressure and can be deduced for all media and compared directly, without difficulties introduced by the use of different concentration units for individual compartments of the environment. Using fugacity as a synoptic descriptor of concentrations in the ecosystem has been applied previously to multi-media concentrations of organochlorines in the Great Lakes (Clark et al. 1988). It is, of course, possible to calculate multimedia equilibrium concentrations using partition coefficients directly, rather than using fugacity as an intermediate, but the equilibrium status of two phases with units such as ng m^{-3} in air and mg kg^{-1} in vegetation may not be obvious.

Ideally, to demonstrate directly the trend of decreasing concentrations, the data should be plotted as a function of distance from source, but because sources are often uncertain and concentrations vary with time as a function of transformation of the material at the location of release, this is rarely possible. The approach adopted here was to compile a distribution of reported concentrations to gain perspective on the range in magnitude of concentrations at various distances from points of release, at least for ecosystems for which sufficient monitoring data have been compiled. Accordingly, Table 4 depicts the distribution of reported concentrations for air, rain, snow, water bodies, soils, sediments, and biota on a decade scale. In some cases, products of transformation are included and in others they were specifically excluded. Some of the data were reported graphically or as ranges, so numerical values were sometimes difficult to establish. Locations for which information was available varied geographically and often lacked information on current and recent meteorology such as wind speed, temperature, and precipitation. Some values reported for each concentration range are approximate because reports gave only

Table 5 Key physical-chemical properties of CPY

Property	Units	Value	Comments
Melting point (mp)	°C	42	
Molar mass	g/mol	350.6	
Fugacity ratio (FR)	–	0.68	Estimated from mp
Vapor pressure (VP) of solid	Pa	0.0023	EPA gives 0.00249
Vapor pressure of sub-cooled liquid	Pa	0.0034	Consistent with FR and solid VP
Solubility of solid in water	$g\ m^{-3}$ or $mg\ L^{-1}$	0.73	EPA gives 1.43
Solubility of sub-cooled solid	$g\ m^{-3}$	1.07	Consistent with FR and solid solubility
Henry's Law constant	$Pa\ m^{-3}\ mol^{-1}$	1.11	VP/Solubility, EPA gives 0.628
Air-water partition coeff. K_{AW}	–	0.00045 Log is −3.35	Calculated from H/RT
Octanol water partition coeff. K_{OW}	–	100,000 Log is 5.0	EPA gives 4.7
Octanol-air partition coeff. K_{OA}	–	2.2×10^8 Log is 8.34	Log is 8.34, $K_{OA} = K_{OW}/K_{AW}$
Organic carbon-water partition coeff. K_{OC}	L/kg	8,500 Log is 3.93	EPA gives 5,860, 4,960, 7,300

Data from Mackay et al. (1997); Muir et al. (2004); Racke (1993); USEPA (2011). Values are at 25 °C unless otherwise stated

minimum, maximum, and mean or median concentrations. Such results are reported as three points. Given these limitations, only an approximate distribution of observed values can be obtained.

Concentrations in air that exceed 20 ng CYP m^{-3} were generally near sources (areas of application), while those in the range 0.01–10 ng CYP m^{-3} were regarded as "regional", corresponding to distances of up to 100 km from sources. Concentrations less than 0.01 ng CYP m^{-3} were considered to be "remote". There is a possibility that lesser concentrations could have been measured close to sources if the prevailing wind direction is not from the source region. Approximately 70% of the data for concentrations in air were in the range of 0.01–1.0 ng CPY m^{-3}. For rain, the greatest frequency (40%) was in the range 1–10 ng CPY L^{-1}. The distribution of concentrations of CPY in snow exhibited similar patterns, but with more concentrations in the range 0.01–0.1 ng CPY L^{-1}.

Physical-chemical properties of chlorpyrifos and chlorpyrifos-oxon. The model developed here was designed to describe transport and fate of CPY from source to remote destinations and thus obtain a semi-quantitative assessment of its LRT characteristics and provide estimates of exposure concentrations at remote locations. Estimates can then be compared with measured concentrations from monitoring programs. The sensitivity of the results to uncertainty in the various input parameters can also be determined. Fundamental to assessing and predicting LRT of CPY and CPYO are reliable values for physical-chemical properties and rates of reaction by different processes that determine partitioning and persistence in the environment. Data from the literature were compiled and critically assessed to obtain consistent values of these physical-chemical properties (Tables 5, 6, and 7).

Table 6 Estimated reaction half-lives of CPY in various media as used in the modeling of LRT

Medium	Value	Comment
•OH radical reaction in air	9.1×10^{-11} cm^3 molecules^{-1} s^{-1}	2nd order rate constant
Half-life in air	1.4 h	•OH radical conc. of 1.5×10^6 molecules cm^{-3}
	3.0 h	•OH radical conc. of 0.7×10^6 molecules cm^{-3}
Half-life in air	3 h	Conservative value assuming lesser concentration of •OH
Half-life in soil	7–30 d	168–720 h
Half-life in surface water	30–50 d	720–1,200 h
Half-life in sediment	50–150 d	1,200–3,600 h

Table 7 Estimated and measured physical-chemical properties of CPYO

Property	Value	Comment
Molar mass	334.6	
Vapor pressure	0.00088 Pa (USEPA 2011)	Goel et al. (2007) give 0.0000062
Solubility in water	26 mg L^{-1} (USEPA 2011)	
Octanol water partition coefficient K_{OW}	776 (USEPA 2011). Log is 2.89	Appears very low compared to K_{OW} for CPY
Half-life in air	11 h	•OH radical conc. of 1.5×10^6 molecules cm^{-3} (Aston and Seiber 1997)
	7.2 h	•OH radical conc. of 1.56×10^6 molecules cm^{-3}
Half-life in air	11 h	Conservative value assuming lesser concentration of •OH
Half-life in soil	9–30 d	
Half-life in water	13.2 d (USEPA 2011)	The more conservative value of 13.2 d was used in the modeling
	40 d, pH = 4, 20 °C	
	4.7 d, pH = 7, 20 °C	
	1.5 d, pH = 9, 20 °C (Tunink 2010)	
Half-life in sediment	132 d	10× half-life in water assumed

Note that some of these values are only illustrative and are subject to considerable error

Selected values presented for CPY in Tables 5 and 6 are well established and judged to be accurate within a factor of approximately 2 but, in some cases, ranges are given to reflect the variability and uncertainty in values. Since CPYO has been less studied, the values presented in Table 7 are subject to more uncertainty than those for CPY and must be treated as tentative.

The vapor pressure and solubility were used only to estimate the air-water partition coefficient K_{AW} and the Henry's Law constant, H. The octanol-water partition coefficient (K_{OW}) was used only indirectly to estimate the organic carbon water partition coefficient (K_{OC}) in the TAPL3 LRT model but, since there are extensive empirical data on K_{OC}, these empirical values were used directly. The octanol-air partition coefficient (K_{OA}) can be used to determine partitioning from air to aerosol particles. Its value is not used directly, but is estimated from the ratio K_{OW}/K_{AW}; however, its relatively low value proves to be less important because monitoring data confirm that CPY does not partition appreciably to aerosol particles in the environment (Yao et al. 2008) or indoors (Weschler and Nazaroff 2010).

From the perspective of LRT, the single most important parameter determining concentrations at any given location and the distance that a chemical can be transported, is transformation half-life in the atmosphere. Results of a study of the atmospheric chemistry of CPY and CPYO at the EUPHORE experimental facility in Spain have been reported showing that the principal process that transforms CPY in the atmosphere is reaction with •OH radicals, although there are also contributions from direct photolysis and reactions with ozone and nitrate radicals (Muñoz et al. 2012). In that study, a second-order rate constant for transformation of 9.1×10^{-11} cm^3 molecules^{-1} s^{-1} was determined. Combining that second order rate constant with a concentration of 1.5×10^6 •OH molecules cm^{-3} gives a first order rate constant of 13.6×10^{-5} s^{-1} which corresponds to a half-life of 1.4 h. Half-lives of CPY, thus depend directly on the assumed concentration of •OH. For CPYO, the corresponding rate constant is less certain ($0.8–2.4 \times 10^{-11}$ cm^3 molecules^{-1} s^{-1}) and was estimated to be a factor of approximately 5.5 slower. Experimental results indicated a 10–30% yield of CPYO from transformation of CPY, which is judged to be relatively small, given the absence of significant yields of other transformation products.

In their assessment of LRT, Muir et al. (2004) used the AOPWIN, structure activity (SAR) program to predict a second-order rate constant for CPY of 9.17×10^{-11} cm^3 molecules^{-1} s^{-1}, a value almost identical to that estimated by Muñoz et al. (2012). Muir et al. used a more conservative concentration of •OH that is tenfold less, which yielded an estimated half-life of 14 h (Muir et al. 2004). The lesser concentration of •OH was selected to account for concentrations of •OH likely to occur in more remote regions and at higher latitudes, for example in Canada. Global concentrations of •OH have been compiled and a concentration of 0.9×10^6 •OH molecules cm^{-3} was reported for April in the Central Valley of California and increasing to 1.46×10^6 in July and decreasing to 0.63×10^6 in October (Spivakovsky et al. 2000). At the latitude of Iowa, USA, concentrations of •OH in summer were approximately 80–85% of the concentrations observed in California. In the assessment of LRT reported here, atmospheric half-lives of 3 and 12 h were selected as being reasonable and conservative daily averages for CPY and CPYO, respectively. The actual half-lives of CPY could be a factor of two shorter, especially during midsummer daylight hours and polluted conditions when concentrations of •OH are greater. Monitoring data suggest that CPYO might have a shorter half-life. Half-lives, based on experimental data for CPY-methyl (CPY-methyl), have been reported to be in the range of 3.5 h for reactions between CPY-methyl and •OH, 15 h for direct photolysis, >8 d for reactions with ozone (O_3) and a half-life of 20 d for transformation of CPY-methyl through reactions with nitrate radicals (Munoz et al. 2011). Given the structural similarity between CPY and CPY-methyl, it is likely that similar proportions apply to both substances for reactions in the atmosphere, but not necessarily in other media such as rainwater and surface water where rates are pH-dependent.

Reported half-lives of CPY in soils vary considerably, which has been attributed to differences in soil organic carbon content, moisture, application rate and microbial activity (Racke 1993). Less data were available for water and sediments. From a critical review of the literature, the half-lives in Table 6 were selected. These are considerably shorter than those predicted by the EPIWIN program and used by Muir et al. (2004). Since these half-lives are uncertain, the selected values must be regarded as tentative, although they are not critical to the determination of potential for LRT because deposited CPY evaporates slowly. These half-lives are, however, important for assessing the extent and duration of exposures in distant water, soil, and sediment ecosystems.

Volatilization. For LRT in the atmosphere, one of the most important parameters is the rate of volatilization from surfaces of leaves and soils. Drift is also important but over shorter distances. The quantity of CPY entering the atmosphere following application is a function of several variables, including the physical-chemical properties of the formulation, whether it is applied as a liquid or granular formulation, the quantity applied, the area to which it is applied, the soil properties where applied, meteorological conditions, spray composition and related parameters and the resulting losses by spray drift. The early period after spraying and particularly 24–48 h after application is critical in determining the fraction of applied CPY that enters the atmosphere and becomes subject to LRT (Racke 1993). Relatively fast initial volatilization of applied CPY is observed in the first 12 h after application. The initial loss rate is hypothesized to result directly from volatilization of the "neat" formulated product. But, as the CPY sorbs to the substratum (e.g., foliage or soil), it becomes subject to photolysis, and the rate of volatilization decreases as a function of time. Photolysis of the formulation occurs on the surface of leaves and soils to form CPYO, which also volatilizes. These assertions are consistent with the results of the study by Zivan (2011), who demonstrated substantial rates of photolysis of CPY to CPYO on various surfaces. In the days subsequent to application, CPY adsorbs more strongly to soil, penetrates more deeply into the soil matrix, becomes less available for volatilization, and becomes subject to biological transformation processes.

The model developed here uses illustrative numerical values of quantities applied and characteristics of the environment to which it is applied. To simulate a desired application, these parameters can be varied to explore the effects of rates and conditions of application on volatilization. Results of pesticide dissipation studies that immediately followed application have been complied and reviewed by several authors (Majewski 1999; van Jaarsveld and van Pul 1999). Results of two experimental field studies are particularly applicable to this LRT study. In the first study, two techniques for direct flux measurement were applied to CPY and CPYO following application of 0.98 kg CPY(a.i.) ha^{-1} to recently cut alfalfa in the Central Valley of California (Rotondaro and Havens 2012). The Aerodynamic method gave a maximum flux of 0.657 μg m^{-2} s^{-1} (2,365 μg CPY m^{-2} h^{-1}) which decreased to 0.002 μg CPY m^{-2} s^{-1} (7.2 μg CPY m^{-2} h^{-1}) by 24 h. The Integrated Horizontal Flux method gave a maximum flux of 0.221 μg CPY m^{-2} s^{-1} (797 μg CPY m^{-2} h^{-1}), which decreased to 0.002 μg CPY m^{-2} s^{-1} (7.2 μg CPY m^{-2} h^{-1}) by 24 h following application. Total loss of CPY mass in the 12–24 h after application ranged from 15.8 to 16.5%, but diurnal variability is expected. CPYO was also observed to evaporate but at a lesser rate of 0.0164 μg m^{-2} s^{-1} for the 3–8 h period after application, which corresponds to 0.85% of the CPY applied. These results confirm that some of the CPY was transformed to CPYO on the surface and/or in the atmosphere immediately above the surface and subsequently entered the atmosphere. The average initial flux was approximately 1,500 μg CPY m^{-2} h^{-1} and decreased by a factor of approximately 200–7.2 μg CPY m^{-2} h^{-1}. In an earlier study, the eddy correlation micro-meteorological technique was used to estimate evaporation fluxes for several pesticides including CPY in the days following application in California (Woodrow and Seiber 1997). For CPY, a flux of 92.3 μg m^{-2} h^{-1} was calculated following application of 1.5 kg CPY ha^{-1}, which is equivalent to 0.15 g CPY m^{-2}. Fluxes of other pesticides were directly correlated with vapor pressure (P, Pa) and inversely proportional to K_{OC} (L kg^{-1}) as well as solubility in water (S, mg L^{-1}). The parameter described by $\frac{P}{(K_{OC}S)}$ is essentially an air/soil partition coefficient analogous to an air/water partition coefficient, thus this correlation has a sound theoretical basis. This flux of 92.3 μg CPY m^{-2} h^{-1} from a site containing 0.15 gm^{-2} corresponds to a loss of a fraction of $92.3 \times 10^{-6}/0.15$ or 615×10^{-6} per hour, which is equivalent to 0.0615% per hour or 1.4% per day. The total flux from an area of 1 ha or 10^4 m^2 is thus predicted to be approximately 0.92 g CPY h^{-1}, with a possible error judged to be a factor of 3.

In summary, it is suggested that, in the 12 h following application of the liquid formulation to the surface, approximately 10–20% of the applied material volatilizes, but variability is expected diurnally, with temperature, rainfall and soil moisture content. Sorption then "immobilizes" the CPY and subsequent volatilization is slower, with a rate of approximately 1% per day that decreases steadily to perhaps 0.1% per day in the subsequent weeks. During these periods on the surface and in the atmosphere, there is direct photolysis of CPY to CPYO. A detailed characterization of the initial 12 h period is given by Rotondaro and Havens (2012), while studies by Woodrow et al. (Woodrow and Seiber 1997; Woodrow et al. 2001) characterized average volatilization during the day or 2 following application. In the context of modeling volatilization losses, the simplest approach is to determine the total applied quantity and area treated, assume an immediate volatilization loss of 10–20% followed by a period of slower volatilization at an approximate initial rate of 1% per day decreasing with a half-life of approximately 3 d to 0.1% after 10 d. Rain and temperature will affect these rates. For illustrative modeling purposes, it was assumed that a typical rate of application is 1.5 kg CPY ha^{-1}, which corresponds to 0.15 g CPY m^{-2} (Woodrow and Seiber 1997) to an illustrative area of 1.0 ha (10^4 m^2).

Concentrations in air. Of primary interest here are concentrations of CPY in the atmosphere following application. A maximum concentration is dictated by the saturation vapor pressure of solid CPY of 0.0023 Pa, which corresponds to approximately 0.00033 g m^{-3} or 330,000 ng m^{-3} in an enclosed ecosystem. It is inconceivable that this concentration could be achieved in the field because of dilution in formulations and mass transfer limitations during evaporation. Concentrations of CPY in air above a potato field in the Netherlands at noon in midsummer ranged from 14,550 to 7,930 ng m^{-3} at 1 and 1.9 m above the crop 2 h after application (Leistra et al. 2005). These declined to a range of 2,950 to 1.84 ng m^{-3} after 8 h and to 26 to 15 ng m^{-3} in the 6 d following application. The initial flux was large (5–9 mg m^{-2} h^{-1}), possibly because of the large surface area of the leaves of this crop. As CPY is not registered for use on potatoes in the U.S., these data were not used in this assessment. Similarly high concentrations of CPY in air following an application of 4.5 kg ha^{-1} to turf were in the range of 1,000–20,000 ng m^{-3} (Vaccaro 1993). This might be a "worst case" in terms of concentrations and represents ~10% of the saturation concentration in air, i.e., the vapor pressure/RT, where RT is the gas constant-absolute temperature group. Immediately after application, concentrations of CPY of approximately 10,000 ng CPY m^{-3} (~3% of saturation) were measured at a height of 1.5 m above an alfalfa crop (Rotondaro and Havens 2012). Concentrations then decreased to approximately 100 ng m^{-3} after the initial more rapid evaporation. The USEPA conducted a modeling study to assess potential exposures of bystanders close to the site of application (USEPA 2013), but these values are not directly relevant to larger distances, in which concentrations would be much smaller because of dilution.

Concentrations of pesticides in air downwind of the site of application can, in principle, be calculated from an estimated flux by assuming a wind-speed, a mixing height, an atmospheric stability class and dimensions of the site. This is most rigorously done by using air dispersion models, such as SCREEN3 (Turner 1994; USEPA 1995). Detailed estimation of near-source concentrations in the atmosphere are beyond the scope of the simulation utilized here, which was focused on transport over distances up to 100s of km. Such estimates are nonetheless useful to estimate the order of magnitude of these "source" concentrations when monitoring data have been obtained in the vicinity of sources. The SCREEN3 model has been used to estimate concentrations in air at ground-level (1.5 m) immediately downwind, such as 10–30 m, from treated crops (Woodrow and Seiber 1997). Measured concentrations of five pesticides were of similar magnitude to predicted concentrations (μg m^{-3}) and similar in magnitude to estimated fluxes (μg m^{-2} s^{-1}), a result that is consistent with the ratio of these two parameters being approximately 1 m s^{-1}. This ratio of flux to concentration can be regarded as an effective wind-speed or mass transfer coefficient into which the evaporated chemical is diluted and is similar to the actual wind-speed of a few meter per second. Measured and simulated concentrations of pesticides in air were in good agreement. Accordingly, using this simple estimation method, ground-level concentrations in air at the site studied by Woodrow et al. are expected to be approximately 92 μg m^{-2} h^{-1} divided by a typical wind-speed of 3,600 m h^{-1}, giving 0.025 μg m^{-3}, which is 25 ng m^{-3}. This result is consistent with the above estimate. Volatilization rates of approximately 1,500 μg CPY m^{-2} h^{-1} (Rotondaro and Havens 2012) yielded a concentration of approximately 500 ng CPY m^{-3}. Concentrations would be expected to be less downwind because of dissipation by vertical and lateral atmospheric dispersion. Concentrations in the range of 100 ng CPY m^{-3} ± a factor of 10 are regarded as typical of areas immediately downwind (~1 km) of application sites, but large variability is expected from differences in rates of application, nature of the crops treated, site area and meteorological conditions, especially temperature and wind-speed.

In support of these concentration ranges, Raina et al. (2010) have reported CPY concentrations at the Canadian agricultural field site at Bratt's Lake SK in 2003 and 2005. Over a 4-d sampling period, concentrations were 1–100 ng CPY m^{-3} with some values as high as 250 ng CPY m^{-3}. These are similar to measured concentrations in the range 4–180 ng CPY m^{-3} adjacent to a citrus orchard at the Lindcove Field Station in California (Aston and Seiber 1997). Concentrations of a variety of pesticides, including CPY, have been measured at locations across Canada (Yao et al. 2008). In the intensive fruit and vegetable growing area of Vineland, Ontario, the greatest concentrations of CPY were 21.9 ng CPY m^{-3} in 2004 and 20.6 ng CPY m^{-3} in 2005. These concentrations suggest that sampling was at a site within a few km of treated areas and possibly during or shortly after application. It has been confirmed that the samples were taken immediately adjacent to the application and were timed to coincide with the application (Personal communication, Dr. T. Harner).

Volatilization from water. It is possible that some CPY enters nearby ponds or streams as a result of spray drift and run-off and subsequently evaporates from these water bodies or flows downstream. To assess the significance of this process a simple kinetic analysis was conducted using the two-resistance or two-film model. If typical water and air mass transfer coefficients (MTCs) for water to air exchange are assumed of 0.05 and 5 m h^{-1} and K_{AW} is 4.5×10^{-4}, respectively, then the water and air phase resistances are $\frac{1}{0.05}$ and $\frac{1}{5\times4.5\times10^{-4}}$ h m^{-1}, respectively, i.e., 20 and 444 h m^{-1} and the overall water phase MTC would be 0.0021 m h^{-1} as follows (1):

$$\frac{1}{(20+444)} = 0.0021\,\mathrm{m\,h^{-1}} \tag{1}$$

The primary resistance to transport thus lies in the air phase. For a water depth of 1 m, the rate constant for evaporation would be 0.0021 h^{-1} and the half-life would be 322 h, which is 13 d. This is similar to the half-lives estimated for transformation of CPY in water, which suggests that both volatilization and transformation are significant pathways of dissipation of CPY in such bodies of water. Partitioning to suspended solids and deposition to bottom sediments are also likely to remove some CPY from solution (Gebremariam et al. 2012) and reduce the volatilization rate. CPY reaching water bodies will thus be subject to other loss processes and relatively slow and delayed evaporation over a period of weeks. It is concluded that secondary volatilization from water bodies is unlikely to be significant compared with the primary volatilization immediately following application.

2.2 *Model of Long Range Transport and Characteristic Travel Distance*

Estimation of mass loss by transformation and deposition. As a parcel of air containing 100 ng CPY m^{-3} is conveyed downwind, the total mass and concentrations of CPY decrease. The mass decreases as a result of transformation processes, primarily reaction with •OH radicals and net deposition. Oxidation primarily results in the formation of CPYO. The rate of the overall process can be represented (2) as follows:

$$V \times C \times k_R \text{ or } 0.693 \times V \times C/t \quad (2)$$

Where: V is volume, C is concentration and k_R and t are the first-order rate constant (0.23 h^{-1}) and half-life (3 h), respectively. There is also loss of mass of CPY by transport from air to the ground, specifically due to deposition in rain or snow, sorption to aerosol particles that subsequently are deposited by wet and dry deposition, and direct sorption to terrestrial and aquatic surfaces as shown by Aston and Seiber (1997), LeNoir et al. (1999) and Bradford et al. (2013). Estimates of these process rates can be made and the overall results can be compared to measured concentrations of CPY. Rates of these processes can be combined into a chemical-specific net mass transfer coefficient or velocity k_M m h^{-1}. The rate of deposition is described in (3):

$$C \times k_M \, \text{g m}^{-2}\text{h}^{-1} \quad (3)$$

and the loss of mass is described by (4):

$$A \times C \times k_M \, \text{g h}^{-1} \quad (4)$$

Where: A is the area and is equivalent to the volume (V) divided by the parcel height (H, expressed in m). Thus, the rate of loss of mass of CPY is described (5) as:

$$\frac{V \times C \times k_M}{H} \text{g h}^{-1} \quad (5)$$

The parameter $\frac{k_M}{H}$ can be regarded as a rate constant. The TAPL3 model, which is discussed later, suggests that this rate constant is approximately 0.0016 h^{-1} for an atmospheric height H of 1,000 m, which is a factor of 144 slower than transformation. For deposition from a lesser atmospheric height such as 100 m, the rate constant is correspondingly greater by a factor of 10, thus there will be greater deposition from a near-ground level plume of higher concentration. The total rate of loss of mass by reaction and deposition is then described (6) as:

$$V \times C \times k_R + \frac{V \times C \times k_M}{H} \text{g h}^{-1} \quad (6)$$

The rate of change of mass M in the parcel is given by the following relationships (7):

$$\frac{dV \times C}{dt} = -\left(V \times C \times k_R + \frac{V \times C \times k_M}{H} \right) = \frac{dM}{dt} \tag{7}$$

Integration from an initial mass M_0 gives the relationship (8):

$$M = M_0^{-\left(\left(k_R + \frac{k_M}{H}\right) \times t\right)} \tag{8}$$

If a constant wind velocity U m h^{-1} is assumed, t can be replaced by $\frac{L}{U}$, where L is distance in m. The CTD (m) is defined as L when the group in the exponent is −1.0, i.e., CTD is:

$$\frac{U}{\left(k_R + \frac{k_M}{H}\right)} \tag{9}$$

The corresponding characteristic travel time (CTT, expressed in h) is:

$$\frac{1}{\left(k_R + \frac{k_M}{H}\right)} \tag{10}$$

This time has the advantage that it applies regardless of the assumed wind velocity. When L equals CTD or CTT equals:

$$\frac{1}{\left(k_R + \frac{k_M}{H}\right)}, \text{ and } \frac{C}{C_0} \text{ is } e^{-1} \tag{11}$$

or 0.368 and 63.2% of the mass is lost by transformation and net deposition. A complication arises when describing the behavior of CPY in that some of the deposited CPY re-evaporates. The actual CTD is thus somewhat longer than that calculated, but, for CPY, this is a relatively small quantity. In practice, this complication is readily addressed by calculating the CTD by an alternative, but equivalent method, which has become standard in LRT calculations. This is done by use of a multimedia, mass balance model to calculate the steady-state mass of chemical in the atmosphere of an evaluative environment which contains water and soil compartments. The only emission is to air and no advective losses from air are included; thus, the only losses from air are degrading reactions and net deposition processes, i.e., deposition and absorption less volatilization. Since the rate of input to air is

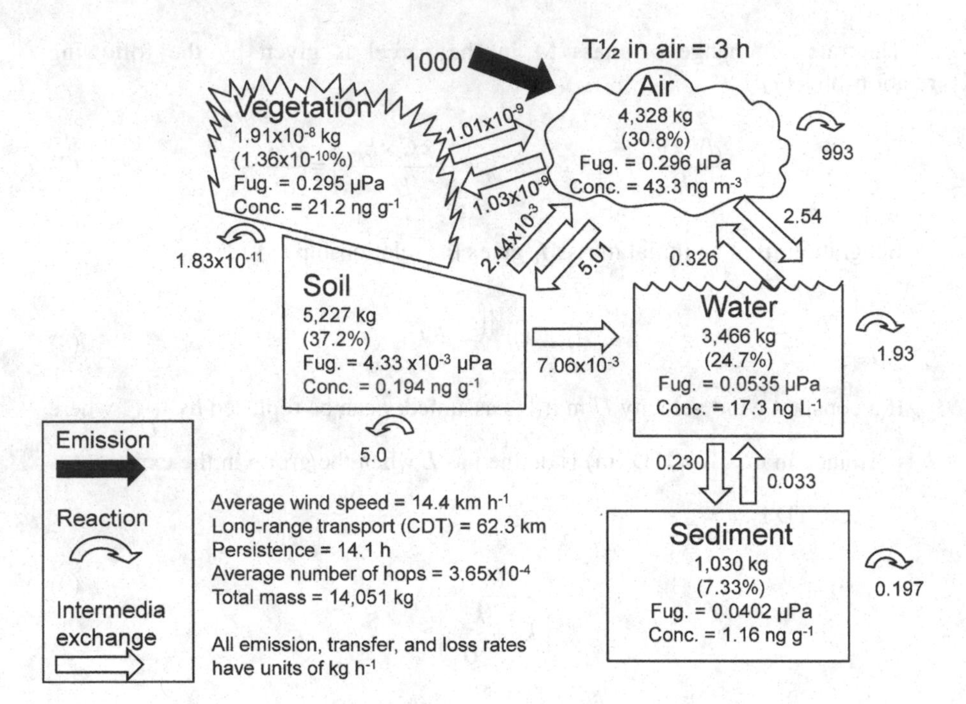

Fig. 2 Mass balance output from the TAPL3 LRT model. A 3 h half life is assumed for atmospheric degradation. The CTD is 62 km. Other parameters are as specified in Tables 5 and 6 and include the upper range (conservative) of half lives in water, soil and sediment in Table 6. The model can be downloaded from www.trentu.ca/cemc

known, that rate must equal the net rate of loss from air. Dividing the calculated mass in air by this rate gives the characteristic time defined above and this can be converted to a distance by multiplying by the wind velocity U, which is conventionally assumed to be 14.4 km h^{-1} or 4 m s^{-1}. Alternatively, the residence time or characteristic travel time (CTT) in air can be calculated as the mass in air divided by the rate of emission. This approach is used in the TAPL3 model (Beyer et al. 2000) and in the similar OECD Tool described by Wegmann et al. (2009)

The output of the TAPL3 simulation model is given diagrammatically for the selected half-life in air of 3.0 h (Fig. 2) and includes the conservative (long) half lives in other media (Table 7). The mass in air is 4,328 kg and the emission rate to air is 1,000 kg/h, thus, the residence time in air and the CTT is 4.3 h and the corresponding rate constant for total loss is 0.231 h^{-1}. The CTD is approximately 62 km, which is the product of 4.3 h and the wind velocity of 14.4 km h^{-1}. The rate of transformation is 993 kg h^{-1} and the net losses by deposition to water, vegetation, and soil total about 7 kg CPY h^{-1}, which corresponds to a rate constant of 0.0016 h^{-1}, and is less than 1% of the rate of degradation. The critical determinant of potential for LRT is the rate of transformation from reactions with •OH radicals

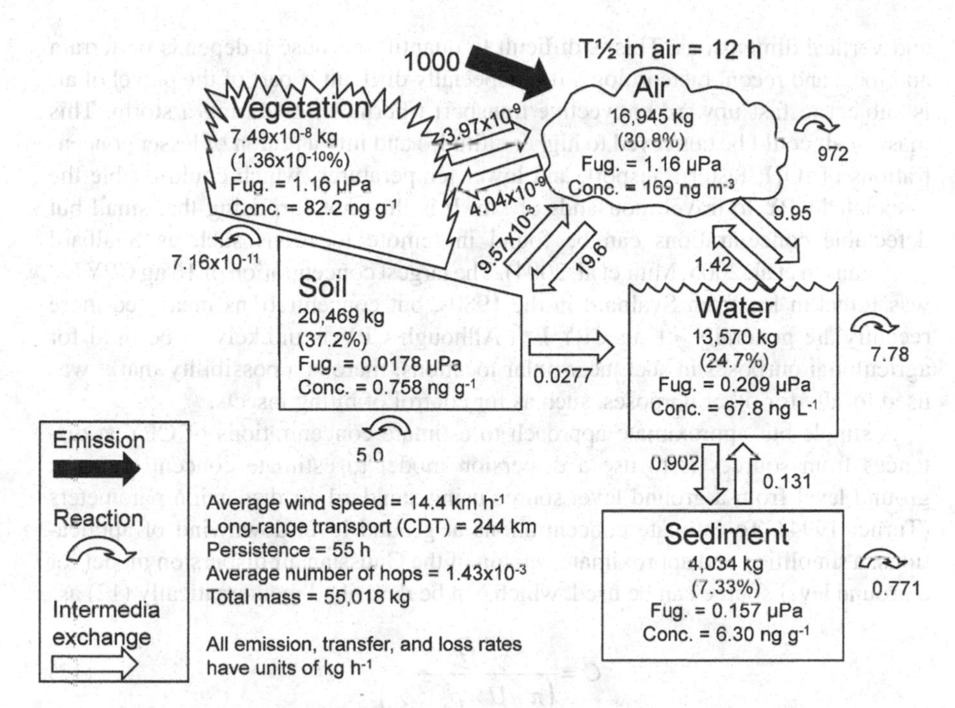

Fig. 3 Mass balance output from the TAPL3 LRT model. A 12 h (conservative) half life is assumed for atmospheric degradation. The CTD is 244 km. Other parameters are as specified in Tables 5 and 6 and include the upper range (conservative) of half lives in water, soil and sediment in Table 6. The model can be downloaded from www.trentu.ca/cemc

in air. If the half-life is increased by an arbitrary factor of 4–12 h as in Fig. 3, the CTD increases to 244 km.

In a similar study, Muir et al. (2004) estimated the CTD of CPY by using two models (TAPL3 and ELPOS) and obtained values of 290 and 283 km for conditions in which the concentration of •OH radicals was smaller, thus yielding a half-life of 14 h. Introducing intermittent rather than continuous precipitation had a negligible effect on predicted concentrations. Their longer CTDs are entirely attributable to their assumed longer half-life, which is a factor of 4.7 greater and is regarded as very conservative, but might be more appropriate for conditions at higher latitudes and during winter. Since the CTD is the distance over which the mass of chemical decreases by a factor of e (2.718), at a distance of two CTDs, the mass would be reduced by a factor of 7.4 and at three CTDs this factor is 20. Under the conditions simulated for CPY, 5% of the initial mass would remain in air at a distance of approximately 180 km if the half-life is assumed to be 3 h. If the half-life is increased to 12 h, the fraction remaining at that distance increases to 47%.

Decreases in concentration caused by dispersion/dilution. In addition to the decrease in concentration corresponding to loss of mass, there is a decrease in concentration attributable to expansion of the volume of the parcel of air in horizontal and vertical dimensions. This is difficult to quantify because it depends on terrain and local and recent meteorology. It is especially difficult if part of the parcel of air is subject to fast upward convective transport (thermals) or during a storm. This mass of air could be conveyed to higher altitudes and into a region of lesser concentrations of •OH, faster transport, and lower temperatures, which could enable the associated CPY to travel thousands of km. It is thus not surprising that small but detectable concentrations can be found in remote locations such as Svalbard (Hermanson et al. 2005; Muir et al. 2004). The largest concentration of 16 ng CPY L^{-1} was found in ice from Svalbard in the 1980s, but concentrations measured more recently are generally <1 ng CPY L^{-1}. Although CPY is unlikely to be used for agricultural purposes in such near-polar locations, there is a possibility that it was used locally for other purposes, such as for control of biting insects.

A simple but approximate approach to estimate concentrations of CPY at distances from sources is to use a dispersion model to estimate concentrations at ground level from a ground level source using standard air dispersion parameters (Turner 1994). To estimate concentrations at ground level downwind of applications, a simplified and approximate version of the Gaussian air dispersion model for a ground level source can be used, which can be described mathematically (12) as:

$$C = \frac{Q}{\left(\pi \times U \times \rho_y \times \rho_z\right)} \tag{12}$$

In (12), C is concentration (g m^{-3}), Q is emission rate (g h^{-1}), π is the mathematical constant that is the ratio of the circumference of a circle to its diameter, U is wind velocity (m h^{-1}) and ρ_y and ρ_z are respectively the horizontal (crosswind) and vertical Pasquill-Gifford dispersion parameters (m) that depend on downwind distance (km) and atmospheric stability class. This equation must be applied with caution because of variation of U as a function of height and topography, but it is used here to suggest the form of an appropriate correlation. Q can be estimated from the total quantity applied and an assumed fraction volatilized during a specified time period. Plots of ρ_y and ρ_z(m) versus downwind distance x (km) have been given (Turner 1994), and can be expressed as correlations for stability class C (13):

$$\rho_y = 100 \times x^{0.91} \text{ and } \rho_z = 61 \times x^{0.91} \tag{13}$$

For example, at 1.0, 10 and 100 km (the maximum distance) ρ_y is 100, 776 and 6,026 m, respectively and corresponding values of ρ_z are 61, 496 and 4,030 m. For an evaporation rate of 1.0 g h^{-1} into a wind of 1 m s^{-1}, the concentrations are 14 ng CPY m^{-3} at 1 km, 0.23 ng CPY m^{-3} at 10 km and 0.0037 ng CPY m^{-3} at 100 km. There is approximately an 8-fold increase in plume width and height from 1 to 10 km, and thus, there is about a 64-fold decrease in concentration. At 100 km, there is a further 61-fold decrease in concentration. For larger areas of application, concentrations of CPY would be correspondingly greater. Under other conditions of moderate atmospheric stability, e.g., categories D or B, the dispersion parameters are smaller or larger respectively by factors such as 1.5–2.0 that can be estimated from the dispersion parameter plots. If the area of application is larger by a factor such as 100, i.e., 1 km^2, then local concentrations downwind of sources would probably be greater. Horizontal dispersion then merely mixes this air and most dilution is by vertical dispersion and the dilution factor discussed above would be of the order of 10 rather than 60. There will also be contributions from evaporation from other soils in the locality that have been subject to prior applications.

Due to uncertainty in calculating concentrations from the volatilization rate Q, it is more convenient and probably more accurate to calculate downwind concentrations from an assumed concentration at, for instance, 1 km from the source i.e., $C_{1\,km}$. By applying the equation for C at 1 km and at x km and taking the ratio, the concentration at a distance x km can be shown to be $\frac{C_{1km}}{x^{1.82}}$. The quantity $x^{1.82}$ can be regarded as a dilution factor. The wind speed cancels when the ratio of concentrations is deduced. In practice, the exponent of x can be lesser, but this gives a reasonable form of the dilution equation. When applied to monitoring data it was determined that an exponent of 1.5 is more appropriate.

Combining the mass loss and the volume expansion gives the concentration downwind as a function of U, L, and the CTT, which is represented by (14):

$$C = \frac{Q}{\left(\pi \times U \times \rho_y \times \rho_z\right)} \times e^{-\left(k_R + \frac{k_M}{H}\right) \times \frac{L}{U}} \text{ or } C = \frac{Q}{\left(\pi \times U \times \rho_y \times \rho_z\right)} \times e^{-\left(k_R + \frac{k_M}{H}\right) \times CTT} \quad (14)$$

Or, more conveniently, by (15):

$$\frac{C_{1km}}{x^{1.82}} \times e^{-\left(k_R + \frac{k_M}{H}\right) \times \frac{L}{U}} \text{ or } \frac{C_{1km}}{x^{1.82}} \times e^{-\left(k_R + \frac{k_M}{H}\right) \times CTT} \quad (15)$$

It is these calculated concentrations (that do not include deposition) that can be compared with monitoring results.

Limitations in predicting concentrations downwind of a source are caused by the uncertainties inherent in the dispersion parameters, Q, U, and L, as well as the possibility that a remote region has experienced CPY transport from multiple sources. Equation (15) does, however, provide a basis for estimating concentrations of CPY in air at more remote locations as far as 100 km from the source. If desired, conservative assumptions can be applied. The effect of wind velocity can also be evaluated. Lesser wind speeds cause an increase in the initial concentration, because as the quotient $\frac{Q}{U}$ increases and transit times $\frac{L}{U}$ increase, the volatilized pesticide is more concentrated in the region of application, there is more transformation locally, and the impact of LRT would be reduced. The equation also enables the relative roles of transformation and dilution by dispersion to be assessed. For example, at relatively short distances downwind, dispersion dominates because the transit time is short relative to the half-life for transformation. At greater distances and longer transit times, transformation is more influential. The equation can also be used to estimate the fraction of the volatilized mass of CPY that will travel a given distance, or be deposited, or the fraction of the applied mass that can reach a specified distance. The quantity of transformation products can also be estimated. Results of the model can also be used to design more targeted monitoring. The model equation for C as a function of $C_{1\ km}$, distance and time is applied later to test agreement with monitoring data.

2.3 *Formation and Fate of Chlorpyrifos-Oxon*

Despite uncertainties in partitioning and reactivity of CPYO, it is possible to estimate CPYO's rate of formation and concentrations in distant atmospheres relative to CPY. These estimates can also be compared with monitoring data. It is assumed for illustrative purposes here, that CPY reacts with •OH to form CPYO in air or on surfaces with a molar yield of 30%; CPYO also reacts by the same mechanism. Half-lives are assumed to be 3 and 12 h for the reactions of CPY and CPYO, respectively. In the later evaluation, we assume a more conservative yield of 70%.

A parcel of air containing M_0 mol of CPY will change in composition with time and distance, forming CPYO, which in turn is degraded. This decay series is analogous to a radioactive decay series. The quantity of CPY (M_1) will follow first order kinetics, which can be described as (16):

$$\frac{dM_1}{dt} = -M_1 \times k_1 \tag{16}$$

This can be integrated to give (17):

$$M_1 = M_0^{(-k_1 \times t)} \tag{17}$$

Where: k_1 is the first order rate constant. The corresponding differential equation for CPYO (M_2) is given by (18):

$$\frac{dM_2}{dt} = Y \times k_1 \times M_1 - k_2 \times M_2 \tag{18}$$

Where: Y is the upper reported molar yield of 0.3, i.e., 30% and k_2 is the transformation rate constant of CPYO. It is likely that Y is larger than is stated above because other transformation products are at lesser yields. Integration of this function gives (19):

$$M_2 = \frac{Y \times k_1}{k_2 - k_1} \times M_0 \times \left(e^{-k_1 \times t} - e^{-k_2 \times t}\right) \tag{19}$$

When $k_1 < k_2$, a "secular" or "transient" equilibrium is established with an approximately constant ratio of the two species. In this case, $k_1 > k_2$ and a "no equilibrium" condition prevails in which the ratio M_2/M_1 increases monotonically with time.

Using the above relationships, half-lives, and yields, the following are the approximate quantities for an initial value of M_0 of 100 mol. After 0.46 h, when 10% of the initial CPY has degraded, 3 mol of CPYO are formed and the ratio CPYO/CPY is 0.033 (3/90). After 3 h, 50% of the initial CPY would have degraded, 14 mol of CPYO would be formed and the ratio CPYO/CPY would increase to 0.28. After 7 h, 80% of the CPY would have reacted and both M_2 and M_1 are 20 and their ratio would reach 1.0. After 10 h, M_1 is 10 and M_2 would reach its maximum value of 21, their ratio becoming 2.1. At longer times, the ratio would continue to increase, because, although M_1 and M_2 would both be decreasing, M_1 would be decreasing faster. For example, at 12 h, the ratio would be 3.2. This behavior results in the possibility that the CPYO/CPY ratio can provide insights into the approximate "age" of the air parcel, although this ratio may be influenced by conversion during sampling and prior to analysis. This ratio was observed to be approximately 1.0 in the summer of 1994 at Lindcove near Fresno CA, which suggests a transit time of ~5 h (Aston and Seiber 1997). A test using SF_6 as a tracer gave comparable transit times. At a more distant location, Ash Mountain in Sequoia National Park, the ratio increased to 7–30, corresponding to a longer transit time. At the even more distant location of Kaweah Canyon (elevation 1,920 m) the CPYO/CPY ratio was 2.7 in June to early July 1994 but later the CPY was less than the LOQ for much of the summer and only CPYO was measurable. Generally, similar results were obtained by LeNoir et al. (1999). The similar concentrations of CPY and CPYO observed in air at Lindcove were also observed in pine needles from the same location. In surface waters in the same region, concentrations of CPY exceeded those of CPYO, possibly because of faster hydrolysis of CPYO or differences in deposition rates and hydrology (LeNoir et al. 1999).

From knowledge of the kinetics or transformation, local meteorology, transit times, and atmospheric deposition characteristics, these results indicate that it is feasible to predict formation and fate of CPYO, and thus, to estimate concentrations in air and other media at distant locations. An implication is that, whereas CPY is the substance of greatest exposure and concern in areas of application, its transformation product CPYO might be of most concern in more distant locations subject to LRT. The absolute quantities of CPY transported to and retained in terrestrial media are small and the concentrations and exposures to aquatic organisms are relatively small, and much smaller than concentrations sufficient to cause toxicity (Aston and Seiber 1997; LeNoir et al. 1999). However, to quantify the risk of impacts on distant terrestrial and aquatic ecosystems, improved information is needed on the properties of CPYO and the parameters required by the simulation models. Seasonally stratified monitoring is also desirable. Concentrations of pesticides in surface water at altitudes greater than 2,040 m in the Sierra Nevada were below detection limits. This result suggests that, because of meteorological constraints, there is less effective transport to higher elevations (LeNoir et al. 1999). Concentrations also become lower because of faster wind speeds at high altitudes. The postulated "cold-condensation" effect, in which low temperatures associated with high elevations cause high deposition rates

Table 8 Estimated Z values of CPY at 25 °C used in fugacity calculations

Environmental phase	Formula	Value (mol m^{-3} Pa)	Comment
Air	1/RT	$Z_A=4.03\times10^{-4}$	R is 8.314
Water	1/H	$Z_W=0.90$	H=1.11
Octanol and lipids	$Z_O=K_{OW}/H$	$Z_O=90{,}000$	$K_{OW}=10^5$
Organic carbon	$Z_{OC}=K_{OC}Z_W\rho_{OC}$	$Z_{OC}=7{,}730$	$K_{OC}=8{,}500$ $\rho_{OC}=1.01$ (density kg L^{-1})
Soils solids of 2% OC	$Z_S=\rho_S Z_{OC} f_{OC}$	$Z_S=371$	ρ_S 2.4 kg L^{-1}, $f_{OC}=0.02$
Sediment solids of 10% OC	$Z_S=\rho_S Z_{OC} f_{OC}$	$Z_S=1{,}855$	$\rho_S=2.4$ kg L^{-1}, $f_{OC}=0.10$
Aerosol particles	$Z_P=0.1\ Z_O$	$Z_P=9{,}000$	Assumes 10% octanol equivalent
Snow Z_N	$Z_N=15\ Z_W$	$Z_N=13.5$	Assuming factor of 15 lesser Henry's Law constant at 0 °C
Biota of 100% lipid equivalent	$Z_B=Z_O$	=90,000	i.e., 100% octanol
Biota of 10% lipid equivalent	$Z_B=0.1\ Z_O$	=9,010	i.e., 10% octanol, 90% water

and greater concentrations in terrestrial and aquatic systems, does not apparently apply to transport of CPY into the Sierra Nevada mountains.

The relationship between CPY and CPYO and their transport in the atmosphere is summarized as follows: Shortly after application, a fraction of the applied CPY volatilizes to the atmosphere where it is dispersed by atmospheric turbulence to lower concentrations estimated to be of the order of 100 ng m^{-3} at a distance of 1 km. It is also subject to transformation to CPYO, which is also subject to dispersion and transport for moderate or long distances. Some CPY will be transported from the plume back to neighboring soils and vegetation by direct gas absorption; however, the resulting concentrations in soils and vegetation will be small and many orders of magnitude less than those in the application area. The vapor pressure and K_{OW} of CPYO are smaller than those of CPY and its solubility in water is greater, thus it has a smaller K_{AW}. As a result, it is subject to faster deposition and there will be enhanced partitioning into water droplets in the air. CPYO is also subject to some gaseous deposition but it is likely to be further degraded in other compartments such as water and moist solid surfaces. Once in water, hydrolysis is rapid (Table 7). This process also explains the very infrequent detection of CPYO in surface waters (Williams et al. 2014). During heavy rainfall immediately following application, local deposition will be maximized. The rates could be estimated but will be speculative and will be difficult to confirm because most locally deposited CPY will result from spray drift and it will be difficult to discriminate between gaseous deposition and spray drift.

Interpretation of measured concentrations of chlorpyrifos in media by use of fugacity. There is an incentive to exploit all the available measured concentrations of CPY for all sampled media, rather than just air. This is feasible by converting all concentrations of CPY to the "common currency" of fugacity as outlined in Tables 8 and 9 (Mackay 2001). Fugacity is the escaping tendency for chemicals to move

Table 9 Concentration-fugacity conversion factors for CPY

Environmental phase	Conversion
Air	1 ng m^{-3} = 1 × 10^{-9}/(350 × 4.03 × 10^{-4}) = 7.1 × 10^{-9} Pa = 7.1 nPa
Water and rain	1 ng L^{-1} = 1 × 10^{-9} × 1,000/(350 × 0.9) = 3.2 × 10^{-9} Pa = 3.2 nPa
Snow	1 ng g^{-1} = 1 × 10^{-9} × 1,000/(350 × 0.9 × 15) = 0.21 × 10^{-9} Pa = 0.21 nPa
Organic carbon	1 ng/g = 1.01 × 10^{6}/(350 × 7,727) = 0.37 nPa
Sediment and soil solids 2% OC	1 ng g^{-1} = 2.4 × 10^{6}/(350 × 371) = 18.4 nPa
Sediment and soil solids 10% OC	1 ng g^{-1} = 2.4 × 10^{6}/(350 × 1,854) = 3.7 nPa
Biota concentrations on a lipid weight basis	1 ng g^{-1} = 10^{6}/(350 × 90,000) = 0.032 nPa
Biota of 10% lipid or octanol equivalent	1 ng g^{-1} = 10^{6}/(350 × 9,010) = 0.32 nPa

from one environmental compartment to another and has the units of pressure. At equilibrium, fugacities of a chemical in all compartments are equal. The relative concentrations in compartments do not change and are defined by the equilibrium partition coefficients, even though individual molecules are still moving between compartments. This conversion requires first that all concentrations (C) be converted to units of mol m^{-3}, which requires that the molar mass and possibly the phase density are known. This concentration is then divided by the appropriate Z value for the medium in which CPY is partitioned. Values of Z, which have units of mol m^{-3} Pa^{-1}, are deduced from partition coefficients. This yields the fugacity, f, as C/Z, of CPY in that medium, thus enabling fugacities in a variety of phases to be compared directly. Essentially, this analysis leads to a characterization of the equilibrium status of CPY in the entire ecosystem.

In many cases, phase fugacities in multi-media environments are similar in magnitude, e.g., water, sediments and small fish might exist at comparable fugacities. An additional advantage of incorporating rain, snow, and terrestrial components in the model is that concentrations of CPY are generally greater in solid and liquid media and can be analyzed more accurately. Concentrations are generally more stable as a function of time. It is with this perspective that considerable effort has been devoted to measuring concentrations of CPY in rain, snow, terrestrial, and aquatic systems in regions of interest. Insights into likely differences in fugacity between air and other media can be obtained by examining ratios of fugacities as predicted by models such as TAPL3. For example, in Fig. 2, the fugacity of CPY in surface water is 12% of that in air, largely because the rate of transformation in water is fast relative to the rate of deposition from air. Z-values and conversion factors are given in Table 6.

Since effects of mixing, transport, and transformation generally cause a decrease in fugacity of CPY as it travels from source to destination, it is expected that measured concentrations and fugacities of CPY will display this trend. In this case, the most convenient units for fugacity are nano Pascals (nPa) i.e., 10^{-9} Pa. The fugacity of liquid CPY as applied is limited by the vapor pressure of 0.002 Pa, (2 × 10^{6} nPa), but it is likely to be smaller because of dilution in carrier fluids or granules. Incorporation of 0.15 g m^{-2} into solid phases of soils to a depth 2.5 cm or 0.025 m gives a bulk soil concentration of 0.017 mol m^{-3} and the corresponding fugacity is 46,000 nPa, a factor of 43 less than that of the applied chemical and is attributable to sorption and dilution.

A concentration of 100 ng m^{-3} in air close to a site of release corresponds to 0.286×10^{-9} mol $m^{-3}$and the fugacity would be 710 nPa. This is a factor of 64 less than the fugacity of the chemical in soil and is from dilution that occurs during evaporation. The total decrease in fugacities of CPY from the point of application is, thus, approximately 64×43 or 2,750. Most measured concentrations of CPY were in the range 0.01–1.0 ng m^{-3}, which corresponds to a range of fugacities of 0.07–7 nPa, a factor of 100–10,000-foldless than that of the initial concentrations of 100 ng m^{-3}. Therefore, CPY undergoes high dilution in the hundreds of km downwind of the source.

The concentration of CPY in rain of approximately 0.4 ng CPY L^{-1} or 400 ng m^{-3} that was reported by Mast et al. (2012) corresponds to approximately 1.1×10^{-9} mol m^{-3} and a fugacity of 1.32 nPa. The corresponding equilibrium concentration in air is 0.18 ng m^{-3} which is typical of concentrations in air in the Sierra Nevada. Fugacities of CPY in air and rain thus appear to be of a similar order of magnitude, which lends support to the use of fugacity as a method of combining and comparing measured concentrations among media.

Conversion of concentrations of CPY in snow to fugacities is more problematic because the Z value for snow is uncertain. This is because the low temperatures and the variable sorption to ice surfaces as distinct from partitioning to liquid water. There might also be greater deposition of aerosols in snow at lower temperatures. Concentrations of CPY in snow were reported to be approximately tenfold greater in snow than in rain (Mast et al. 2012). This result is consistent with the greater Z value, which is due to the lesser Henry's Constant and vapor pressure of CPY. The enthalpy of vaporization, which has been reported to be 73 kJ mol^{-1} for CPY (Goel et al. 2007) corresponds to a 15-fold decrease in vapor pressure from 25 to 0 °C. The value of Z for snow appears to be a factor of 10–20-fold greater than that of water. For this reason, rates of deposition of CPY associated with snow are expected to be greater than those in rain from a similar atmospheric concentration. Snow concentrates and integrates CPY more than does rain and can be useful for monitoring the presence of CPY, but using this information quantitatively is problematic because of uncertainties in translating concentrations of CPY in air to those in snow, especially for more intense snow-fall events when extensive scavenging of chemicals from the atmosphere occurs.

Concentrations in biota such as zooplankton, tadpoles, lichen, and pine needles can also be converted to fugacities by assuming a content of lipid, or more correctly an equivalent content of octanol. If data are reported on a lipid weight basis, conversion to fugacity involves division by the Z value of lipid or octanol. The average CPY lipid-based concentration in tadpoles from the Sierra Nevada in 2008–2009 has been reported to be 22.2 ng CPY g^{-1} (Mast et al. 2012). The corresponding fugacity of CPY is 0.7 nPa, which is similar in magnitude to the fugacities of air and rain.

Concentrations of pesticides were measured in frogs at 7 high elevation sites in the Sierra Nevada in 2009 and 2010 (Smalling et al. 2013). Although CPY was one of the most heavily used pesticides in the area, it was not detected in frog tissues above the LOD of 0.5 ng g^{-1}. In comparison, *p,p'*-DDE was widely detected with a 75th centile of 40 ng g^{-1} and the fungicide, tebuconazole was detected with a 75th centile of 120 ng g^{-1}. Concentrations of CPY in zooplankton in lakes in Ontario, expressed on a wet weight basis, have been reported to be 0.004 ng CPY g^{-1}wwt, but concentrations as great as 0.08 ng CPY g^{-1} can occur (Kurt-Karakus et al. 2011). Corresponding concentrations, normalized to the fraction of lipid (2%) in zooplankton results in a range of concentrations of 0.2 and to 4 ng CPY g^{-1} lipid in lakes distant from points of application of CPY. Corresponding fugacities for this range of concentrations are 0.0064 and 0.13 nPa. Bioconcentration factors (BCF) are considerably smaller than would be predicted from the octanol-water partition coefficient (K_{OW}) or from estimations based on simulation models such as BCFWIN. These lesser values for site-specific BCF calculated from measured concentrations are likely attributable to biotransformation. Aston and Seiber (1997) obtained pine needle/air bioconcentration factors of 9,800 of CPY that might be a function of the octanol/air partition coefficient and the quantity of lipid-like material in the cuticle. In summary, fugacity can act as a bridge between monitored concentrations in biota, air, and precipitation in regions subject to LRT in the atmosphere. The corollary is that estimated concentrations in air can be used to estimate concentrations in biota and possibly contribute to assessments of risk of adverse effects.

2.4 *Long-Range Atmospheric Transport of Chlorpyrifos and Its Oxon*

It is useful to present a perspective on the relevant distances in regions of the U.S. that have been monitored for CPY and CPYO. Much of the available data have been collected from the Central Valley of CA and adjacent National Parks in the Sierra Nevada. The Parks are 50–100 km from the areas of application in the Central Valley and have altitudes from 600 to 4,000 m. The region is approximately 50 km west of the border between California and Nevada, but the meteorology at higher elevations is complex and simple estimates of concentration versus distance are impossible. In Eastern and Midwest regions of the U.S., distances relative to application areas are less defined and are probably several hundreds of km. For example, the distance from central Iowa to the U.S. East Coast is approximately 1,000 km.

An example of monitored concentrations along a transect from source to destination is the work of Aston and Seiber (1997), who measured concentrations of CPY in June 1994 over a transect from Lindcove, CA (elevation 114 m) to Ash Mountain 22 km distant (elevation 533 m) and to Kaweah a further 10 km distant (elevation 1,920 m). Concentrations decreased from approximately 100 ng CPY m^{-3} at Lindcove to 0.1–0.5 ng CPY m^{-3} at Ash Mountain and to 0.1–0.3 ng CPY m^{-3} at Kaweah.

Those authors also present tracer data for SF_6 that suggest a dilution factor of 100 from a source 9 km SW of Lindcove to Ash Mountain, i.e., a distance of approximately 31 km. It is monitoring data of this type that can provide quantitative information on LRT and assist in calibrating models.

A semi-quantitative interpretation of measured concentrations of CPY and CPYO, assisted by use of the fate and transport model developed in this study is provided here including the effects of transformation, transport, and dispersion/dilution processes on downwind concentrations. A half-life of CPY of 3 h in air (Table 6) is assumed, but to test the sensitivity of the results to this half-life, the effect of a value of 12 h is also used. A wind speed of 15 km h^{-1} (4.16 m s^{-1}) is assumed for estimating the CTD. In the model, the concentration at a distance downwind C_L and distance x km can be estimated from the concentration $C_{1\,km}$ at 1 km by (20) to give C_L as:

$$\frac{C_{1km}}{x^N} \times e^{-\left(k_R + \frac{k_M}{H}\right) \times \frac{L}{U}} \text{ or } \frac{C_{1km}}{x^N} \times e^{-\left(k_R + \frac{k_M}{H}\right) \times CTT} \tag{20}$$

The exponent N is assigned an illustrative value of 1.5, $\frac{k_M}{H}$ is assigned a value of 0.002 h^{-1}, which is small in comparison to the reaction rate constant of 0.231 h^{-1}. $\frac{L}{U}$ is the transit time in h. The rate constant for CPYO reaction was increased to 0.139 h^{-1}, i.e., a half-life of 5 h and the yield of CPYO from CPY was increased to 70%. The CTD of CPY occurs when $\left(-\left(k_R + \frac{k_M}{H}\right) \times \frac{L}{U}\right)$ is 1.0, or equivalently when L is $\frac{U}{\left(k_R + \frac{k_M}{H}\right)}$.

These parameter assignments were selected by comparing available monitoring data to predicted values from the simulation model and adjusting parameters by hand until the selected input parameters resulted in simulated results that were comparable to measured concentrations. The objective was not to rigorously calibrate the model, but rather to test the feasibility of developing and applying the LRT model to estimate concentrations of CPY and CPYO at locations remote from site of application. The results of applying the model developed in this study are summarized for CPY (Table 10) and illustrated for CPY and CPYO (Fig. 4). Near the area of application, such as at a distance of 1 km and assuming a 0.1 h air transit time, air concentrations ($C_{1\,km}$) were assigned a value of 100 ng CPY m^{-3} (~700 nPa). At these short transit times, relatively little of the CPY would have been transformed, although there might be transformation to CPYO on the surface and adjacent atmosphere if conditions are sunny and favor greater concentrations of •OH. Concentrations of CPY are primarily controlled by rates of evaporation and dispersion rather than reactions with •OH.

At a distance of 120 km and 8.4 h transit time, which is equivalent to two CTDs, 84% of the volatilized CPY would have been transformed and 16% would remain.

Table 10 Estimates of the transformation of CPY and concentrations in air at various distances downwind of an application

Distance km	Transit time h	F (reacted)	Conc. ng m^{-3}	Comment
1 3	0.1–0.2	<0.05	20–100	Application area
10	0.67	0.14	5	Local
30	2	0.38	0.7	Regional
60	4	0.62	0.15	One CTD
120	8	0.84	0.022	Two CTDs
180	12	0.94	0.005	Three CTDs
240	16	0.98	0.001	Four CTDs
300	20	0.99	0.0003	Five CTDs
1,000	67	>0.999	<0.0001	Fifteen CTDs

A wind speed of 14.4 km h^{-1} is assumed. The fraction reacted, F, is calculated assuming a half-life of 3 h as $e^{-0.231 \times t}$ where t is the transit time and the transformation rate constant is 0.231 h^{-1}

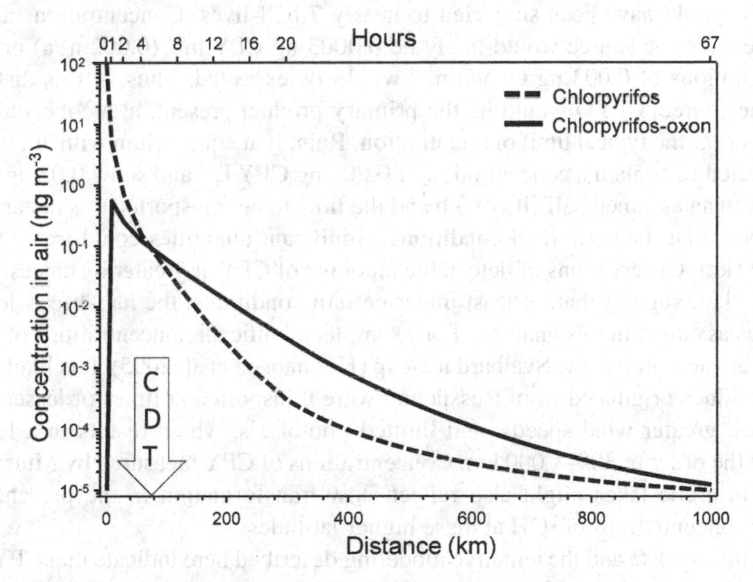

Fig. 4 Concentrations of CPY and CPYO modelled at various times and distances downwind from an application

Concentrations of CPY in air would have decreased to 0.022 ng CPY m^{-3} (0.16 nPa). At this distance, transformation would have become a greater proportion of the total dissipation, and concentrations of CPYO would be expected to exceed those of CPY by a factor of 2, but may be affected by differing deposition rates. At steady state, rain water would be predicted to have a concentration of 0.1 ng CPY L^{-1} and snow a concentration of 1.5 ng CPY L^{-1}. If a very conservative CPY half-life of 12 h were assumed, the fraction of CPY transformed would be only 38% and much greater concentrations are expected. At a distance of 180 km and 12 h transit time, that is equivalent to three CTDs, 94% of CPY would be predicted to have been transformed with only 4% remaining. Concentrations of CPY would be approximately 0.005 ng m^{-3} (0.035 nPa). Approximately 70% of the concentrations measured in air are in the range of 0.01–1.0 ng CPY m^{-3} and probably correspond to distances from sources of 30–200 km. Predicted concentrations in rain at steady state would be 0.02–2.0 ng CPY L^{1} and those in snow would be 0.3–30 ng CPY L^{-1}, with some 39% of the reported concentrations in snow being in this range. Most of these data are restricted to one region, the Sierra Nevada Mountains in the U.S. Predicted fugacities and concentrations in snow are speculative since the air/snow partition coefficient is uncertain and concentrations are undoubtedly influenced by timing of the snowfall relative to applications. Heavier snowfall, such as occurs in the Sierra Nevada might result in dilution in the precipitation and near-total scavenging of CPY from the atmosphere.

At a distance of 300 km and about 20 h transit time, which is equivalent to approximately five CTDs, 1.0% of the initial mass of CPY would remain because the CPY would have been subjected to nearly 7 half-lives. Concentrations at this distance from the source would likely be 0.0003 ng CPY m^{-3} (0.002 nPa) or less. Concentrations of 0.003 ng CPYO m^{-3} would be expected. Thus, at this distance from the source, CPYO would be the primary product present, at a concentration which is near the typical limit of quantitation. Rain, if at equilibrium with air, would be expected to contain a concentration of 0.001 ng CPY L^{-1} and snow 0.02 ng CPY L^{-1}. Given an assumed half-life of 3 h and the time to be transported this distance, it is unlikely that, under normal conditions, significant quantities could travel more than 300 km. Observations of detectable amounts of CPY at greater distances, such as 1,000 km, suggest that, at least under certain conditions, the half-life is longer than was assumed in this analysis. For example, significant concentrations of CPY have been measured in the Svalbard ice-cap (Hermanson et al. 2005). It is likely that these residues originated from Russia and were transported at times of lesser temperatures, greater wind speeds, and limited photolysis, which results in a longer CTD of the order of 300–1,000 km. Concentrations of CPY measured by Muir et al. (2004) in arctic lakes might also reflect slow transformation in the presence of smaller concentrations of •OH at these higher latitudes.

Monitoring data and the tentative modeling described here indicate that CPY and CPYO are detectable in air at concentrations exceeding 0.1 ng m^{-3} at distances of up to 60 km from the source and at 0.01 ng m^{-3} at distances up to 200 km, except in the Sierra Nevada where there are meteorological constraints on flows of air masses. There will be corresponding concentrations in rain, snow, and in terrestrial media such as pine needles and biota. There is an incentive to monitor these media because of the greater concentrations and increased analytical reliability. The "zone of potential influence" of LRT in this case is one to two CTDs or up to 60–120 km from the point of application. Reactivities of CPY and CPYO are such that concerns about LRT are much more localized than for organochlorines, which are more persistent and thus might have CTDs of thousands of km. The results of the analysis presented here suggest that it is feasible to extend assessments of LRT beyond the mere estimation of CTD and CTT to address the magnitude of the concentrations and fugacities along a typical LRT transect and to estimate absolute multi-media concentrations and deposition rates. There is also a need to focus more on the transformation products such as CPYO, but major uncertainties exist about the formation rates and properties of transformation products which preclude full interpretation of monitoring data and modeling. It is likely that any risks associated with LRT are attributable more to CPYO than to CPY; however, the concentrations predicted in air and water are much smaller than toxicity values for either of these compounds (Giddings et al. 2014) and risks are *de minimis*. The proposed model can also be applied to gain an understanding of the likely effects of the various parameters such as wind speed and temperature.

3 Fate in Water

The fate of CPY in water was extensively reviewed by Racke (1993), and data are provided in (Solomon et al. 2013a); key points are summarized here with a focus on information that has become available since 1993. As discussed above, there are significant differences between dissipation and degradation of CPY in water, but earlier studies did not always distinguish between dissipation and degradation. In the laboratory, and in the absence of modifiers such as methanol, reported half-lives (DT_{50deg}) for hydrolysis in distilled and natural waters ranged from 1.5 to 142 d (SI Table 1) at pH values between 5 and 9 (Racke 1993), which are considered to represent realistic field values. The mean half-life of these values was 46 d and the geometric mean was 29 d. At pH <5, reported half-lives were generally longer (16–210 d) and at pH >9, shorter (0.1–10 d). The presence of copper (Cu^{++}) resulted in shorter half-lives (<1 d), even at pH <5 (Racke 1993). In studies published since 2000, similar half-lives have been reported (SI Table 1). A DT_{50deg} of 40 d for CPY was reported in distilled water but DT_{50deg} (120 to 40 d) varied in sterile natural waters from rivers flowing into Chesapeake Bay. Concentration of Cu^{++} was a major driver of rate of hydrolysis, although other factors such as salinity were also identified (Liu et al. 2001). Concentrations of total suspended solids (TSS) greater than 10 mg L^{-1} resulted in lesser rates of hydrolysis of CPY, but dissolved organic carbon did not affect the rate. In water, CPY has been shown to bind strongly with variable strength and reversibility to Ca-saturated reference smectites but strongly and with poor reversibility to Ca-saturated humic acid (from Aldrich) (Wu and Laird 2004). The binding to suspended clays might explain the effect of TSS on hydrolysis rate observed by Liu et al. Half-lives from the newer laboratory-studies ranged from 1.3 to 126 d with a mean and geometric mean of 23 and 13 d, respectively (SI Table 1). The overall mean and geometric mean were 37 and 21 d, respectively (SI Table 1).

Under field conditions, it is difficult to separate degradation from dissipation and the half-lives measured are normally based on the latter (DT_{50dis}). A number of reports have noted relatively rapid dissipation of CPY in microcosms. DT_{50dis} of 9.6–6.1 d in microcosms treated with 0.005–5 μg L^{-1} were reported in small laboratory-based studies conducted in mesocosms in the Netherlands (Daam and Van den Brink 2007). However, smaller DT_{50dis} values (<4 d) were reported for outdoor mesocosms treated with 1 μg L^{-1} in Thailand (Daam et al. 2008). Using small (70-L) open-air estuarine microcosms to investigate dissipation of ^{14}C CPY, a DT_{50dis} of ~5 d was reported under tropical conditions with loss to air a major driver of dissipation (Nhan et al. 2002). In studies conducted in flowing, outdoor mesocosms, a DT_{50dis} was reported to be <1 d, probably as a result of hydraulic dilution. However, in still-water-only laboratory mesocosms, DT_{50dis} ranged from 10 to 18 d (Pablo et al. 2008). Dissipation in small (2.4-L) laboratory microcosms with water and gravel was biphasic with a phase-1 DT_{50dis} of 2.25–3 d and a phase-2 DT_{50dis} of 14–18 d (Pablo et al. 2008). DT_{50dis} of CPY in microcosms was reported to be ~5 d from water (Bromilow et al. 2006). Overall, dissipation of CPY in natural waters under field conditions was rapid with the range of DT_{50dis} s from 4 to 10 with a geometric mean of 5 d (SI Table 1).

4 Fate in Soils and Sediments

Studies on the fate of CPY in soils and sediments were summarized in the review by Racke (1993) and discussed in the context of adsorption and desorption in a detailed review in 2012 (Gebremariam et al. 2012). Most of the half-lives in soil (DT_{50dis} and DT_{50deg}) summarized from laboratory studies in Racke (1993) were in the range of 1.9–120 d for rates of application associated with agricultural uses, with most in the range of 7–30 d (Table 6). Longer half-lives (DT_{50deg}) were reported for rates of application for the now-cancelled use for control of termites in soil. Half-lives in soil were dependent on temperature (a doubling in rate of degradation for a 10 °C increase in temperature) and soil pH, with faster rates at greater pH (0.0025 d^{-1} at pH 3.8 to 0.045 d^{-1} at pH 8) (Racke 1993). Mean and geometric mean values for all data (SI Table 2) were 82 and 32 d, respectively.

Generally, dissipation (DT_{50dis}) of CPY in soils under field conditions was reported to be more rapid than in the laboratory. The DT_{50dis} was reported to range from <2 to 120, with mean and geometric means of 32 and 22 d, respectively (SI Table 3); most values were in the range of 7–30 d (Table 6). Comparison of rates of dissipation of CPY from soils from Brazil under laboratory conditions suggested a tenfold greater rate of dissipation in the field than in the laboratory (Laabs et al. 2002).

Half-lives (DT_{50deg}) in sediments were reported to range from 6 to 223 d (SI Table 4), with longer times likely reflecting more anaerobic conditions. Some more recent studies have reported dissipation of CPY from sediments in microcosms, a more realistic scenario. The DT_{50dis} values for CPY were reported to range from 68 to 144 d in wetland sediments under flooded conditions (Budd et al. 2011). Measurements of dissipation of CPY from sediments collected in San Diego and Bonita Creeks (Orange County, CA, USA) gave DT_{50dis} values of 20 and 24 d under aerobic and 223 and 58 d under anaerobic conditions, respectively (Bondarenko and Gan 2004). DT_{50dis} of CPY in microcosms was reported to be 15–20 d from sediment (Bromilow et al. 2006). The DT_{50dis} value measured in sediment in a laboratory-based marine microcosm study was approximately 6 d under tropical conditions

(Lalah et al. 2003), but was likely overestimated because metabolites were not separated from the ^{14}C CPY. The DT_{50dis} values measured in pore-water ranged from 7 to 14 d in water-gravel laboratory-based microcosms that were treated with 0.2–20 μg CPY L^{-1} (Pablo et al. 2008). The mean and geometric mean DT_{50dis}s for CPY in laboratory and microcosm tests were 68 and 39 d, respectively (SI Table 4).

5 Fate in Organisms

The fate of CPY in organisms is a function of absorption, distribution, metabolism and excretion (ADME) and has been well studied in mammals (Testai et al. 2010). Observations have also been recorded for other animals such as fish (Racke 1993; Barron and Woodburn 1995), aquatic organisms (Giesy et al. 1999) and birds (Solomon et al. 2001). The focus in this paper is on newer studies, and only key information from older studies will be addressed. Integration of the processes of ADME in organisms at quasi-equilibrium is described by several factors, which are ratios between abiotic and biotic compartments. These include bioconcentration factors (BCF), bioaccumulation factors (BAFs), biota/sediment accumulation factors (BSAFs) and, in the case of movement in the food web, biomagnification (BMFs) or trophic magnification factors (TMFs) (Gobas et al. 2009).

Several studies have been conducted in aquatic organisms to measure concentrations of CPY in fish and other organisms during uptake, at equilibrium, and during dissipation. These have been used to calculate various magnification factors. Bioconcentration factors (BCFs) reported from laboratory studies reviewed by Racke (1993) and Barron and Woodburn (1995) in 17 species of freshwater (FW) and saltwater fish exposed to CPY at concentrations <10 μg/L for ≥26 d ranged from 396 to 5,100 with a mean of 1,129 and a geometric mean of 848 (SI Table 5). Similar values were observed in several studies conducted in microcosms or ponds under field conditions, which also have been reviewed in Racke (1993). Here the mean BCF was 1,734 and geometric mean 935 (SI Table 5). Assuming a K_{OW} of 100,000 and a lipid content of 5% suggests an equilibrium BCF of 5,000, but lower than equilibrium values can be expected as a result of metabolic conversion and slow uptake.

Several studies on uptake of CPY from water and sediments have been reported since 2000 (Table 11). Results of several other recently-published studies were not usable. Two studies of marine clams were conducted using ^{14}C-CPY but results were only reported as percentages (Kale et al. 2002; Nhan et al. 2002) and BCFs could not be calculated. Uptake of CPY from water by the fish, hybrid red tilapia, was measured by gas-chromatography (Thomas and Mansingh 2002) but a BCF could not be calculated. A study of uptake and depuration of ^{14}C-CPY reported BCFs for 15 species of FW aquatic invertebrates (Rubach et al. 2010). Unfortunately, the BCFs were based on total ^{14}C in the organisms and, because the ^{14}C-label was in the di-ethyl-phosphorothiol moiety of the CPY molecule, radioactivity measured in the organisms did not represent only CPY, but included other phosphorylated proteins such as AChE, BuChE, and paraoxonase. Therefore, as has been pointed

Table 11 Bioaccumulation factors for CPY in various biota from studies conducted after 2000

Species	Taxon	BCF/BAF/BSAF	Time of exposure (d)	Exposure-concentration	Reference	Comment
Crassostrea virginica	Mollusk	BCF ± SD 565 ± 172	28-exposure, 14-depuration	0.61 μg/L	Woodburn et al. (2003)	Determined using [2,6-pyridine ^{14}C]-CPY
Lumbriculus variagatus	Oligochaete	BAF 1.8 13.9 17.2 57.3 BSAF 67.1 99.2 34.9 6.2	12	385 54 25.4 29.9 μg/kg dwt in sediment	Jantunen et al. (2008)	Determined using ^{14}C-pyridine-labelled CPY. TOC in sediments = 21.5, 3.5, 1.7, 0.13, resp. BSAF normalized for lipid and OC
Aphanius iberus	Fish	BCF 3.1	3	3.2 μg/L	Varo et al. (2000)	Approaching equilibrium at 72 h. Analysis by GLC
Danio rerio eleutheroembryos	Fish	BCF 3,548 6,918 Geometric mean = 4,954	2-exposure 3-depuration	1 μg/L 10 μg/L	El-Amrani et al. (2012)	Determined using GLC analysis. Results might overestimate BCFs—metabolism not fully developed
Aphanius iberus	Fish	BMF ± SD 0.30 ± 0.2	32	94 ± 41 ng/g lipid wt	Varo et al. (2002)	Lipid corrected based on conc. in food, analysis by GLC

out previously (Ashauer et al. 2012), these data were unusable. Uptake of CPY was rapid in *Gammarus pulex* (Ashauer et al. 2012), with equilibrium reached in less than 1 d. Formation of an unidentified metabolite and CPYO were rapid with rate constants of 3.5 and 0.132 d^{-1}, respectively. The elimination rate constant CPYO in *G. pulex* was 0.298 d^{-1}. Because the ^{14}C-label was in the Et-O moieties in the molecule, a BCF for CPY could not be calculated. In a study in laboratory-based marine microcosms, BCFs of 89–278 and 95–460 were reported in oysters and fish, respectively (Lalah et al. 2003), but were likely overestimated as the metabolites were not separated from the ^{14}C-CPY. Studies with usable results demonstrate that in most cases, BCFs, BAFs, and BSAFs are small (<2,000) and not indicative of bioaccumulation or toxicologically significant exposures to predators via the food chain (Table 11). The two new reports of BCFs from fish (Table 11) were based on very short exposures (≤3 d), and thus, cannot be compared or combined with the studies reviewed by Racke (1993), which were conducted for ≥28 d. The one report of a BMF (0.32 in the fish *Aphaniusiberus*) was based on an exposure of 32 d and is not indicative of biomagnification (Varo et al. 2002). One study in eleutheroembryos of *Danio rerio* reported a BCF value greater than 2,000 (El-Amrani et al. 2012), most likely because metabolic capacity in this early a stage of development is not fully developed.

6 Assessment of Chlorpyrifos as a POP or PBT

The Stockholm convention (United Nations Environmental Programme 2001) and the UN-ECE POP Protocol (United Nations Economic Commission for Europe 1998) was established to identify and manage organic chemicals that are persistent, bioacumulative, and toxic (PBT), in that they have the potential to exceed the threshold for toxicity, and to be transported to remote regions (persistent organic pollutants, POPs). Classification criteria for POPs were developed from the physical, chemical, biological, and environmental properties of the so-called "dirty dozen" (Ritter et al. 1995a, b) and are based on trigger values for persistence (P) bioaccumulation (B), toxicity (T), and propensity for long range transport (LRT) (Table 12). Several other initiatives to assess chemicals for properties that might confer P, B, and T have been put in place. These are the Convention for the Protection of the Marine Environment of the North-East Atlantic (OSPAR 1992), the Toxic Substances Management Policy (Environment Canada 1995), the Toxics Release Inventory Reporting (USEPA 1999a), the New Chemicals Program (USEPA 1999b), REACH (European Community 2011). These initiatives exclude pesticides but EC regulation No. 1107/2009 (European Community 2009) is specifically directed towards pesticides and is the focus of further discussion here. EC regulation No. 1107 uses classification criteria similar to those of the POPs (Table 12), but these are somewhat more conservative for P and B. The criteria used to classify the PBT character of pesticides under EC regulation No. 1107/2009 are simple (Table 12); the process is basically a hazard assessment that does not make full use of the rich

Table 12 Criteria for the categorization of compounds as POPs or PBT

Persistent (P)	Bioaccumulative (B)	Toxicity (T)	Potential for long-range transport (LRT)
POP (Stockholm Convention)			
Water: DT_{50}> 2 months Sediment: DT_{50}> 6 months Soil: DT_{50}> 6 months Other evidence of persistence	BCF >5,000 or Log K_{OW}>5 Other, e.g., very toxic or bioaccumulation in nontarget species. *Trigger values for BMF, BAF, and BSAF not available*	No specific criteria other than "significant adverse effects"	Air: DT_{50} >2 d or monitoring modeling or data that shows long-range transport
PBT (EC No. 1107/2009)			
Marine water: t½ >60 d Fresh water t½ >40 d Marine sediment: t½ >180 d Freshwater sediment: t½ >120 d Soil: t½ >120 d	BCF >2,000 in aquatic species. *Trigger values for BMF, BAF, and BSAF not available*	Chronic NOEC <0.01 mg/L or is a carcinogen, mutagen, or toxic for reproduction, or other evidence of toxicity. *Trigger values for non-aquatic species not available*	None

From European Community (2009); United Nations Economic Commission for Europe (1998); United Nations Environmental Programme (2001). Author comments in *italics*

set of data available for risk assessment of pesticides (Solomon et al. 2013). In addition, as has been pointed out previously, classification criteria are inconsistent for PBT among regulations in various jurisdictions (Moermond et al. 2011) and, in some cases, appropriate criteria and/or guidance are not provided (Solomon et al. 2013). Under EC regulation No. 1107/2009, exceeding trigger values for P, B, and T results in a ban and, exceeding two of three, results in being listed for substitution with pesticides that are less P, B, and/or T. In the sections below, we assess CPY as a POP and PBT based on criteria for POPs (Stockholm) and PBT (EC regulation No. 1107/2009). To our knowledge, neither CPY nor CPYO are officially being considered for classification as POPs or PBTs, although, some have suggested that they be considered (Watts 2012).

In assessing P and B for chemicals, the concern is for the general environment, not for a particular local scenario. Because extreme values that are observed in specific situations are not representative of all locations, it is best to use mean values. Moreover, because many P or B processes are driven by first-order kinetics, the geometric mean value is most appropriate for comparing triggers for classification. Accordingly, these were used in the following sections.

Classification as a POP. CPY does not meet the criteria for P, B, and LRT for classification as a POP. Persistence in water, sediment, and soil (Table 6) is less than the trigger values (Sects. 3 and 4, above), and there is no evidence to suggest ecologically significant persistence in the environments of use. The geometric mean of half-life values in water tested in the laboratory was 21 d. Half-lives in water in the presence of sediments in the field were even smaller (geometric mean of 5 d). These are less than the trigger values of 60 d for POPs. Geometric mean half-lives in soil tested under laboratory and field conditions had geometric means of 32 and 22 d, respectively, both of which are less than the 180 d trigger. The geometric mean DT_{50dis}s for CPY in sediments tested in the laboratory and microcosms was 39 d, which is less than the trigger value of 180 d.

Geometric means of values for BCF, BAF, and BSAF measured in the laboratory (assumed to be equivalent, Sect. 5, and Table 11) and the field were 848 and 935, respectively, all of which were less than the trigger value of 5,000. Studies of trophic magnification of CPY in the field were not found in the literature but, based on food-chain magnification measured in model ecosystems with ^{14}C-labelled material (Metcalf and Sanborn 1975), CPY does not magnify to the same extent as any of the currently identified POPs that were also tested.

The criterion for toxicity, "significant adverse effects", used to classify chemicals a POP, is somewhat vague (Solomon et al. 2009) in that specific numerical criteria are not provided. All pesticides are toxic to some organisms; otherwise they would not be used. However, in the context of POPs and LRT, adverse effects are more properly interpreted as ecologically significant outcomes on survival, growth, development, and reproduction in organisms well outside the boundaries of the site of application. Based on the conclusions of several of the companion papers, CPY does not exceed the trigger for "significant adverse effects" in or outside the regions of use (Cutler et al. 2014; Giddings et al. 2014; Moore et al. 2014).

The half-life of CPY in the atmosphere of 1.4 d does not exceed the trigger value of 2 d for LRT. CPY is found at distances from areas of application, and even in remote locations (Sect. 2.1); but the concentrations in air (Table 1), rain and snow (Table 2), aquatic and terrestrial media (Table 3), and biota (Table 4) are small and less than the threshold of toxicity for aquatic organisms and birds (Giddings et al. 2014; Moore et al. 2014). Even assuming a longer half life of 3 d as was done earlier, the concentrations at remote locations are low and do not approach toxicity thresholds.

Assessment of CPYO as a POP is complicated by the fact that it is a degradation product of CPY and is usually present with the parent material in the environment as well as during tests of effects of CPY. By itself, CPYO also does not exceed the triggers for POPs (Table 12) with respect to persistence in water, soil, and sediment (Table 7). No data were available for BCF of CPYO, but studies with ring-labelled CPY provide equivalency for CPYO and it does not trigger the criterion for B. Toxicity for CPYO is subsumed in that of CPY and it does not trigger "significant adverse effects". The half-life in air of approximately 11 h (Table 7) is less than 25% of the LRT trigger of 2 d. In addition, replacement of the =S with =O in CPYO increases polarity; CPYO is about 25-fold more water soluble, and has a K_{OW} that is 100-fold smaller than that of CPY (Table 7). Thus, CPYO will partition more into water in the atmosphere (precipitation) and will be more likely to rain-out into surface water or snow. Because of the greater electronegativity of the P-atom, CPYO is more reactive than CPY and will undergo hydrolysis more rapidly than CPY; the half-life in water (Table 7) is approximately half that of CPY (Table 6). Thus, CPYO will partition out of air into water, where it is less persistent. The overall persistence in these two media is not suggestive of the characteristics of a POP. Concentrations in surface waters in remote locations (Sect. 2.4 and Fig. 4) are less than amounts that would cause toxicity (Giddings et al. 2014). In addition, adverse effects in aquatic organisms have not been linked to exposures to CPY (Datta et al. 1998; Davidson et al. 2012), and, by extension to the oxon that might be formed. Thus, neither CPY nor CPYO trigger the criteria for POPs and LRT.

Classification as a PBT. In terms of assessment of PBT under EC regulation No. 1107/2009, there is no guidance for using multiple values for P and B. However, because of the multiple uses of CPY over a large number of agricultural sites, the geometric mean is the most appropriate value to compare to the trigger value. In addition, values derived under field conditions can be used for validation. For surface waters, the geometric mean of laboratory-based half-lives was 21 d but, under more realistic conditions in microcosms, half-lives were less than 10 d (Sect. 3). These are less than the trigger of 40 d. Based on these values, the trigger value for P is not exceeded. Geometric mean half-lives in soil tested under laboratory and field conditions were 32 and 22 d, respectively. These are less than the trigger of 120 d. The geometric mean DT_{50dis} for CPY in sediments tested in the laboratory and microcosms was 39 d, less than the trigger value of 120 d.

The geometric mean values for BCF or BAF tested in the laboratory and the field were 848 and 935, respectively (Sect. 5). These are less that the trigger value of 2,000. Moreover, CPY does not trigger the criterion for Pv or Bv. In addition, CPYO also does not trigger the criteria for P and B or Pv or Bv.

The trigger of 10 $\mu g\ L^{-1}$ for T is exceeded for CPY; the most sensitive NOEC reported for aquatic organisms is 0.005 $\mu g\ L^{-1}$ for *Simocephalus vetulus* in a microcosm experiment (Daam and Van den Brink 2007). Since CPY is an insecticide, toxicity to arthropods is expected, however; the key question is the relevance of this to the exposures in the general environment and, as discussed in the companion papers (Giddings et al. 2014; Williams et al. 2014), this is not indicative of significant ecotoxicological risks in the North-American environment, even in areas close to where it is applied. In aquatic organisms, CPYO has similar toxicity to CPY (Giddings et al. 2014), but it is only infrequently found in surface waters and then only at very small concentrations (Williams et al. 2013). This is consistent with the greater reactivity of CPYO and its rapid hydrolysis in the environment. Thus, although formed from CPY in the atmosphere, CPYO is not persistent enough to present a risk to aquatic organisms, although it does trigger the T criterion under EC 1107/2009.

7 Summary

The fate and movement of the organophosphorus insecticide chlorpyrifos (CPY; CAS No.2921-88-2) and its metabolite chlorpyrifos-oxon (CPYO; CAS No.5598-15-2) determine exposures in terrestrial and aquatic environments.

Detectable concentrations of the organophosphorus insecticide CPY in air, rain, snow and other environmental media have been measured in North America and other locations at considerable distances from likely agricultural sources, which indicates the potential for long range transport (LRT) in the atmosphere. This issue was addressed by first compiling monitoring results for CPY in all relevant environmental media. As a contribution to the risk assessment of CPY in remote regions, a simple mass balance model was developed to quantify likely concentrations at locations ranging from local sites of application to more remote locations up to hundreds of km distant. Physical-chemical properties of CPY were reviewed and a set of consistent values for those properties that determine partitioning and reactivity were compiled and evaluated for use in the model. The model quantifies transformation and deposition processes and includes a tentative treatment of dispersion to lesser atmospheric concentrations. The model also addressed formation and fate of CPYO, which is the major transformation product of CPY. The Characteristic Travel Distance (CTD) at which 63% of the original mass of volatilized CPY is degraded or deposited-based on a conservative concentration of •OH radicals of 0.7×10^6 molecules cm^{-3} and a half-life of 3 h, was estimated to be 62 km. At lesser concentrations of •OH radical, such as occurs at night and at lesser temperatures, the CTD is proportionally greater. By including monitoring data from a variety of media, including air, rain, snow and biota, all monitored concentrations can be converted to the equilibrium criterion of fugacity, thus providing a synoptic assessment of concentrations of CPY and CPYO in multiple media. The calculated fugacities of CPY in air and other media decrease proportionally with increasing distance from sources, which can provide an approximate prediction of downwind concentrations and fugacities in media and can contribute to improved risk assessments for CPY and especially CPYO at locations remote from points of application, but still subject to LRT. The model yielded estimated concentrations that are generally consistent with concentrations measured, which suggests that the canonical fate and transport processes were included in the simulation model. The equations included in the model enable both masses and concentrations of CPY and CPYO to be estimated as a function of distance downwind following application. While the analysis provided here is useful and an improvement over previous estimates of LRT of CPY and CPYO, there is still need for improved estimates of the chemical-physical properties of CPYO.

Based on the persistence in water, soils, and sediments, its bioconcentration and biomagnification in organisms, and its potential for long-range transport, CPY and CPYO do not trigger the criteria for classification as a POP under the Stockholm convention or a PB chemical under EC 1107/2009. Nonetheless, CPY is toxic at concentrations less than the trigger for classification as T under EC1107/2009; however, this simple trigger needs to be placed in the context of low risks to non-target organisms close to the areas of use. Overall, CPY and CPYO are judged to not trigger the PBT criteria of EC 1107/2009.

Acknowledgements The authors wish to thank the anonymous reviewers of this paper for their suggestions and constructive criticism. Prof. Giesy was supported by the Canada Research Chair program, a Visiting Distinguished Professorship in the Department of Biology and Chemistry and

State Key Laboratory in Marine Pollution, City University of Hong Kong, the 2012 "High Level Foreign Experts" (#GDW20123200120) program, funded by the State Administration of Foreign Experts Affairs, the P.R. China to Nanjing University and the Einstein Professor Program of the Chinese Academy of Sciences. Funding for this study was provided by Dow AgroSciences.

References

Ashauer R, Hintermeister A, O'Connor I, Elumelu M, Hollender J, Escher BI (2012) Significance of xenobiotic metabolism for bioaccumulation kinetics of organic chemicals in *Gammarus pulex*. Environ Sci Technol 46:3498–3508

Aston LS, Seiber JN (1997) Fate of summertime airborne organophosphate pesticide residues in the Sierra Nevada Mountains. J Environ Qual 26:1483–1492

Barron MG, Woodburn KB (1995) Ecotoxicology of chlorpyrifos. Rev Environ Contam Toxicol 144:1–93

Beyer A, Mackay D, Matthies M, Wania F, Webster E (2000) Assessing long-range transport potential of persistent organic pollutants. Environ Sci Technol 34:699–703

Bondarenko S, Gan J (2004) Degradation and sorption of selected organophosphate and carbamate insecticides in urban stream sediments. Environ Toxicol Chem 23:1809–1814

Bradford DF, Heithmar EM, Tallent-Halsell NG, Momplaisir G-M, Rosal CG, Varner KE, Nash MS, Riddick LA (2010a) Temporal patterns and sources of atmospherically deposited pesticides in alpine lakes of the Sierra Nevada, California, U.S.A. Environ Sci Technol 44: 4609–4614

Bradford DF, Stanley K, McConnell LL, Tallent-Halsell NG, Nash MS, Simonich SM (2010b) Spatial patterns of atmospherically deposited organic contaminants at high elevation in the southern Sierra Nevada mountains, California, USA. Environ Toxicol Chem 29:1056–1066

Bradford DF, Stanley KA, Tallent NG, Sparling DW, Nash MS, Knapp RA, McConnell LM, Massey-Simonich SL (2013) Temporal and spatial variations of atmospherically deposited organic contaminants at high elevations in Yosemite National Park, California, USA. Environ Toxicol Chem 32:517–525

Bromilow RH, de Carvalho RF, Evans AA, Nicholls PH (2006) Behavior of pesticides in sediment/water systems in outdoor mesocosms. J Environ Sci Hlth B 41:1–16

Budd R, O'Geen A, Goh KS, Bondarenko S, Gan J (2011) Removal mechanisms and fate of insecticides in constructed wetlands. Chemosphere 83:1581–1587

Clark T, Clark K, Paterson S, Norstrom R, Mackay D (1988) Wildlife monitoring, modelling, and fugacity. They are indicators of chemical contamination. Environ Sci Technol 22:120–127

Cutler GC, Purdy J, Giesy JP, Solomon KR (2014) Risk to pollinators from the use of chlorpyrifos in the United States. Rev Environ Contam Toxicol 231:219–265

Daam MA, Van den Brink PJ (2007) Effects of chlorpyrifos, carbendazim, and linuron on the ecology of a small indoor aquatic microcosm. Arch Environ Contam Toxicol 53:22–35

Daam MA, Crum SJ, Van den Brink PJ, Nogueira AJ (2008) Fate and effects of the insecticide chlorpyrifos in outdoor plankton-dominated microcosms in Thailand. Environ Toxicol Chem 27:2530–2538

Datta S, Hansen L, McConnell L, Baker J, LeNoir JS, Seiber JN (1998) Pesticides and PCB contaminants in fish and tadpoles from the Kaweah River Basin, California. Bull Environ Contam Toxicol 60:829–836

Davidson C, Stanley K, Simonich SM (2012) Contaminant residues and declines of the Cascades frog (*Rana cascadae*) in the California Cascades, USA. Environ Toxicol Chem 31:1895–1902

El-Amrani S, Pena-Abaurrea M, Sanz-Landaluze J, Ramos L, Guinea J, Camara C (2012) Bioconcentration of pesticides in zebrafish eleutheroembryos (Danio rerio). Sci Tot Environ 425:184–190

Environment Canada (1995) Toxic Substances Management Policy—Persistence and Bioaccumulation Criteria. Final report of the ad hoc Science Group on Criteria. Environment Canada, Ottawa, ON, Canada. En40-499/1-1995

European Community (2009) Regulation (EC) No 1107/2009 of the European Parliament and of the Council of 21 October 2009 Concerning the Placing of Plant Protection Products on the Market and Repealing Council Directives 79/117/EEC and 91/414/EEC (91/414/EEC). Off J Eur Commun 52:1–50

European Community (2011) Regulation (EU) No 253/2011 of 15 March 2011 Amending Regulation (EC) No 1907/2006 of the European Parliament and of the Council on the Registration, Evaluation, Authorisation and Restriction of Chemicals (REACH) as Regards Annex XIII. Off J Eur Commun 54:7–12

Foreman WT, Majewski MS, Goolsby DA, Wiebe FW, Coupe RH (2000) Pesticides in the atmosphere of the Mississippi River Valley, part II–air. Sci Tot Environ 248:213–226

Gebremariam SY, Beutel MW, Yonge DR, Flury M, Harsh JB (2012) Adsorption and desorption of chlorpyrifos to soils and sediments. Rev Environ Contam Toxicol 215:123–175

Giddings JM, Williams WM, Solomon KR, Giesy JP (2014) Risks to aquatic organisms from the use of chlorpyrifos in the United States. Rev Environ Contam Toxicol 231:119–162

Giesy JP, Solomon KR, Coates JR, Dixon KR, Giddings JM, Kenaga EE (1999) Chlorpyrifos: ecological risk assessment in North American aquatic environments. Rev Environ Contam Toxicol 160:1–129

Gobas FA, de Wolf W, Burkhard LP, Verbruggen E, Plotzke K (2009) Revisiting bioaccumulation criteria for POPs and PBT assessments. Integr Environ Assess Manag 5:624–637

Goel A, McConnell LL, Torrents A (2005) Wet deposition of current use pesticides at a rural location on the Delmarva Peninsula: impact of rainfall patterns and agricultural activity. J Agric Food Chem 53:7915–7924

Goel A, McConnell LL, Torrents A (2007) Determination of vapor pressure-temperature relationships of current-use pesticides and transformation products. J Environ Sci Hlth B 42:343–349

Hageman KJ, Simonich SL, Campbell DH, Wilson GR, Landers DH (2006) Atmospheric deposition of current-use and historic-use pesticides in snow at national parks in the western United States. Environ Sci Technol 40:3174–3180

Hermanson MH, Isaksson E, Teixeira C, Muir DC, Compher KM, Li YF, Igarashi M, Kamiyama K (2005) Current-use and legacy pesticide history in the Austfonna Ice Cap, Svalbard, Norway. Environ Sci Technol 39:8163–8169

Jantunen AP, Tuikka A, Akkanen J, Kukkonen JV (2008) Bioaccumulation of atrazine and chlorpyrifos to *Lumbriculus variegatus* from lake sediments. Ecotoxicol Environ Safety 71:860–868

Kale SP, Sherkhane PD, Murthy NB (2002) Uptake of ^{14}C-chlorpyrifos by clams. Environ Technol 23:1309–1311

Kuang Z, McConnell LL, Torrents A, Meritt D, Tobash S (2003) Atmospheric deposition of pesticides to an agricultural watershed of the Chesapeake Bay. J Environ Qual 32:1611–1622

Kurt-Karakus PB, Teixeira C, Small J, Muir D, Bidleman TF (2011) Current-use pesticides in inland lake waters, precipitation, and air from Ontario, Canada. Environ Toxicol Chem 30:1539–1548

Laabs V, Amelung W, Fent G, Zech W, Kubiak R (2002) Fate of ^{14}C-labeled soybean and corn pesticides in tropical soils of Brazil under laboratory conditions. J Agric Food Chem 50:4619–4627

Lalah JO, Ondieki D, Wandiga SO, Jumba IO (2003) Dissipation, distribution, and uptake of ^{14}C-chlorpyrifos in a model tropical seawater/sediment/fish ecosystem. Bull Environ Contam Toxicol 70:883–890

Leistra M, Smelt JH, Weststrate JH, van den Berg F, Aalderink R (2005) Volatilization of the pesticides chlorpyrifos and fenpropimorph from a potato crop. Environ Sci Technol 40:96–102

LeNoir JS, McConnell LL, Fellers GM, Cahill TM, Seiber JN (1999) Summertime transport of current-use pesticides from California's Central Valley to the Sierra Nevada mountain range, USA. Environ Toxicol Chem 18:2715–2722

Liu B, McConnell LL, Torrents A (2001) Hydrolysis of chlorpyrifos in natural waters of the Chesapeake Bay. Chemosphere 44:1315–1323

Mackay D, Shiu W-Y, Ma KC (1997) Illustrated handbook of physical-chemical properties and environmental fate for organic chemicals volume V pesticide chemicals. Lewis Publishers, Boca Raton, FL, p 812

Mackay D (2001) Multimedia environmental models: the fugacity approach. Lewis Publishers, Boca Raton, FL, pp 261

Majewski MS, Foreman WT, Goolsby DA, Nagakaki N (1998) Airborne pesticide residues along the Mississippi River. Environ Sci Technol 32:3689–3698

Majewski MS (1999) Micrometeorologic methods for measuring the post-application volatilization of pesticides. Water Air Soil Pollut 115:83–113

Mast MA, Alvarez DA, Zaugg SD (2012) Deposition and accumulation of airborne organic contaminants in Yosemite National Park, California. Environ Toxicol Chem 31:524–533

Metcalf RL, Sanborn JR (1975) Pesticides and environmental quality in Illinois. Ill State Nat Hist Surv Bull 31:381–436

Moermond C, Janssen M, de Knecht J, Montforts M, Peijnenburg W, Zweers P, Sijm D (2011) PBT assessment using the revised Annex XIII of REACH—a comparison with other regulatory frameworks. Integr Environ Assess Manag 8:359–371

Moore DRJ, Teed RS, Greer C, Solomon KR, Giesy JP (2014) Refined avian risk assessment for chlorpyrifos in the United States. Rev Environ Contam Toxicol 231:163–217

Muir DC, Teixeira C, Wania F (2004) Empirical and modeling evidence of regional atmospheric transport of current-use pesticides. Environ Toxicol Chem 23:2421–2432

Munoz A, Vera T, Sidebottom H, Mellouki A, Borras E, Rodenas M, Clemente E, Vazquez M (2011) Studies on the atmospheric degradation of chlorpyrifos-methyl. Environ Sci Technol 45:1880–1886

Muñoz A, Vera T, Ródenas M, Borras E, Vázquez M, Sánchez P, Ibañez A, Raro M, Alacreu F (2012) Gas phase photolysis and photooxidation of Chlorpyrifos and Chlorpyrifos Oxon. Dow AgroScience (Unpublished Report), Indianapolis, IN, USA

Nhan DD, Carvalho FP, Nam BQ (2002) Fate of ^{14}C-chlorpyrifos in the tropical estuarine environment. Environ Technol 23:1229–1234

OSPAR (1992) Convention for the protection of the marine environment of the North-East Atlantic. OSPAR Comission, London, UK. http://www.ospar.org/html_documents/ospar/html/ospar_convention_e_updated_text_2007.pdf

Pablo F, Krassoi FR, Jones PR, Colville AE, Hose GC, Lim RP (2008) Comparison of the fate and toxicity of chlorpyrifos—laboratory versus a coastal mesocosm system. Ecotoxicol Environ Safety 71:219–229

Peck AM, Hornbuckle KC (2005) Gas-phase concentrations of current-use pesticides in Iowa. Environ Sci Technol 39:2952–2959

Racke KD (1993) Environmental fate of chlorpyrifos. Rev Environ Contam Toxicol 131:1–151

Raina R, Hall P, Sun L (2010) Occurrence and relationship of organophosphorus insecticides and their degradation products in the atmosphere in Western Canada agricultural regions. Environ Sci Technol 44:8541–8546

Ritter L, Solomon KR, Forget J, Stemeroff M, O'Leary C (1995a) Persistent organic pollutants. Report for the Intergovernmental Forum on Chemical Safety. Canadian Network of Toxicology Centres, Guelph, ON, Canada. PCS/95.38 http://www.chem.unep.ch/pops/ritter/en/ritteren.pdf

Ritter L, Solomon KR, Forget J, Stemeroff M, O'Leary C (1995b) A review of selected persistent organic pollutants. DDT-Aldrin-Dieldrin-Endrin-Chlordane-Heptachlor-Hexachlorobenzene-Mirex-Toxaphene-Polychlorinated biphenyls-Dioxins and Furans. Review prepared for IPCS.

International Programme on Chemical Safety (IPCS) of the United Nations, Geneva. Number PCS/95.39 http://www.who.int/ipcs/assessment/en/pcs_95_39_2004_05_13.pdf
Rotondaro A, Havens PL (2012) Direct flux measurement of chlorpyrifos and chlorpyrifos-oxon emissions following applications of Lorsban Advanced Insecticide to Alfalfa. Dow AgroScience (Unpublished Report), Indianaoplis, IN, USA
Rubach MN, Ashauer R, Maund SJ, Baird DJ, Van den Brink PJ (2010) Toxicokinctic variation in 15 freshwater arthropod species exposed to the insecticide chlorpyrifos. Environ Toxicol Chem 29:2225–2234
Smalling KL, Fellers GM, Kleeman PM, Kuivila KM (2013) Accumulation of pesticides in pacific chorus frogs (*Pseudacris regilla*) from California's Sierra Nevada Mountains, USA. Environ Toxicol Chem 32:2026–2034
Solomon KR, Giesy JP, Kendall RJ, Best LB, Coats JR, Dixon KR, Hooper MJ, Kenaga EE, McMurry ST (2001) Chlorpyrifos: ecotoxicological risk assessment for birds and mammals in corn agroecosystems. Human Ecol Risk Assess 7:497–632
Solomon KR, Dohmen P, Fairbrother A, Marchand M, McCarty L (2009) Use of (eco)toxicity data as screening criteria for the identification and classification of PBT/POP compounds. Integr Environ Assess Manag 5:680–696
Solomon KR, Williams WM, Mackay D, Purdy J, Giddings JM, Giesy JP (2014) Properties and uses of chlorpyrifos in the United States. Rev Environ Contam Toxicol 231:13–34
Solomon KR, Matthies M, Vighi M (2013) Assessment of PBTs in the EU: a critical assessment of the proposed evaluation scheme with reference to plant protection products. Environ Sci EU 25:1–25
Spivakovsky CM, Logan JA, Montzka SA, Balkanski YJ, Foreman-Fowler M, Jones DBA, Horowitz LW, Fusco AC, Brenninkmeijer CAM, Prather MJ, Wofsy SC, McElroy MB (2000) Three-dimensional climatological distribution of tropospheric OH: update and evaluation. J Geophys Res 105:8931–8980
Testai E, Buratti FM, Consiglio ED (2010) Chlorpyrifos. In: Krieger RI, Doull J, van Hemmen JJ, Hodgson E, Maibach HI, Ritter L, Ross J, Slikker W (eds) Handbook of pesticide toxicology, vol 2. Elsevier, Burlington, MA, pp 1505–1526
Thomas CN, Mansingh A (2002) Bioaccumulation, elimination, and tissue distribution of chlorpyrifos by red hybrid Tilapia in fresh and brackish waters. Environ Technol 23:1313–1323
Tunink A (2010) Chlorpyrifos-oxon: determination of hydrolysis as a function of pH. Dow AgroScience (Unpublished Report), Indianaoplis, IN, USA
Turner DB (1994) Workbook of atmospheric dispersion estimates. Lewis Publ. CRC Press, Boca Raton, FL, pp 192
United Nations Economic Commission for Europe (1998) Protocol to the 1979 convention on long-range transboundary air pollution on persistent organic pollutants. United Nations Economic Commission for Europe, Geneva, Switzerland. http://www.unece.org/env/lrtap/full%20text/1998.POPs.e.pdf
United Nations Environmental Programme (2001) Stockholm convention on persistent organic pollutants. Secretariat of the Stockholm Convention, Geneva, Switzerland. http://chm.pops.int/Convention/tabid/54/language/en-US/Default.aspx
USEPA (1995) SCREEN3 Model User's Guide. United States Environmental Protection Agency, Office of Air Quality Planning and Standards Emissions, Monitoring, and Analysis Division, Research Triangle Park, NC, USA. EPA-454/B-95-004 http://www.epa.gov/scram001/userg/screen/screen3d.pdf
USEPA (1999a) Persistent Bioaccumulative Toxic (PBT) Chemicals; Final Rule. 40 CFR Part 372. Fed Reg 64:58665–587535
USEPA (1999b) Category for persistent, bioaccumulative, and toxic new chemical substances. Fed Reg 64:60194–60204
USEPA (2011) Revised Chlorpyrifos Preliminary Registration Review Drinking Water Assessment. United States Environmental Protection Agency, Office of Chemical Safety and Pollution Prevention, Washington, DC, USA. PC Code 059101 http://www.epa.gov/oppsrrd1/registration_review/chlorpyrifos/EPA-HQ-OPP-2008-0850-DRAFT-0025%5B1%5D.pdf

USEPA (2013) Chlorpyrifos; preliminary evaluation of the potential risks from volatilization. United States Environmental Protection Agency, Office of Chemical Safety and Pollution Prevention. Washington, DC, USA. Numbers 399484 and 400781

Vaccaro JR (1993) Risks associated with exposure to chlorpyrifos formulation components. In: Racke KD, Leslie AR (eds) Pesticides in urban environments: fate and significance, vol 522, ACS Symposium Series. American Chemical Society, Washington, DC, pp 197–396

van Jaarsveld JA, van Pul WAJ (1999) Modelling of atmospheric transport and deposition of pesticides. Water Air Soil Pollut 115:167–182

Varo I, Serrano R, Pitarch E, Amat F, Lopez FJ, Navarro JC (2000) Toxicity and bioconcentration of chlorpyrifos in aquatic organisms: *Artemia parthenogenetica*(Crustacea), *Gambusia affinis*, and *Aphanius iberus* (Pisces). Bull Environ Contam Toxicol 65:623–630

Varo I, Serrano R, Pitarch E, Amat F, Lopez FJ, Navarro JC (2002) Bioaccumulation of chlorpyrifos through an experimental food chain: study of protein HSP70 as biomarker of sublethal stress in fish. Arch Environ Contam Toxicol 42:229–235

Watts M (2012) Chlorpyrifos as a Possible Global POP. Pesticide Action Network North America, Oakland, CA. www.pan-europe.info/News/PR/121009_Chlorpyrifos_as_POP_final.pdf

Wegmann F, Cavin L, MacLeod M, Scheringer M, Hungerbuhler K (2009) The OECD software tool for screening chemicals for persistence and long-range transport potential. Environ Model Software 24:228–237

Weschler CJ, Nazaroff WW (2010) SVOC partitioning between the gas phase and settled dust indoors. Atmos Environ 44:3609–3620

Williams WM, Giddings J, Purdy J, Solomon KR, Giesy JP (2014) Exposures of aquatic organisms resulting from the use of chlorpyrifos in the United States. Rev Environ Contam Toxicol 231:77–118

Woodburn KB, Hansen SC, Roth GA, Strauss K (2003) The bioconcentration and metabolism of chlorpyrifos by the eastern oyster, *Crassostrea virginica*. Environ Toxicol Chem 22:276–284

Woodrow JE, Seiber JN (1997) Correlation techniques for estimating pesticide volatilization flux and downwind concentrations. Environ Sci Technol 31:523–529

Woodrow JE, Seiber JN, Dary C (2001) Predicting pesticide emissions and downwind concentrations using correlations with estimated vapor pressure. J Agric Food Chem 49:3841–3846

Wu J, Laird DA (2004) Interactions of chlorpyrifos with colloidal materials in aqueous systems. J Environ Qual 33:1765–1770

Yao Y, Harner T, Blanchard P, Tuduri L, Waite D, Poissant L, Murphy C, Belzer W, Aulagnier F, Sverko E (2008) Pesticides in the atmosphere across Canadian agricultural regions. Environ Sci Technol 42:5931–5937

Zivan O (2011) Atmospheric levels and transport of organophosphates pesticides and their derivatives at agricultural settlements. Environmental Engineering, Vol. M.Sc., Technion—Israel Institute of Technology, Haifa, IL, pp 56

Exposures of Aquatic Organisms to the Organophosphorus Insecticide, Chlorpyrifos Resulting from Use in the United States

W. Martin Williams, Jeffrey M. Giddings, John Purdy, Keith R. Solomon, and John P. Giesy

1 Introduction

Chlorpyrifos (*O*,*O*-diethyl *O*-(3,5,6-trichloro-2-pyridinyl) phosphorothioate) is an organophosphorus insecticide that has been detected in surface waters of the United States (CDPR 2012a; Martin et al. 2011; NCWQR 2012; WDOE 2012). The potential for chlorpyrifos (CPY) to occur in surface water is governed by complex interactions of factors related to application, agronomic practices, climatological conditions during and after application, soil pedology and chemistry, hydrologic responses of drainage systems, and its physicochemical properties that affect mobility and persistence under those environmental settings. These conditions vary among patterns of use such as the crop to which it is applied within the different regions of the country that have different soil types and climates. CPY use and registrations have changed over time as a result of market forces and product stewardship, including the ban of retail use and

The online version of this chapter (doi:10.1007/978-3-319-03865-0_4) contains supplementary material, which is available to authorized users.

W.M. Williams (✉)
Waterborne Environmental Inc., Leesburg, VA, USA
e mail: williamsm@waterborne-env.com

J.M. Giddings
Compliance Services International, Rochester, MA, USA

J. Purdy
Abacus Consulting, Campbellville, ON, Canada

K.R. Solomon
Centre for Toxicology, School of Environmental Sciences, University of Guelph, Guelph, ON, Canada

J.P. Giesy
Department of Veterinary Biomedical Sciences and Toxicology Centre, University of Saskatchewan, 44 Campus Dr., Saskatoon, SK S7N 5B3, Canada

J.P. Giesy and K.R. Solomon (eds.), *Ecological Risk Assessment for Chlorpyrifos in Terrestrial and Aquatic Systems in the United States*, Reviews of Environmental Contamination and Toxicology 231, DOI 10.1007/978-3-319-03865-0_4, © The Author(s) 2014

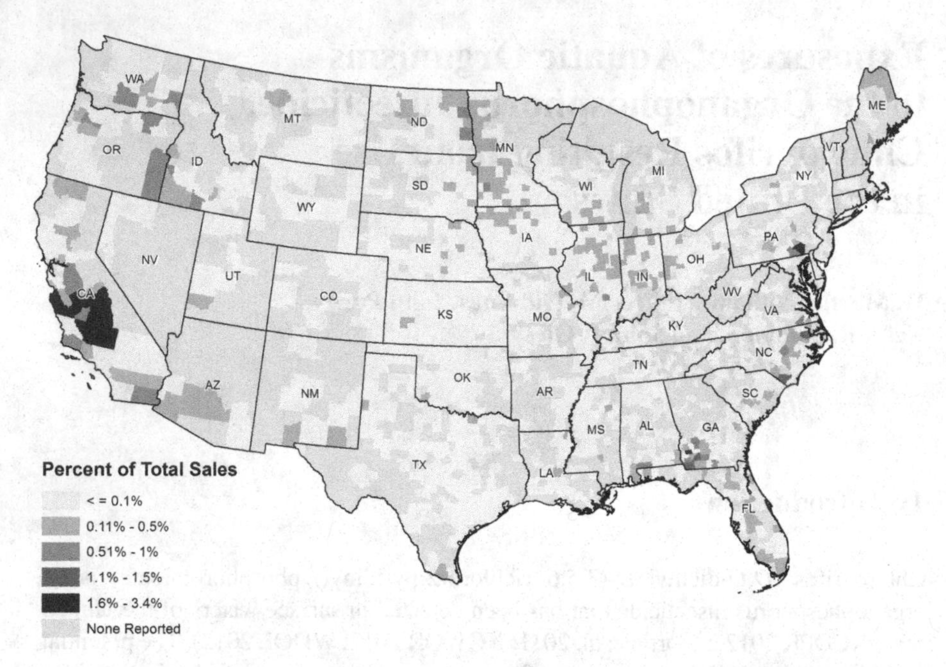

Fig. 1 Geographical distribution of chlorpyrifos use in the United States from 2010 to 2011. Derived from confidential sales data (from Dow AgroSciences, 2012; see SI for a color version of this Figure)

the implementation of other label changes for environmental stewardship that were implemented in 2001. The objective of this study was to characterize likely exposures of aquatic organisms to CPY in the U.S. by evaluating patterns of use, environmental chemistry, available monitoring data, and via simulation modeling. The results of the data analyses and simulation modeling are a key component of the CPY risk assessment described in the companion paper (Giddings et al. 2014).

1.1 Distribution of Use of Chlorpyrifos

CPY is one of the insecticides most widely used throughout the world to limit insect and mite damage to a number of important crops, including soybeans, corn, tree nuts, alfalfa, wheat, citrus, peanuts, and vegetables, among others (Solomon et al. 2014). Regions in the U.S. with the largest use of CPY in 2010–2011, expressed as percent of insecticide, include the Central Valley of California; the Snake River basin in Oregon and southwestern Idaho; parts of Minnesota, Iowa, Wisconsin, Illinois, Indiana, Ohio and Michigan in the central and eastern corn belt; and areas in Georgia and North Carolina along the Atlantic and Gulf coastal plain (Fig. 1). Soybeans, corn, tree nuts (almonds, pecans and walnuts), apples, alfalfa, wheat, and sugar beets accounted for approximately 80% of the use of CPY in 2007 (Fig. 2).

CPY is available as a granular product for soil treatment, or several flowable formulations (all formulations that are sprayed) that can be applied to foliage, soil,

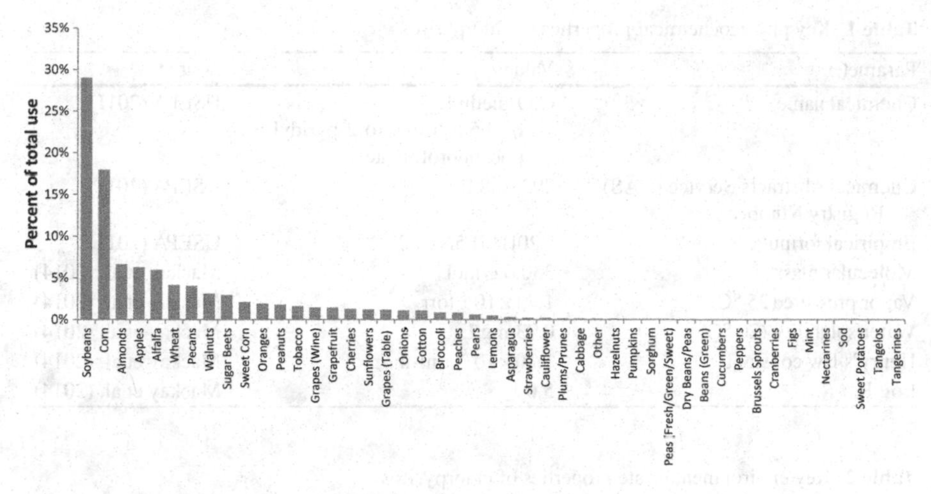

Fig. 2 Use of chlorpyrifos by crop (2007) (data from Gomez 2009)

or dormant trees (Solomon et al. 2014). Application can occur pre-plant, at-plant, post-plant or during the dormant season using aerial equipment, chemigation, ground boom or air-blast sprayers, tractor-drawn spreaders, or hand-held equipment.

1.2 Environmental Fate Properties

CPY has short to moderate persistence in the environment as a result of several pathways of dissipation, including volatilization, photolysis, abiotic hydrolysis and microbial degradation that can occur concurrently. Dissipation by volatilization from foliage, as controlled by the physical and chemical properties of CPY (Tables 1 and 2) is the dominant process during the first 12 h after application, but decreases as the formulation adsorbs to foliage or soil (Mackay et al. 2014). During the days following application, CPY strongly adsorbs to soil and penetrates the soil profile to become less available for volatilization, and consequently other degradation processes become important. The magnitude of CPY adsorption varies substantially between different soils, with a spread of two orders of magnitude in calculated Kd coefficients (SI Table A4).

Key factors affecting degradation of CPY in soil have been reviewed previously (Racke 1993) and these factors have been substantiated by additional recent research cited herein. CPY can be degraded by UV radiation, dechlorination, hydrolysis, and microbial processes. However, hydrolysis of CPY is the primary mechanism and results in formation of 3,5,6-trichloropyridinol (TCP). This step can be either abiotic or biotic, and the rate is 1.7- to 2-fold faster in biologically active soils. The rate of abiotic hydrolysis is pH-dependent and occurs more rapidly under alkaline conditions. It is also faster in the presence of catalysts such as certain types of clay (Racke 1993). Degradation of TCP is dependent on biological activity, and leads to

Table 1 Key physicochemical properties of chlorpyrifos

Parameter	Value	Source
Chemical name	O,O-diethyl o-(3,5,6-trichloro-2-pyridyl phosphorothioate	USEPA (2011)
Chemical Abstracts Service (CAS) Registry Number	2921-88-2	USEPA (2011)
Empirical formula	C20H17F5N2O2	USEPA (2011)
Molecular mass	350.6 g/mol	Mackay et al. (2014)
Vapor pressure (25 °C)	1.73×10^{-5} torr	Mackay et al. (2014)
Water solubility (20 °C)	0.73 mg/L	Mackay et al. (2014)
Henry's law constant	1.10×10^{-5} atm-m^3/mol	Mackay et al. (2014)
Log K_{ow}	5.0	Mackay et al. (2014)

Table 2 Key environmental fate properties of chlorpyrifos

Parameter	Value	Source
Hydrolysis (t½)	pH 5: 73 d pH 7: 72 d pH 7: 81 d pH 9: 16 d	USEPA (2011)
Aqueous photolysis (t½)	29.6 d	USEPA (2011)
Aerobic soil metabolism (t½)[a]	1.9–1576 d	See Table SI A1
Aerobic aquatic metabolism (t½)	22–51 d	See Table SI A5
Anaerobic soil metabolism (t½)	15 and 58 d	USEPA (2011)
Anaerobic aquatic metabolism (t½)	39 and 51 d	USEPA (2011)
Soil adsorption coefficient K_{OC}	973–31,000 cm^3 g^{-1}	See Table SI A4
Terrestrial field dissipation (t½)	1.3–120 d	See Table SI A3

t½ = half-life
[a]All values <335 d except one

formation of bound residues and reversible formation of 3,5,6-trichloro-2-methoxypyridinol (TMP). Under aerobic conditions, the major terminal degradation product of CPY is CO_2. Since TCP and TMP are not considered to be residues of concern (USEPA 2011), they were not included in the model simulations presented below. Finally, while photolysis and oxidation are known to form the chlorpyrifos oxon (CPYO) in air and on foliar surfaces (Mackay et al. 2014), this route is either insignificant in soil or the oxon degrades as quickly as it is formed. CPYO has not been reported in radiotracer soil degradation studies in the laboratory or field. Concentrations of CPYO were not included in simulations because it has not been observed in soils.

Under standardized laboratory conditions, rates of degradation of CPY in soil, expressed as the half-life, have ranged from 1.3 to 1,575 d (all values except one are <335 d; see SI, Table A1). Although half-lives of 7–120 d are considered typical, this range narrows for medium textured soils across several states where reported half-lives range from 33 to 56 d (USEPA 1999). This variability in reported half-lives has been attributed to differences in soil organic carbon and moisture contents, prior CPY application rate, and microbial activity at the time of sampling, but no quantitative relationships have been reported (Racke 1993). Rates of degradation are inversely proportional to rates of application, possibly because concentrations in soil water reach the solubility limit of approximately 1 μg CPY L^{-1}. Generally, when applied as a granular product, CPY dissipation is slower than when applied as a liquid (Racke 1993). Dissipation under field conditions is also variable, with half-lives ranging from 1.3 to 120 d (SI, Table A2).

Results of laboratory aerobic degradation studies with CPY exhibit bi-phasic behavior in some soils. Initial rates of degradation are greater than overall rates by factors of 1.1 to 2.9 (Racke 1993). This behavior of CPY is not as apparent for some soils for which half-lives were calculated by use of simple, first-order kinetics (de Vette and Schoonmade 2001). Some half-lives reported in the literature (SI Table A1) have been derived by assuming first-order kinetics for degradation, which can overestimate the environmental persistence of CPY. This artifact is discussed in greater detail in the second paper of this series (Solomon et al. 2013).

CPY rapidly dissipates from plant surfaces, primarily from volatility and secondarily from photolysis, with most reported dissipation half-lives being on the order of several days (SI, Table A3). In a field study of CPY loss to air conducted in California, maximum fluxes via volatilization occurred in the first 8 h after application to recently cut alfalfa (Rotondaro and Havens 2012). The total loss of mass was calculated from fluxes determined by the Aerodynamic (AD) and Integrated Horizontal Flux (IHF) methodologies and ranged from 15.8 to 16.5% of the applied mass of CPY.

Based on reported water-sediment adsorption coefficients normalized to fraction of organic carbon in sediments (K_{OC}) of 973–31,000 $cm^3\ g^{-1}$ (mean 8216 $cm^3\ g^{-1}$, SI Appendix X, Table A4), CPY has moderate to high potential to adsorb to soil. Uptake by roots, translocation, and metabolism of CPY in plants are negligible and thus CPY is non-systemic, although metabolism of foliar-applied CPY does occur (Racke 1993).

In aquatic systems, abiotic degradation from aqueous hydrolysis of CPY has been reported to occur with half-lives of 73, 72, and 16 d at pH 5, 7, and 9, respectively at 25 °C (Racke 1993). An aqueous hydrolysis half-life of 81 d at pH 7 has been reported (USEPA 2011). Half-lives of 22–51 d have been reported from studies of aerobic metabolism in aquatic systems (SI Table A5, Kennard 1996; Reeves and Mackie 1993). A half-life of 29.6 d was observed in a aqueous photolysis study performed with CPY under sterile conditions at pH 7 in phosphate-buffered solution under natural sunlight (Batzer et al. 1990).

Transport of CPY off-site following application has been extensively examined across a range of field conditions as affected by several factors: antecedent soil moisture, soil physical and chemical properties, soil erosion, plant canopy coverage, plant development stage, time intervals of 2-h to 7-d between application and rainfall events, and a range of rainfall events with return frequencies as little as 1-in-833 yr (Cryer and Dixon-White 1995; McCall et al. 1984; Poletika and Robb 1994; Racke 1993). CPY mass in runoff ranged from 0.003 (McCall et al. 1984; Poletika and Robb 1994)) to 4.4% (McCall et al. 1984; Poletika and Robb 1994) of applied mass. A study conducted in Iowa under record high rainfall conditions concluded that the majority of compound was transported attached to eroded sediment (Cryer and Dixon-White 1995). However, based on a combination of low erosion and availability of large amounts of residues on cotton foliage for wash-off during storms simulated soon after application of CPY, Poletika and Robb (1994) suggested that the majority of the CPY was transported in the dissolved phase of the runoff in Mississippi. Thus, both dissolved and adsorbed fractions need to be considered as transport pathways to surface water.

2 Measurements of Chlorpyrifos in Aquatic Environments

2.1 *Chlorpyrifos in Surface Water*

The most comprehensive dataset of pesticide concentrations has been compiled from the USGS National Water-Quality Assessment (NAWQA) Program and the National Stream Quality Accounting Network (NASQAN). These represent concentrations measured from 1992 to 2010 (Martin and Eberle 2009; Martin et al. 2011). Both NAWQA and NASQAN programs utilized similar methods to collect and process samples. Pesticide concentrations were determined by the USGS National Water Quality Laboratory (NWQL) by using gas chromatography/mass spectroscopy (GC/MS) in selective ion monitoring (SIM) mode. The USGS determined and applied a consistent minimum concentration as a bias correction to account for changes in recovery and limit of detection (LOD)[1] during the sample collection period. A consistent method of rounding was applied to concentration values and quality control (QC) samples were removed from the data file before analysis. To allow for trend analysis, USGS added an attribute to the database to allow users to create a subset of the data that had no more than one sample per calendar week to avoid weighting the analysis toward periods of more frequent sampling.

The dataset reported by Martin et al. (2011) was characterized based on the percent of samples that contained detectable CPY. Characterization also included calculating 90th, 95th, and 99th centile concentrations and maximum concentrations for all samples. Data were categorized by year, Farm Resource Region[2] (FRR; 9 total), and drainage basin land-use class (4 total). The basin land-use classes were agricultural (>50% agricultural and ≤5% urban), undeveloped (≤25% agricultural and ≤5% urban); urban (>25% urban and ≤25% agricultural); and mixed (all other combinations of urban, agricultural, and undeveloped land). To examine the effect of the 2001 ban on retail sales of CPY on measured concentrations, data for 1992–2001 and 2002–2010 were characterized separately.

[1] Level of detection (LOD) level of quantitation (LOQ) and method detection limit (MDL) are used as defined by MacDougall and Crummet (1980).

[2] A map and explanation of USDA's Economic Research Service (ERS) Farm Resource Regions is available at http://ageconsearch.umn.edu/bitstream/33625/1/ai000760.pdf.

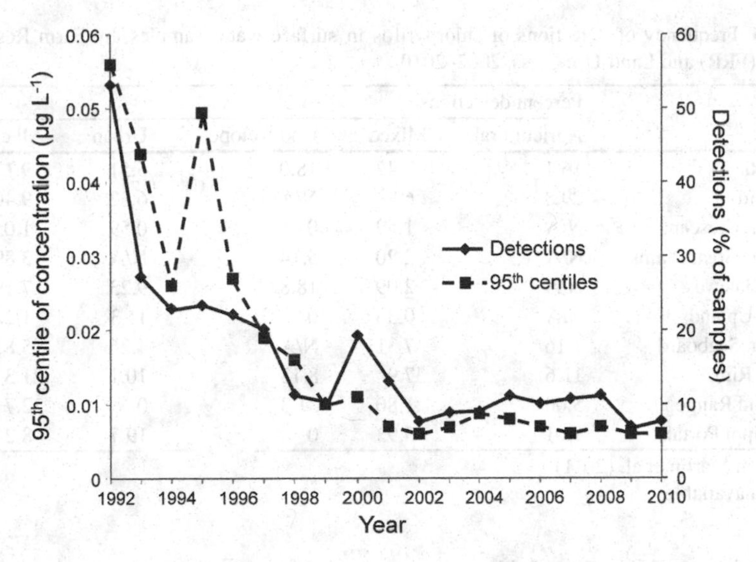

Fig. 3 Detections of chlorpyrifos and 95th centile concentrations in U.S. surface water, 1992–2010 (data from Martin et al. 2011)

Frequencies of detection and 95th centile concentrations decreased more than five-fold between 1992 and 2010 (Fig. 3). Detections in 1992–2001 ranged from 10.2 to 53.1%, while 2002–2010 detections ranged from 7.0 to 11.2%. The 95th centile concentrations ranged from 0.007 to 0.056 μg L^{-1} in 1992–2001 and 0.006–0.008 μg L^{-1} in 2002–2010. Localized intensive studies of CPY in surface waters also indicated the dramatic decrease in detection of CPY between 2001 and 2002, for example in Texas (Banks et al. 2005).

Farm Resource Regions (USDA-ERS 2000) with the greatest percent detections (Table 3) and the greatest 95th centile concentrations (Table 4) were the Fruitful Rim region and the Basin and Range region. Farms in the Basin and Range and Fruitful Rim FRRs include cattle, wheat, sorghum, fruit, vegetable, nursery and cotton farms in the western third of the U.S. and southern Texas, Florida, and southeastern Georgia and South Carolina (Martin et al. 2011).

The greatest frequency of detections (Table 3) and 95th centile concentrations (Table 4) occurred in undeveloped and agricultural land-use classes. The two land-use classes with the most urban land (urban and mixed) had the smallest frequency of detections and 95th centile concentrations, consistent with the cessation of CPY-based product sales for most homeowner uses in December 2001 (Johnson et al. 2011).

Databases in California and Washington were also examined to identify additional data on concentrations of CPY in surface waters. The California Department of Pesticide Regulation Surface Water Database (CA-SWD) was developed in 1997 to collect and provide information on pesticides in California surface waters (CDPR 2012a). Samples were taken from rivers, creeks, urban streams, agricultural drains, the San Francisco Bay delta region, and runoff of urban storm-water from August

Table 3 Frequency of detections of chlorpyrifos in surface water samples, by Farm Resource Region (FRR) and Land-Use Class, 2002–2010

	Percent detections				
Region	Agricultural	Mixed	Undeveloped	Urban	All classes
All FRRs	16.1	7.42	18.0	5.51	9.11
Heartland	29.8	6.83	N/A[a]	6.32	9.40
Northern Crescent	N/A	1.59	0	0.59	1.05
Northern Great Plains	N/A	2.20	5.14	N/A	3.89
Prairie Gateway	N/A	2.09	18.8	9.23	7.19
Eastern Uplands	N/A	0.33	0	11.3	1.22
Southern Seaboard	1.16	7.31	N/A	4.26	5.81
Fruitful Rim	11.6	37.9	8.13	10.1	21.3
Basin and Range	5.00	1.80	31.3	0	22.7
Mississippi Portal	9.01	5.93	0	19.7	8.24

Data from Martin et al. (2011)
[a]No data available

Table 4 Ninety-fifth centiles of chlorpyrifos concentrations in surface water samples, by Farm Resource Region (FRR) and Land-Use Class, 2002–2010

	95th centiles of measured concentrations (μg L^{-1})				
Region	Agricultural	Mixed	Undeveloped	Urban	All classes
All FRRs	0.0097	0.006	0.008	0.0067	0.007
Heartland	0.018	0.005	N/A[a]	0.0076	0.007
Northern Crescent	N/A	0.005	0.005	0.005	0.005
Northern Great Plains	N/A	0.005	0.005	N/A	0.005
Prairie Gateway	N/A	0.005	0.009	0.008	0.0067
Eastern Uplands	N/A	0.005	0.005	10.0	0.005
Southern Seaboard	0.005	0.006	N/A	0.005	0.005
Fruitful Rim	0.0086	0.022	0.005	0.0098	0.015
Basin and Range	0.005	0.005	0.01	0.005	0.009
Mississippi Portal	0.0065	0.006	0.005	0.012	0.007

Data from Martin et al. (2011)
[a]No data available

1990 through July 2010. From 2002 to 2010, the percent detections for CPY in the CA-SWD ranged from 7.6 to 27%, the 95th centile concentrations ranged from 0.010 to 0.1 μg L^{-1}, and maximum concentrations ranged from 0.15 to 3.9 μg L^{-1}. Percent detections in the CA-SWD from 2002 to 2010 was in the same range as for the Martin et al. (2011) data, while the maximum and 95th centile concentrations were generally greater.

The majority of the applicable surface water monitoring data available in the Washington Department of Ecology's Environmental Information Management (WA-EIM) database (WDOE 2012) was generated as part of the Surface Water Monitoring Program for Pesticides in Salmonid-Bearing Streams. The monitoring program was specifically designed "to address pesticide presence in Endangered Species Act (ESA)-listed, salmonid-bearing streams during typical pesticide use periods," with weekly monitoring from March through October each year (WDOE 2012). As with the CA-SWD, the WA-EIM monitoring data were not suitable for trend analysis, but comparisons to the data sets of Martin et al. (2011) and CA-SWD (CDPR 2012a) could be made. From 2002 to 2011, percent detections in the WA-EIM database ranged from 4.9 to 32%, rates similar to those observed by Martin et al. (2011) and CA-SWD (CDPR 2012a). The 95th centile concentrations ranged from 0.033 to 0.3 μg L^{-1}, and maximum concentrations ranged from 0.35 to 0.59 μg L^{-1}. In general, CPY concentrations in the EIM dataset were greater than those reported by Martin et al. (2011), but less than those reported by CA-SWD (CDPR 2012a).

Another source of information on concentrations of CPY in surface water used in this assessment was compiled by the tributary monitoring program operated by the National Center for Water Quality Research (NCWQR) at Heidelberg University (NCWQR 2012). Concentrations of nutrients, sediment, and pesticide were included in the NCWQR monitoring program database. Concentrations of CPY were reported for 6,301 samples from ten stations within the Lake Erie Basin and Ohio from 2002 to 2011. CPY was detected in 6 samples (<1%), with concentrations ranging from 0.02 to 0.37 μg L^{-1}. The percent of detections was less in the NCWQR dataset than in the other datasets.

Overall, the database with the greatest number of samples (more than 10,000) and broadest geographical representation (Martin et al. 2011) shows that CPY was detected in 9% of samples analyzed between 2002 and 2010 and that 95% of the samples contained CYP at less than 0.007 μg L^{-1}, and the maximum was 0.33 μg L^{-1}. The regional databases, which were more focused on areas of pesticide use than the USGS database, had more frequent detections (13–17%) and greater concentrations (95th centiles 0.010–0.3 μg L^{-1}). Even in the WA-EIM database, less than 1% of samples analyzed since 2007 exceeded 0.1 μg L^{-1}.

2.2 *Chlorpyrifos in Sediment*

The NAWQA (NAWQA 2012), CA-SWD (CDPR 2012a), and WA-EIM (WDOE 2012) databases also contained data on concentrations of CPY in sediments. The NAWQA database included results of 76 analyses for CPY in sediments between 2007 and 2010 (NAWQA 2012). CPY was detected in two sediments (detection rate of 2.6%). One sediment sample had an estimated concentration of 1 μg kg^{-1} and the other had a measured concentration of 58.6 μg kg^{-1}. The LOD was 2 μg kg^{-1} for all samples. Data for sediment from the CA-SWD (CDPR 2012a) were only available from one study that was conducted in 2004 (Weston et al. 2005). Of 24 sediments analyzed, CPY was detected in nine, or a detection rate of 37.5%. Concentrations ranged from 1.5 to 19 μg kg^{-1} and the limit of quantitation (LOQ) was 1 μg kg^{-1} for all samples. The WA-EIM database (WDOE 2012) had only one detection for CPY in 23 sediment samples (detection rate of 4.4%) post-2001; the estimated concentration was 1 μg kg^{-1}.

2.3 *Chlorpyrifos Marine Monitoring Data*

Only the WA-EIM database (WDOE 2012) contained marine monitoring data. No CPY was detected in any of the 42 samples of water collected after 2001, which had method reporting limits of 0.002–0.0023 μg L^{-1}. No sediment samples were analyzed.

2.4 *Occurrence of Chlorpyrifos Oxon*

Though not a listed USEPA contaminant (USEPA 2012), the NAWQA (NAWQA 2012), NASQAN (NASQAN 2012), CA-SWD (CDPR 2012a), and WA-EIM (WDOE 2012) databases also contained data on detections of CPYO and concentrations that provide some insight into the possible earlier distribution of CPY. The NAWQA database included results of 7,098 analyses for CPYO in surface waters between 1999 and 2012. CPYO was detected in 16 samples, a rate of 0.23%. These detections, however, were all estimated concentrations, which were less than the LOQ, which ranged from 0.011 to 0.054 μg L^{-1}. The LODs were 0.007 to 0.34 μg L^{-1}. In instances when CPY was sampled at the same time as CPYO was detected, concentrations of CPY were less than the MDLs of 0.005 to 0.006 μg L^{-1} in 75% of samples (i.e., 9 of 12).

In the NASQAN database, CPYO was an analyte in 2,025 surface water samples collected between 2001 and 2012, and was detected in 9 samples-a detection rate of 0.44%. Estimated concentrations (<LOQ) ranged from 0.011 to 0.036 μg L^{-1} and the detection limits were from 0.016 to 0.34 μg L^{-1}. In all cases when samples were analyzed for CPY at the same time as CPYO (nine samples), concentrations of CPY were less than the MDL of 0.005 to 0.006 μg L^{-1}. Neither the CA-SWD (CDPR 2012a) nor the EIM (WDOE 2012) databases contained any detections for CPYO in surface waters. The CA-SWD included results of 288 analyses from 2005 to 2010 (LOQ = 0.05 to 0.06 μg L^{-1}), while the EIM database contained results of 964 analyses (MDL = 0.048 to 0.1 μg L^{-1}) from 2009 to 2011.

2.5 *Exposures in Relation to Changes in Use-Pattern*

Three different United States Geological Survey (USGS) studies used the same pesticide concentration dataset (Martin and Eberle 2009) to examine trends in pesticide concentrations in corn-belt streams (Sullivan et al. 2009), urban streams (Ryberg et al. 2010), and streams of the western U.S. (Johnson et al. 2011). The dataset containing concentrations of insecticides was compiled in the same manner as the Martin et al. (2011) dataset, which included data only from 1992 to 2008.

Pesticide concentration trends that commonly occur in streams and rivers of the corn belt of the U.S were assessed Sullivan et al. (2009) and the relative applicability and performance several statistical methods for trend analysis were evaluated. Temporal trends in concentrations of 11 pesticides (including CPY) with sufficient data were assessed at as many as 31 stream sites for two time periods: 1996–2002 and 2000–2006. Most sites had too few concentrations that were greater than the MDLs to determine meaningful trends in CPY concentrations. All the 11 sites for which flow-adjusted trends could be analyzed during 1996–2002 exhibited downward trends in concentrations of CPY, including significant downward trends at two sites and highly significant downward trends at five sites. Only three sites could be analyzed during 2000–2006. One site had a statistically significant downward trend, the second site had a non-significant upward trend, and the third site had a highly significant upward trend. Overall, the results indicate downward trends in concentrations of pesticides in general. This included decreasing CYP concentrations in Corn Belt streams and rivers during 1996–2006, which were explained largely by the corresponding decreases in annual use due to regulatory actions or market forces.

The second USGS study (Ryberg et al. 2010) used the Martin and Eberle (2009) dataset to assess trends of pesticide concentrations in 27 urban streams in the Northeastern, Southern, Midwestern and Western regions of the U.S. Three partially-overlapping 9-yr periods (1992–2000, 1996–2004, and 2000–2008) were examined for eight herbicides, five insecticides, and three degradation products. The data were analyzed for trends in concentrations by use of a parametric regression model. Due to small and declining frequencies detection, trends in concentration of CPY were not assessable at most sites, particularly during 2000–2008. Most of the streams for which adequate data were available exhibited significant downward trends in concentrations (i.e., 10 of 11 sites during 1996–2004 and two of five sites during 2000–2008). Between 2000 and 2008, most measured concentrations of CPY were less than the MDLs and only five sites had more than 10 detections. Concentrations of CPY at two of those sites continued to decline significantly, but there were no significant downward trends at the remaining three sites during the latter period. The downward trends of CPY concentrations in urban streams were consistent with the regulatory phase-out of residential uses of CPY between 1997 and 2001.

The most recent USGS study to use the Martin and Eberle (2009) dataset assessed trends in concentrations of two insecticides and five herbicides in 15 streams in California, Oregon, Washington, and Idaho from 1993 to 2005 (Johnson et al. 2011). Because of changes in amounts applied and methods of application in their associated catchments, pesticide concentration trends were estimated by using a parametric regression model to account for flow, seasonality, and antecedent hydrologic conditions. Short-term models of trends were developed for all sites for the period 2000–2005, while long-term models of trends were developed at 10 of the 15 study sites: two small urban sites (1996–2005), three small agricultural sites (1993–2005), and five large mixed land-use sites (1993–2005). Of the seven sites that had a sufficient number of uncensored (>MDL) concentrations of CPY for short-term trend analysis of flow-adjusted concentrations (2000–2005), only one site (Yakima River, WA; a large mixed land-use site) had a significant upward trend. For the long-term trend models, six sites had a sufficient number of uncensored concentrations of CPY for analysis of flow-adjusted concentrations. Three of the six sites had significant downward trends, one had a significant upward trend (Zollner Creek, OR; a small agricultural site), and the other two sites had no significant trends. Downward trends in concentrations of CPY at the small urban and California agricultural sites appear to reflect the phase-out of CPY and/or reduction in its use. Upward trends in the Yakima River, WA and Zollner Creek, OR might be due to increases in use from an increase in the planted acreage of certain crops (e.g., corn) and/or restrictions on the use of other organophosphate insecticides.

Overall, these three USGS analyses of the Martin and Eberle (2009) data generally indicate trends of decreasing pesticide concentrations, including CPY, in corn belt streams and rivers (Sullivan et al. 2009), urban streams in four regions of the U.S. (Ryberg et al. 2010), and in small urban and agricultural sites in California and the Pacific Northwest (Johnson et al. 2011).

3 Modeling of Chlorpyrifos in Aquatic Environments

Collectively, the CPY monitoring data provided useful and relevant insight towards quantifying the range of concentrations expected in surface waters. However, relatively few monitoring programs have sampled at a frequency sufficient to quantify the time-series pattern of exposure. Therefore, numerical simulations were used to characterize CPY concentrations in water and sediment for three representative high exposure environments in the United States. The environments were selected by parallel examination of use intensity across the U.S., susceptibility of CPY to runoff with respect to soil and weather variability across the U.S., and a sensitivity analysis of CPY runoff potential for various patterns of use. From the analyses, three geographical regions, each defined as several contiguous counties, were identified as having greater potential exposure to CPY. These regions were in central California, southwestern Georgia, and the Leelanau peninsula of Michigan. A small watershed, defined as having a 3rd order stream outlet, was selected from each region based on having a high density of cropland eligible for receiving CPY applications according to registered uses. Models were configured for each watershed and simulations conducted for up to 30 yr of consecutive CPY use using historical weather records for each region modeled. Daily mean concentrations of CPY in water and sediment from runoff, erosion, and drift sources were predicted at the watershed outlets.

3.1 *Selection and Justification of the Models*

Several models were used, the number of which depended on the level of detail needed for the specific phase of the assessment. Models were selected for their ability to represent the key fate and transport processes of CPY. Based on use practices and chemical properties discussed in Sect. 1.1, above, potential transport pathways to aquatic systems include water and soil erosion from rainfall and irrigation and spray drift during application.

The Pesticide Root Zone Model, PRZM, was selected to evaluate pesticide runoff potential because of its ability to account for pertinent environmental processes at an appropriate spatial scale and time step for chemical dissipation. PRZM is a dynamic, compartmental model developed for simulating movement of water and chemical in unsaturated soil systems within and below the plant root zone (Carousel et al. 2005). The model simulates time-varying hydrologic behavior on a daily time step, and includes physical processes of runoff, infiltration, erosion, and evapotranspiration. The chemical transport component of PRZM calculates pesticide uptake by plants, volatilization, surface runoff, erosion, decay, vertical movement, foliar loss, dispersion and retardation. PRZM includes the ability to simulate transport of metabolites, irrigation, and hydraulic transport below the root zone. PRZM is the standard model used for ecological and drinking water pesticide risk assessments by the U.S. Environmental Protection Agency's Office of Pesticide Programs (USEPA 2009). The model has undergone extensive validation, in which the results were compared with measured concentrations from numerous studies of field-scale runoff and leaching, conducted for pesticides in the United States (Carousel et al. 2005; Jones and Russell 2001). Moreover, PRZM has been integrated into several watershed assessment models in the U.S. (Parker et al. 2007; Snyder et al. 2011).

Two versions of PRZM were used here. The majority of simulations were conducted using PRZM version 3.12.2; the version that is incorporated into the pesticide registration review process of the USA (USEPA 2009). Simulations for California were conducted using a version of WinPRZM (FOCUS 2012) that was modified (Hoogeweg et al. 2012) to simulate pesticide losses in furrow and flood irrigation tail-water, which is a significant source of pesticide loadings in many areas of California and elsewhere that utilize these irrigation practices. WinPRZM contains additional enhancements that are not available in PRZM 3.12.2 that have been added for pesticide registration evaluation in Europe, including the ability to simulate soil adsorption using the Freundlich isotherm, temperature and soil-moisture-dependent degradation, and non-equilibrium sorption to soil (FOCUS 2012). WinPRZM produces identical exposure concentrations to PRZM version 3.12.2 when those options are not used.

The Exposure Analysis Modeling System, version 2.98.04 (EXAMS), was used to evaluate the relative effects of use practices (i.e., CPY labels) on exposure concentrations. Simulations utilized standard scenarios developed by USEPA's Office of Pesticide Programs for pesticide registration review, which are configured to run with EXAMS (USEPA 2009). The scenarios represent a 10-ha field draining into a 1-ha × 2 m deep pond. EXAMS combines a chemical fate and transport model with a steady-state hydraulic model to simulate the following processes: advection, dispersion, dilution, partitioning between water, biota, and sediment, and degradation in water, biota, and sediment (Burns 2004).

The regulatory version 2.05 of the AgDRIFT® model (Teske et al. 2002) was selected to estimate spray drift deposition onto aquatic water bodies. This version of

AgDRIFT® contains the Tier I levels for ground and orchard air blast spraying, and the Tier I, II and III levels for aerial spraying. Tier II and III ground and orchard air blast spraying screens have not been developed. The Tier I predictions for this assessment were used for all application methods.

The 2005 version of the Soil and Water Assessment Tool (SWAT) model was chosen to link and route CPY transport in the three focus watersheds. SWAT is a semi-distributed model developed by the U.S. Department of Agriculture to predict effects of land management practices on water, sediment and agricultural chemical yields in large complex watersheds that have varying soils, land use, and management conditions over long periods of time (Neitsch et al. 2010). Model components include weather, surface runoff, return flow, percolation, evapo-transpiration (ET), transmission losses, pond and reservoir storage, crop growth and irrigation, groundwater flow, reach routing, nutrient and pesticide loading, and water transfer. PRZM was used to simulate CPY in runoff and erosion, because SWAT is unable to simulate losses of CPY due to volatilization and runoff from furrow/flood irrigation tail-waters.

The Risk Assessment Tool to Evaluate Duration and Recovery (RADAR) was used to evaluate SWAT model output. RADAR is a software program commissioned by the Ecological Committee for FIFRA Risk Assessment Methods (ECOFRAM) in the late 1990s to conduct evaluations of exposure events from time-series data (ECOFRAM 1999) and to assess pulse-dose study designs. RADAR relies on the concept of defining a threshold value (or trigger value) for the water column concentration believed by the user to be of some interest in interpreting exposure. The software identifies each occurrence, in which CPY residues exceed this threshold and defines such occasion as an "event." Once an event has been triggered, the program calculates the duration that residues continue to be above this level (event duration) and then how long before the concentration again exceeds this threshold value (post-event interval).

3.2 Chemical Input Parameter Values, Source, and Rationale

The chemical properties used in models used in this assessment are summarized in Table 5. Selection of chemical properties began by evaluating the properties used by the U.S. Environmental Protection Agency for a dietary exposure assessment conducted in 2011 (USEPA 2011). USEPA's assessment was designed to be a conservative screening evaluation of pesticide fate and transport. However, this USEPA assessment did not adequately represent the behavior of CPY at the level of resolution and accuracy needed for the present assessment. Therefore, an expanded database on environmental fate properties was assembled and reviewed to better characterize relevant fate and transport processes.

Physicochemical properties. Available molecular weight, water solubility, and vapor pressure values were used for other elements of this risk assessment as needed (Mackay et al. 2014). PRZM simulates volatilization by using a dimensionless expression of Henry's law constant. The value provided in Table 5 was obtained

Table 5 Environmental properties used in model simulations

Property	Value used in model simulations
Molecular weight	350.6 g mol^{-1}
Water solubility	0.73 mg L^{-1}
Vapor pressure	1.87×10^{-5} torr
Henry's law constant	
PRZM model	0.5×10^{-5} (dimensionless)
EXAMS model	6.2×10^{-6} atm m^3 mol^{-1}
Soil adsorption / desorption, K_{OC}	8,216 cm^3 g^{-1} (mean K_{OC} from 37 data points)
Aerobic soil metabolism (t½)	Run 1: 28.3 d[a]
	Run 2: 96.3 d[a]
Aqueous hydrolysis (t½)	pH 5: 73 d
	pH 7: 81 d
	pH 9: 16 d
Aqueous photolysis (t½)	29.6 d
Aerobic aquatic metabolism (t½)	50.8 d
Anaerobic aquatic metabolism (t½)	63 d
Foliar degradation (t½)	3.28 d
Foliar washoff (cm^{-1})	0.1
Plant uptake	0.0

[a]In the text, these numbers are rounded to whole numbers; the table values were used in the modeling

by calibration of PRZM to achieve 10–15% loss of CPY in the first 2 d and 20–25% maximum after several weeks, values comparable to that reported in Sect. 1.2 and by Mackay et al. (2014). This fitted value is approximately two orders of magnitude smaller than the dimensionless values for Henry's law constants cited by the U.S. Department of Health and Health Services (USDHHS 1997) and by (Fendinger and Glotfelty 1990; Glotfelty et al. 1987; Suntio et al. 1987). Other parameters required for simulating volatilization include the diffusion coefficient of pesticide in air (DAIR) and the enthalpy of vaporization (ENPY). Values of 4,188 cm^2 d^{-1} and 14.3 kcal mol^{-1} were obtained for these parameters from Carousel et al. (2005).

Soil-water partition coefficients. Soil-water partition coefficients (Kd) were calculated by multiplying the average K_{OC} of 8,216 cm^3 g^{-1} times the organic carbon content (percent) of the soil series used in a specific model simulation. The K_{oc} value was derived as the mean value from the 37 studies presented in SI Table A4.

Soil degradation of CPY. To bracket the expected environmental behavior, model scenarios were evaluated with two soil degradation rates. As discussed in Sect. 1.2 and in Solomon et al. (2014), aerobic soil metabolism appears to be biphasic. The 28-d and 96-d half-lives represent conservative estimates for each phase. Rate constants were not adjusted for individual soils (i.e., to account for effects of hydrolysis on pH or photolysis).

Foliar processes. Degradation on foliar surfaces was represented by a half-life of 3.28 d. This half-life represents the upper 90th centile confidence bound on the mean half-life calculated from the 18 studies of accumulation of CPY (SI Table A3). If a range of values was listed, the largest value was used in the calculation for a study (SI Table A3). If the half-life value given was expressed as being less than a given value, that value was used as a conservative estimate in the calculation. The upper 90th centile confidence bound on the mean half-life was calculated (1), (USEPA 2009) and used in the model.

$$t_{input} = -t_{1/2} + \left[\frac{\left(t_{90,n} \times s \right)}{n^{1/2}} \right] \quad (1)$$

where, t_{input} = half-life input value (*time*) $t_{1/2}$ = mean of sample half-lives (*time*), s = sample standard deviation (*time*), n = number of half-live values available (–), and $t_{90,n-1}$ = one-sided Student's t value at $\alpha = 0.1$.

A foliar wash-off fraction of 0.1 cm^{-1} rainfall for CPY was used in the PRZM model (Carousel et al. 2005). This value was further supported by the results of dislodgeable foliar residue studies and foliar wash-off measurements (Poletika and Robb 1994; Racke 1993). The review by Racke (1993) indicates that the proportion of applied CPY that can be washed off foliage decreases with time after application, not unexpectedly, given the large values of K_{OC} and Log K_{OW} for CPY. Because dislodgeable residues diminish within hours after application and sampling times likely differ between studies, reported results have high variability. Relevant values of the wash-off fraction could not be derived from many of the older studies, since this would require assumptions about both the proportion of dislodgeable residue released by rain and the amount of rainfall required to release it. For example, Hurto and Prinster (cited in Racke 1993), reported dislodgeable residues on turfgrass of 0.03 μg cm^{-2}, and total residues of 0.53 to 0.68 μg cm^{-2} within the first 1–2 h after application, giving a wash-off fraction of 0.044 to 0.057. Since not all dislodgeable residues as measured in the laboratory (released by immersion in detergent and water) are washed off in rain, this is likely to be an overly conservative estimate. Although no information existed on the amount of rain needed to achieve such a reduction, it is likely to be substantial.

The best direct measure of foliar CPY wash-off was obtained from a field study on cotton (Poletika and Robb 1994). The results of this study (Fig. 4) show that the foliar wash-off fraction of CPY is strongly time-dependent, and has a maximum value of ~0.08 approximately 2 h after application. Other results that were based on irrigation of the treated area immediately after application were considered to be unrealistic. Under actual use conditions, CPY products are not to be applied when rainfall is anticipated within 2 h after application, since this could diminish the efficacy against foliar pests (except where the product is intentionally watered in to provide efficacy against pest in soil or thatch).

Finally, uptake of CPY by plants from the soil was considered negligible and this parameter was set to 0.0 in the model. CPY is not systemic and tends not to enter plants or move within the plant vascular system (Racke 1993).

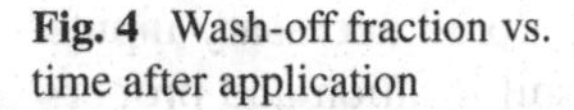

Fig. 4 Wash-off fraction vs. time after application

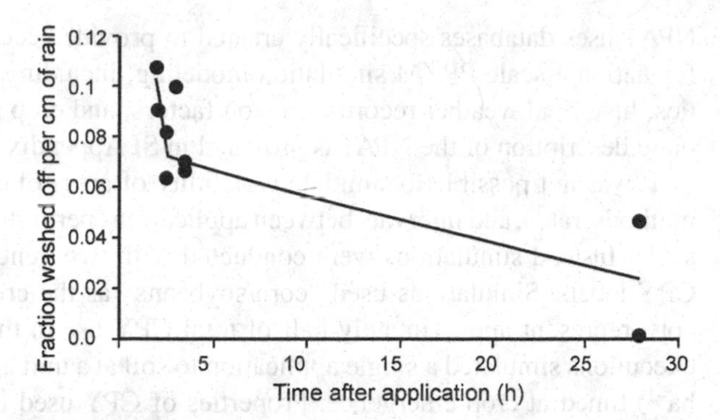

Properties of CPY in water. For the sensitivity analysis of patterns of use of CPY, degradation in the EXAMS model was regarded to occur through aerobic aquatic metabolism, aqueous hydrolysis, and photolysis. Aerobic aquatic metabolism was represented by the upper 90th centile confidence bound on the mean half-life ($t_{½}$=50.8 d) calculated from (1) by using the data in SI Table A5. Aqueous hydrolysis of CYP at pH 5, 7 and 9 has been reported with half-life values of 73, 81, and 16 d, respectively. The longer 81-d value was used for the modeling scenarios. The half-life for aqueous photolysis was taken as 29.6 d in the simulations. Volatilization was also simulated using the Henry's law constant of 6.2×10^{-6} atm m^3 mol^{-1}. The calibrated dimensionless value discussed above is not applicable for aquatic media. Degradation in sediment was assumed to occur via anaerobic aquatic metabolism (63-d half-life), which was calculated from (1) by using the reported laboratory half-life values of 39 and 51 d.

3.3 Selection of Exposure Scenarios

Distribution of CPY across the U.S. Intensity of use was a factor in the selection of watersheds. County sales data (Fig. 1) cannot be used to determine the precise location of product use, but it does provide a general indication of the spatial distribution of CPY use. Areas in the U.S. with the greatest intensity of use from 2010 to 2011 (depicted in blue in Fig. 1, see SI for color map) include Kern, Tulare, Santa Cruz, Fresno counties in central California; Lancaster County in southeastern Pennsylvania; and Calhoun, Decatur, and Mitchell counties in southeast Georgia. Other areas of high use at a lesser density are depicted by the green shading in Fig. 1 (see SI for color map).

National vulnerability assessment of CPY runoff. A national vulnerability assessment was conducted to characterize the potential for CPY to be transported beyond a treated field in runoff water and eroded sediment throughout the conterminous United States. The assessment involved use of the National Pesticide Tool (NPAT) to simulate the relative runoff potential of CPY for all major agricultural soil types.

NPAT uses databases specifically created to provide access to all necessary inputs for national-scale PRZM simulation modeling, including soil location and properties, historical weather records, erosion factors, and crop parameters. A more complete description of the NPAT is provided in SI Appendix B.

It was not possible to simulate variability of areas of crops planted, application methods, rates, and intervals between applications permitted on labels at the national scale. Instead simulations were conducted with two generalized interpretations of CPY labels. Simulations used "corn/soybeans" as the crop because soybeans and corn represent approximately half of total CPY use in the U.S. One of the NPAT executions simulated a single application to soil at a unit application rate (1.0 kg a.i. ha^{-1}) timed at crop emergence. Properties of CPY used in the simulation are provided in Table 5 except that foliar and aquatic properties were not used in simulations and only the predominant 28-d half-life was simulated for aerobic soil metabolism. The second simulation represented two applications of CPY at 0.5 kg a.i. ha^{-1} at 21 and 35 d post-emergence. Properties of CPY used in the simulation are consistent with those of the soil-applied application, with the exception that foliar properties are utilized; however, volatilization from soil was not simulated because it is not as significant with foliar applications. Combined, the simulations performed included more than 64,000 combinations of soil and weather conditions for each of the two types of application considered. Each simulation was conducted for 30 consecutive years of CPY application, and used 30 yr of historical weather data representative of the geographical location of the simulation. Daily predictions of runoff and erosion losses of CPY were estimated for each simulation and expressed as annual loads (kg ha^{-1} yr^{-1}).

The annual load calculated for the 90th centile year was used for assessment purposes. Statistics were performed using an Extended Reach File 1 (ERF) polygon basis (USEPA 2006) for mapping watersheds (Figs. 5 and 6). There are approximately 61,000 ERF polygons in the conterminous U.S.A. Each model simulation associated with a soil polygon that spatially intersects an ERF polygon was assigned a relative weight in the distribution, based on the estimated area of the soil that resides within the ERF (2).

$$\text{Prob}\left(Sim_x\right) = A_x / \sum_{i=1,n} \left(A_i\right) \tag{2}$$

Where, *Prob* (Sim_x) is the area weighted probability associated with simulation "*x*", A_x is the area associated with simulation "*x*" that resides within the ERF, and $\sum_{i=1,n}$ (A_i) is the area sum of the areas of all simulations that reside within the ERF.

The results of soil and foliar executions of NPAT are shown in Figs. 5 and 6, respectively. These figures present the combined runoff and erosion losses of CPY expressed as the percent of applied active ingredient at the ERF resolution. Similar patterns can be seen in Figs. 5 and 6 with the greatest losses occurring in the high summer rainfall regions on relatively heavy soils in the U.S., such as Mississippi Delta regions of Louisiana, Mississippi, and Arkansas. Relatively large losses are also concentrated in northern Missouri and bordering areas. This latter region of

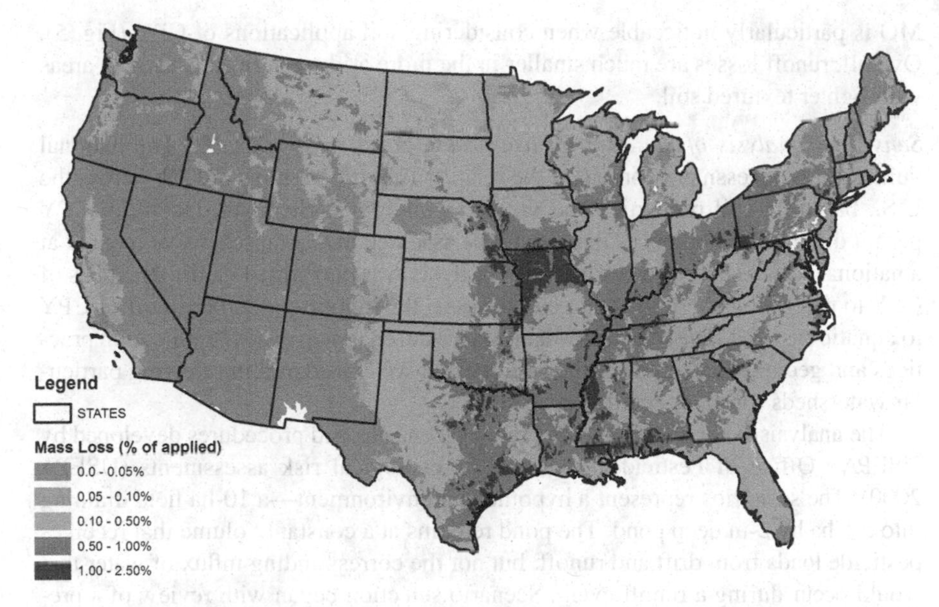

Fig. 5 Relative runoff potential of chlorpyrifos in runoff and erosion with soil applications (see SI for a color version of this Figure)

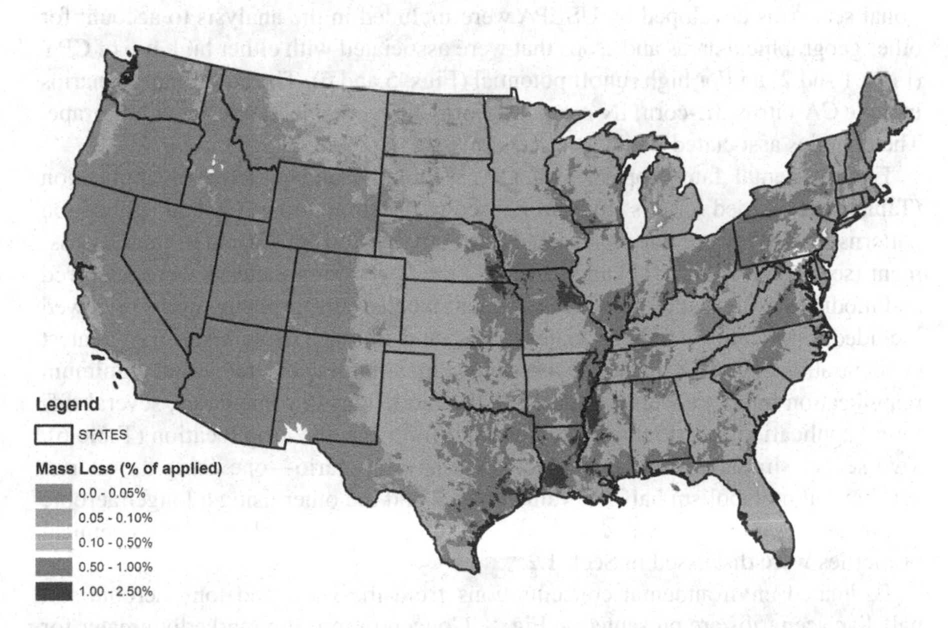

Fig. 6 Relative runoff potential of chlorpyrifos in runoff and erosion with foliar applications (see SI for a color version of this Figure)

MO is particularly noticeable when considering soil applications of CPY (Fig. 5). Overall, runoff losses are much smaller in the more arid western states and in areas with lighter textured soil.

Sensitivity analysis of the effect of use practices on runoff of CPY. The national vulnerability assessment compared the relative runoff potentials of CPY across the U.S., based on soil properties and weather conditions. However, labels for CPY permit different use practices that cannot be assessed in a comprehensive manner at a national scale. Therefore, a sensitivity analysis was performed on use practices of CPY to determine conditions that can result in the highest potential runoff of CPY to aquatic systems. The sensitivity analysis was used to narrow the application practices and geographical areas of the country that we considered in selecting particular watersheds for more detailed analyses.

The analysis was conducted using model scenarios and procedures developed by USEPA's Office of Pesticide Programs for ecological risk assessments (USEPA 2000). The scenarios represent a hypothetical environment—a 10-ha field draining into a 1-ha by 2-m deep pond. The pond remains at a constant volume that receives pesticide loads from drift and runoff, but not the corresponding influx of water that would occur during a runoff event. Scenario selection began with review of a preliminary drinking water assessment of CPY conducted by USEPA (USEPA 2011). The scenarios associated with the highest exposures were selected for the sensitivity analysis: CA-grape, PA-turf, GA-pecans, MI-cherries, and FL-citrus. Several additional scenarios developed by USEPA were included in the analysis to account for other geographical areas and crops that were associated with either high use of CPY (Figs. 1 and 2) and/or high runoff potential (Figs. 5 and 6). The additional scenarios include CA-citrus, IL-corn, IN-corn, NC-corn, NE-corn, NC-apples, and NY-grape. The counties associated with these scenarios are depicted (Fig. 7).

Environmental fate properties of CPY (Table 5) and patterns of application (Table 6) developed for this study were used in the simulations. Certain application patterns for CPY were not represented correctly by USEPA (2011) in their assessment (see Racke et al. 2011) and therefore, the application patterns were reviewed and modified as necessary to better represent labeled uses. Specific items reviewed included: application methods, dates, rates, and timing. To simulate the greatest concentrations of CPY in aquatic systems, maximum label rates and minimum reapplication intervals were evaluated in the modeling. In some cases, several different application practices were represented for a specific crop location (Table 6). Two sets of simulations were conducted for each scenario—one using the shorter aerobic soil metabolism half-life value of 28 d and the other using a longer aerobic soil metabolism half-life value of 96 d. The source and rationale for these chemical properties were discussed in Sect. 1.2.

Estimated environmental concentrations from the short and long aerobic soil half-live scenarios are presented in Fig. 8. Concentrations are markedly greater for certain scenarios (e.g., GA-pecan1, FL-citrus2, MI-cherries1, and IL-corn3). For other scenarios (e.g., CA-citrus3 and FL-citrus2) differences were negligible. Two scenarios (GA-pecans1 and MI-cherries1) resulted in the highest estimates of exposure concentrations and became a significant factor in selecting watersheds for the exposure assessment.

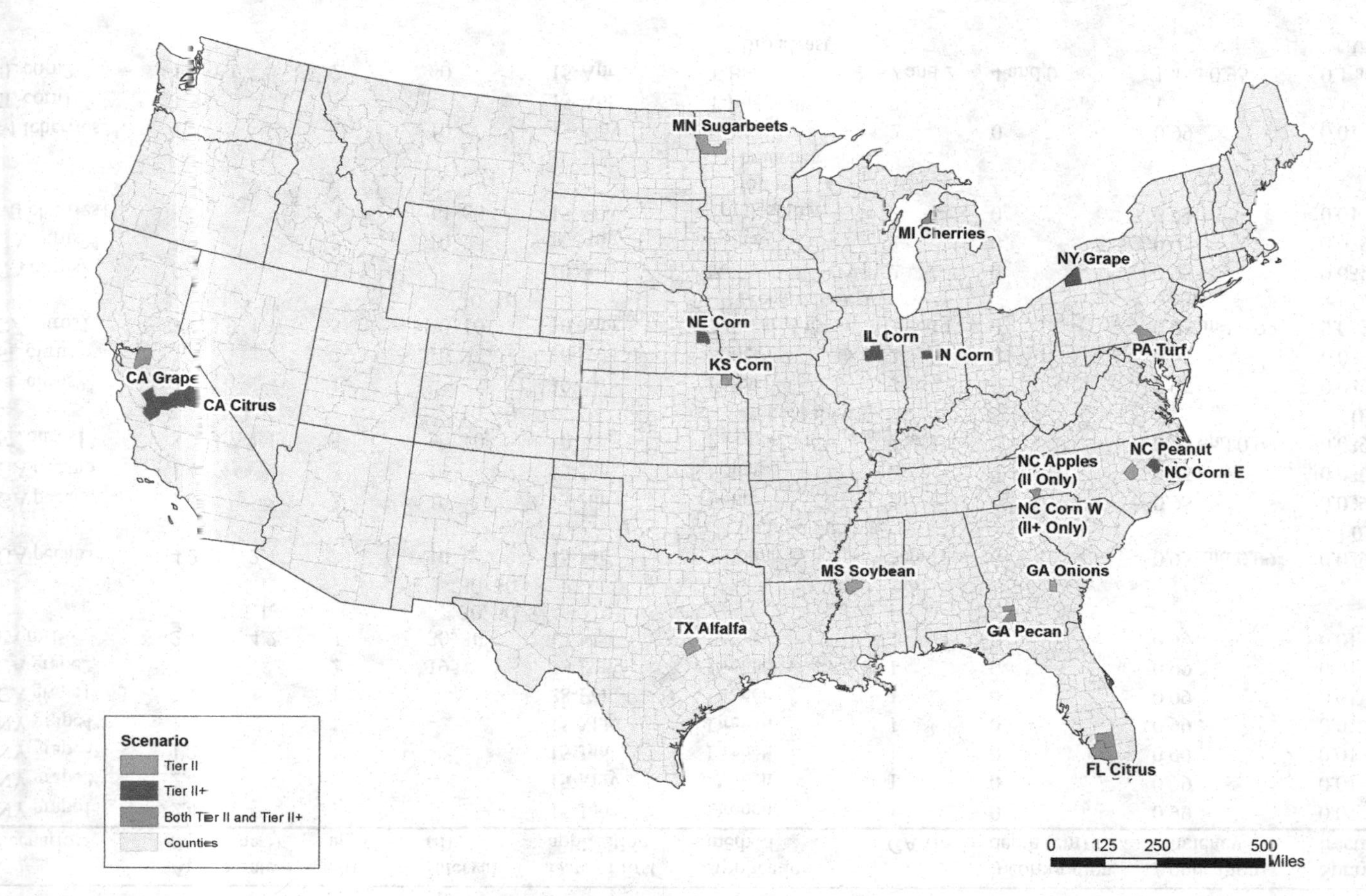

Fig. 7 Locations of scenarios for crop/label sensitivity analysis (see SI for a color version of this Figure)

Table 6 Scenarios and chlorpyrifos application parameters for sensitivity-analysis of use-pattern

Scenario	Applic. rate (kg a.i. ha^{-1})	No. apps.	Interval(s) (d)	Date of first application	Application method	CAM	Incorporation depth (cm)	Application efficiency	Spray drift fraction
NY grape1[a]	2.4	1	–	15-Jun	Drench	1	0	0.99	0.01
NY grape2[a]	2.4	1	–	15-May	Drench	1	0	0.99	0.01
NY grape3[a]	1.2	1	–	15-Jun	Drench	1	0	0.99	0.01
NY grape4[a]	1.2	1	–	15-May	Drench	1	0	0.99	0.01
CA grape1[b]	2.1	1	–	28-Feb	Drench	1	0	0.99	0.01
CA grape2[b]	1.1	2	195	28-Feb	Drench	1	0	0.99	0.01
PA turf[c]	2.1, 4.2, 4.2, 1.1, 1.1, 1.1, 1.1	7	30, 30, 30, 60, 30, 30	14-Mar	Foliar	2	0	0.99	0.01
GA pecan1[d]	4.2, 2.1, 2.1	3	10	15-Jun	1 Foliar, 2 orchard floor	2 and 1	0	0.95 and 0.99	0.039 and 0.01
GA pecan2[e]	1.0	3	10	15-Jun	Foliar	2	0	0.95	0.039
GA pecan3[e]	1.4	3	10	15-Jun	Foliar	2	0	0.95	0.039
FL citrus1[d]	3.7, 3.7, 1.1, 1.1, 1.1	5	30, 10, 10, 10	10-Aug	2 Foliar, 3 to orchard floor	2 and 1	0	0.95 and 0.99	0.039 and 0.01
FL citrus2[d]	3.7	1	–	10-Aug	Foliar	2	0	0.95	0.039
FL citrus3[d]	3.7	2	10	10-Aug	Foliar	2	0	0.95	0.039
CA citrus1[d]	3.9, 3.9, 1.1, 1.1, 1.1	5	30, 10, 10, 10	10-Aug	2 Foliar, 3 to orchard floor	2 and 1	0	0.95 and 0.99	0.039 and 0.01
CA citrus2[d]	6.3	1	–	10-Aug	Foliar	2	0	0.95	0.039
CA citrus3[d]	3.9	2	10	10-Aug	Foliar	2	0	0.95	0.039
MI cherries1[d]	3.16	3	14, 60	15-May	Trunk spray/ lower branches	2	0	0.99	0.01
M Icherries2[d]	1.7	2	10	15-May	Foliar	2	0	0.99	0.01
IL corn16[f]	1.5	1	–	15-Apr	T-Band	7	4	1	0.1
IL corn2[f]	1.5, 1.1	2	60	15-Apr	T-Band, Broadcast	7 and 2	4 and 0	1 and 0.95	0.1 and 0.039

IL corn3[f]	1.1	3	10	15-Apr	Broadcast	2	0	0.95	0.039
IN corn1[f]	1.5	1	–	1-May	T-Band	7	4	1	0.1
IN corn2[f]	1.5, 1.1	2	60	1-May	T-Band, Broadcast	7 and 2	4 and 0	1 and 0.95	0.1 and 0.039
IN corn3[f]	1.1	3	10	1-May	Broadcast	2	0	0.95	0.039
KS corn1[f]	1.5	1	–	20-Apr	T-Band	7	4	1	0.1
KS corn2[f]	1.5, 1.1	2	60	20-Apr	T-Band, Broadcast	7 and 2	4 and 0	1 and 0.95	0.1 and 0.039
KS corn3[f]	1.1	3	10	20-Apr	Broadcast	2	0	0.95	0.039
NC cornE1[f]	1.5	1	–	1-Apr	T-Band	7	4	1	0.1
NC cornE2[f]	1.5, 1.1	2	60	1-Apr	T-Band, Broadcast	7 and 2	4 and 0	1 and 0.95	0.1 and 0.039
NC cornE3[f]	1.1	3	10	1-Apr	Broadcast	2	0	0.95	0.039
NC cornW1[f]	1.5	1	–	10-Apr	T-Band	7	4	1	0.1
NC cornW2[f]	1.5, 1.1	2	60	10-Apr	T-Band, Broadcast	7 and 2	4 and 0	1 and 0.95	0.1 and 0.039
NC cornW3[f]	1.1	3	10	10-Apr	Broadcast	2	0	0.95	0.039
NE corn1[f]	1.5	1	–	10-May	T-Band	7	4	1	0.1
NE corn2[f]	1.5, 1.1	2	60	10-May	T-Band, Broadcast	7 and 2	4 and 0	1 and 0.95	0.1 and 0.039
NE corn3[f]	1.1	3	10	10-May	Broadcast	2	0	0.95	0.039
MS soybean[e]	1.1	3	10	15-May	Broadcast	2	0	0.95	0.039
TX alfalfa[e]	1.1, 1.1, 1.1, 0.56	4	10	17-May	Broadcast	2	0	0.95	0.039
MN sugarbeet[e]	1.1	3	10	28-May	Broadcast	2	0	0.99	0.01

[a]From Cornell University Cooperative Extension (2012a)
[b]From Lorsban Advanced Supplemental Labeling (Dow AgroSciences 2009)
[c]From Cornell University Cooperative Extension (2012b)
[d]From Lorsban Advanced Specimen Label (Dow AgroSciences 2010)
[e]From (USEPA (2011)
[f]From Gomez (2009)

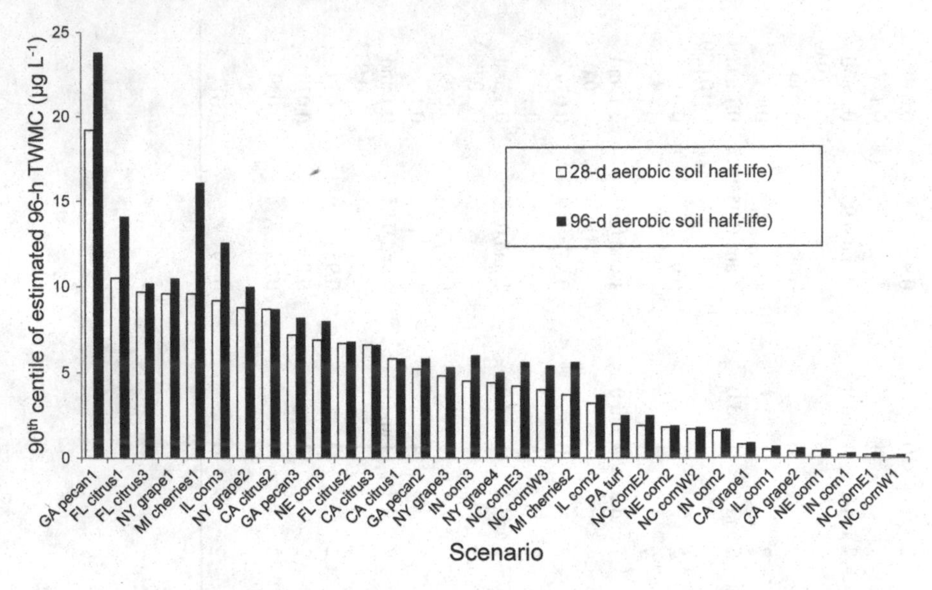

Fig. 8 Influence of aerobic soil metabolism half-life on 90th centile 96-h time-weighted mean estimated environmental concentrations for crop/label sensitivity analysis

Assessment of vulnerability for runoff of CPY in California. An assessment of vulnerability specific to California was conducted for several reasons. First, because of the large volume of CPY used in that state (Fig. 1); second, the historical detections of CPY reported from monitoring studies conducted in the state (CDPR 2012a; CEPA 2011a, b; USGS 2013); third, the availability of detailed records on pesticide applications from the California Department of Pesticide Regulation's (CDPR) Pesticide Use Reporting (PUR) database (CDPR 2012b); and fourth, because of the intensive use of furrow and flood irrigation on crops registered for use with CPY in the state (Orang et al. 2008). Losses in furrow and flood irrigation tail-water have been identified as a major cause of transport of pesticides, including CPY, to non-target aquatic systems (Budd et al. 2009; Gill et al. 2008; Long et al. 2010; Starner et al. 2005). This pathway could not be simulated with the National Pesticide Assessment (NPAT) model.

The assessment of vulnerability for California was conducted using the Co-occurrence of Pesticides and Species Tool (CoPST), a modeling framework developed by the California Department of Water Resources (CDWR) to evaluate the potential spatial and temporal co-occurrence of pesticides with threatened or endangered aquatic and semi-aquatic species (Hoogeweg et al. 2011). The framework integrates a number of databases necessary for temporal and spatial analysis, including historical records of pesticide use and soil properties, weather records, agronomic data, and species habitat for the Sacramento River, San Joaquin, and the San Francisco Bay Delta Estuary watersheds (SI Appendix C). Using the modeling framework, daily pesticide runoff and erosion losses in the Central Valley of CA were estimated for historical CPY applications to agricultural fields. Approximately 47,860 historical applications of CPY were simulated for the period 2000–2008.

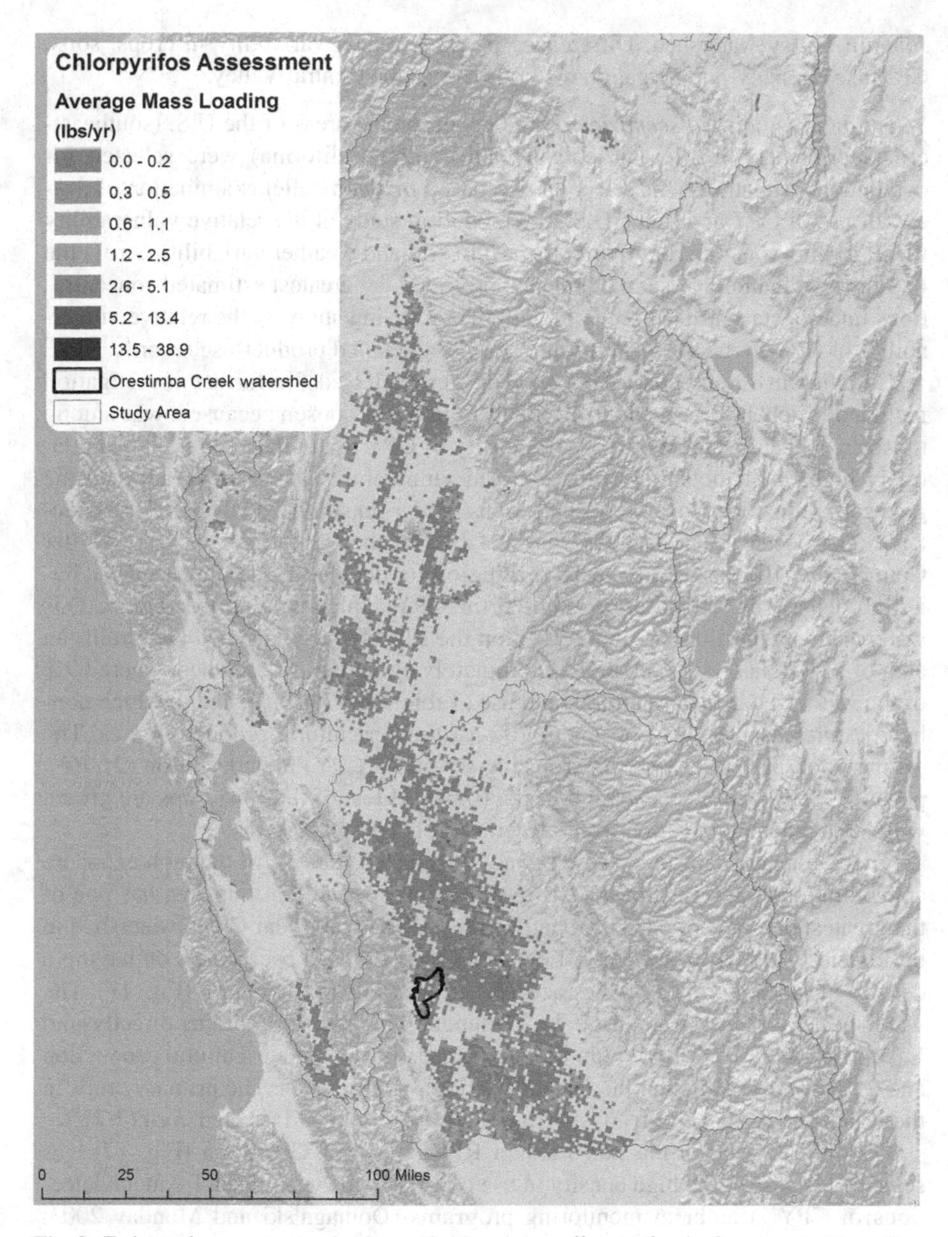

Fig. 9 Estimated average annual chlorpyrifos loss in runoff water for the Sacramento River, San Joaquin River, and Bay Delta estuary watersheds of California (see SI for a color version of this Figure)

Average annual mass loadings (Fig. 9) ranged from as little as 0.45 g to 18.5 kg per Public Land Survey Section (PLSS) per year. The largest annual mass loadings from runoff and erosion were predicted in southern Yuba County of the Sacramento Valley watershed and San Joaquin, Madera, and Stanislaus counties of the San

Joaquin Valley watershed. Differences in loads reflect variability in crops, soils, CPY use intensity, and irrigation methods across the Central Valley.

Selection of watershed scenarios. Three geographical areas of the U.S. (southeastern Georgia, northwestern Michigan, and central California) were selected for detailed investigation. This selection was based on the parallel examination of density of use of CPY across the U.S.; the modeling study of the relative vulnerability to CPY being found in runoff with respect to soil and weather variability across the U.S.; the evaluation of use- patterns that provided the greatest estimated concentrations under a standardized scenario; and the modeling study of the relative vulnerability of CPY to runoff in California, based on detailed product-use records.

Counties in southeastern Georgia that were selected for further investigation included Mitchell, Baker, and Miller. This region was chosen because it was among those areas that sustained the highest density of CPY use in the country, it had relatively high runoff potential in the national vulnerability assessment, and had the greatest predicted exposure concentrations in the sensitivity analysis of use practices (GA-pecans). The Dry Creek watershed in the Flint River basin of Miller County (Fig. 10) was selected from this region for detailed modeling because it has a high density of labeled crops eligible for CPY applications in the region. This watershed is a third-order tributary within the Flint River watershed and drains an area of 12,322 ha. Dry creek is a predominately agricultural watershed, where 69% of the watershed is in agricultural land use (Table 7 and Fig. 10), and in which non-hay/pasture crops account for about 70% of the agricultural production area. The primary non-pasture crops (i.e., crops registered for CPY) include cotton (21.7%), peanut (16.3%), corn (5.5%), and pecan (0.6%). Most of the row crops are grown with supplemental irrigation from center pivot irrigation systems.

The Leelanau peninsula of Michigan was selected because of the high exposure concentrations estimated for the MI-cherry scenario, and that this area has one of the greatest densities of cherry orchards in the USA. The Cedar Creek watershed in the Betsie-Platte River system in Leelanau County was chosen based on having a density of labeled crops eligible for CPY applications in the region (Fig. 11). The watershed drains 7,077 ha into Lake Leelanau, which, in turn, drains directly into Lake Michigan. Land-use within the watersheds is 35% in agricultural production and 23% of the area in non-hay/pasture agriculture (Table 7). The primary crops in the watershed are hay (12.0%), cherries (11.1%), alfalfa (8.1%), and corn (2.7%)

Orestimba Creek in the San Joaquin River basin of California (Fig. 12) was selected because of the high density of use of CPY in the watershed, frequent detections of CPY in ambient monitoring programs (Domagalski and Munday 2003; Ensminger et al. 2011; Zhang et al. 2012), and because an existing model setup of the watershed was available from the University of California (Luo et al. 2008; Luo and Zhang 2010). The watershed drains 55,998 ha from the mountainous area in the west, where brush and scrubland is predominant, into the agriculturally intensive valley floor. Overall, agriculture accounts for 22% of the total watershed area with the primary crops being walnuts, almonds, corn and alfalfa (Table 7). In the intensely cropped valley floor, agriculture accounts for over 80% of land use. In the time period from 2000 to 2008, 11,658 kg of CPY were applied.

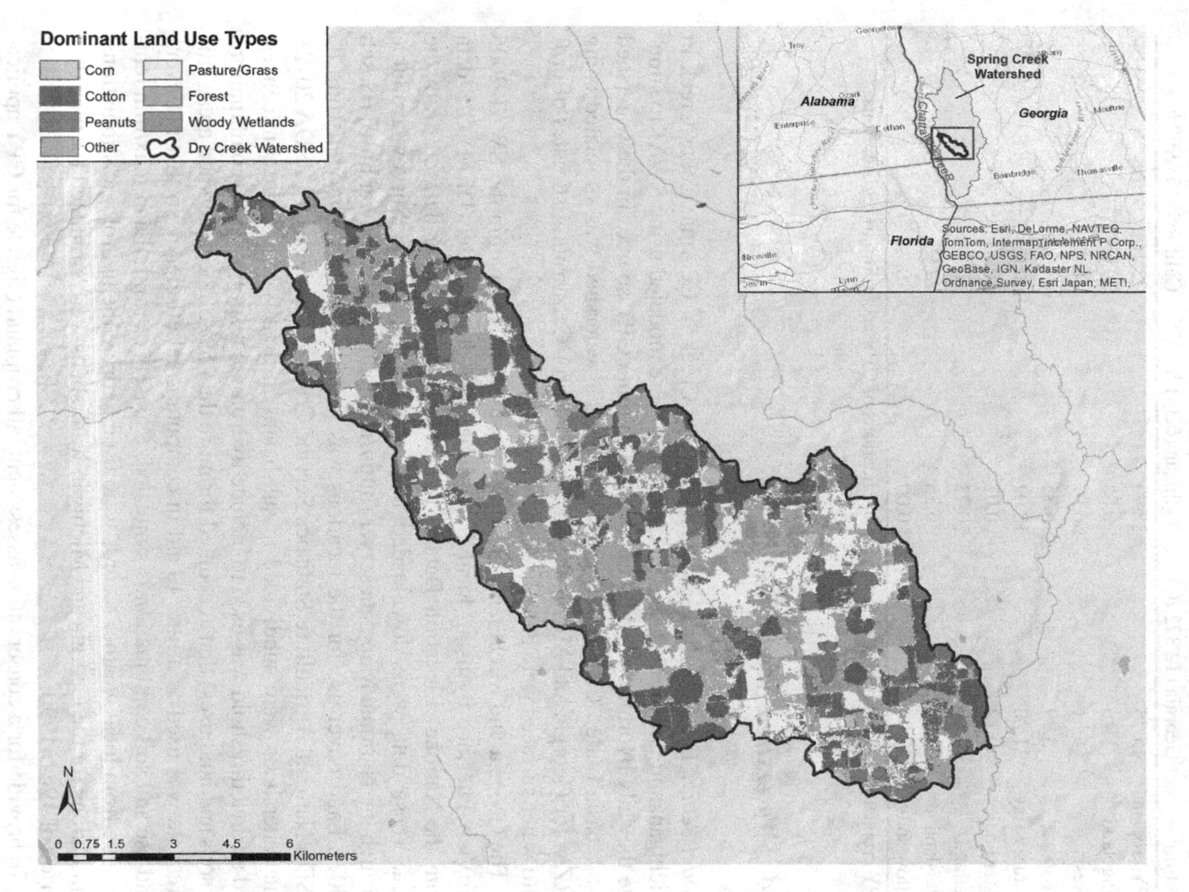

Fig. 10 Distribution of land use in Dry Creek watershed, GA (see SI for a color version of this Figure)

Table 7 Land use distributions in Dry Creek watershed in Georgia, Cedar Creek watershed in Michigan, and Orestimba Creek watershed in California

	Land use distribution (%)		
Aggregated land-use	Dry Creek watershed, Georgia (12,322 ha)	Cedar Creek watershed, Michigan (6,381 ha)	Orestimba Creek watershed, California (55,997 ha)
CPY cropland	45.75	22.72	22.16
Rangeland	1.46	0.88	29.63
Forests	16.31	36.97	14.88
Urban	3.99	6.75	3.58
Pasture / Hay	21.26	12.02	0.00
Water bodies	0.12	0.07	0.15
Wetlands	6.35	20.23	0.91
Grassland	2.73	0.36	27.95
Fallow soils	2.04	0.00	0.74

CPY cropland = crops eligible for receiving applications of CPY according to registered uses

3.4 Watershed Scenarios

Model Setup. The three focus watersheds (Dry Creek in Georgia, Cedar Creek in Michigan, and Orestimba Creek in California) were modeled using SWAT to simulate the daily hydrology and hydraulics and to route CPY sources in the watershed to the basin outlet. CPY runoff mass within each watershed was simulated using PRZM for Georgia and Michigan and winPRZM for California. Estimates of drift onto water surfaces were calculated from AgDrift.

The Georgia and Michigan watersheds were delineated within the SWAT 2005 ArcView interface by using local 10 m digital elevation model (DEM) GIS data from the National Elevation Dataset (NED) and hydrography datasets from the area. These data were downloaded from the US Geological Survey available through the National Map data server (http://nationalmap.gov/viewer.html) (Gesch 2007). Land cover/use for the locations was mapped within the model from the USDA National Agriculture Statistics Service Cropland Data Layer (USDA 2012). The watersheds were subdivided in sub-basins, primarily at stream confluences, and an area threshold was used to include hydrologic response units (combinations of soils and land uses) that accounted for more than 5% of the watershed area. This threshold was used as a filter to aid in computational efficiency so that not every land use and soil combination is simulated. Datasets of rainfall and max/min temperature from local weather records were used to drive the hydrologic simulations. Simulations for the Georgia and Michigan watersheds were simulated for the 30-yr period of 1961–1990.

To provide for a conservative assessment, all cropland eligible for CPY applications according to product labels in the Dry Creek and Cedar Creek watersheds were represented as "GA-pecan1" or "MI-cherries1". For example, areas designated as pecan, cotton, sorghum, corn, peanut, and peaches in the Dry Creek watershed were represented as GA-pecan1, with respect to soil, crop, and CPY application

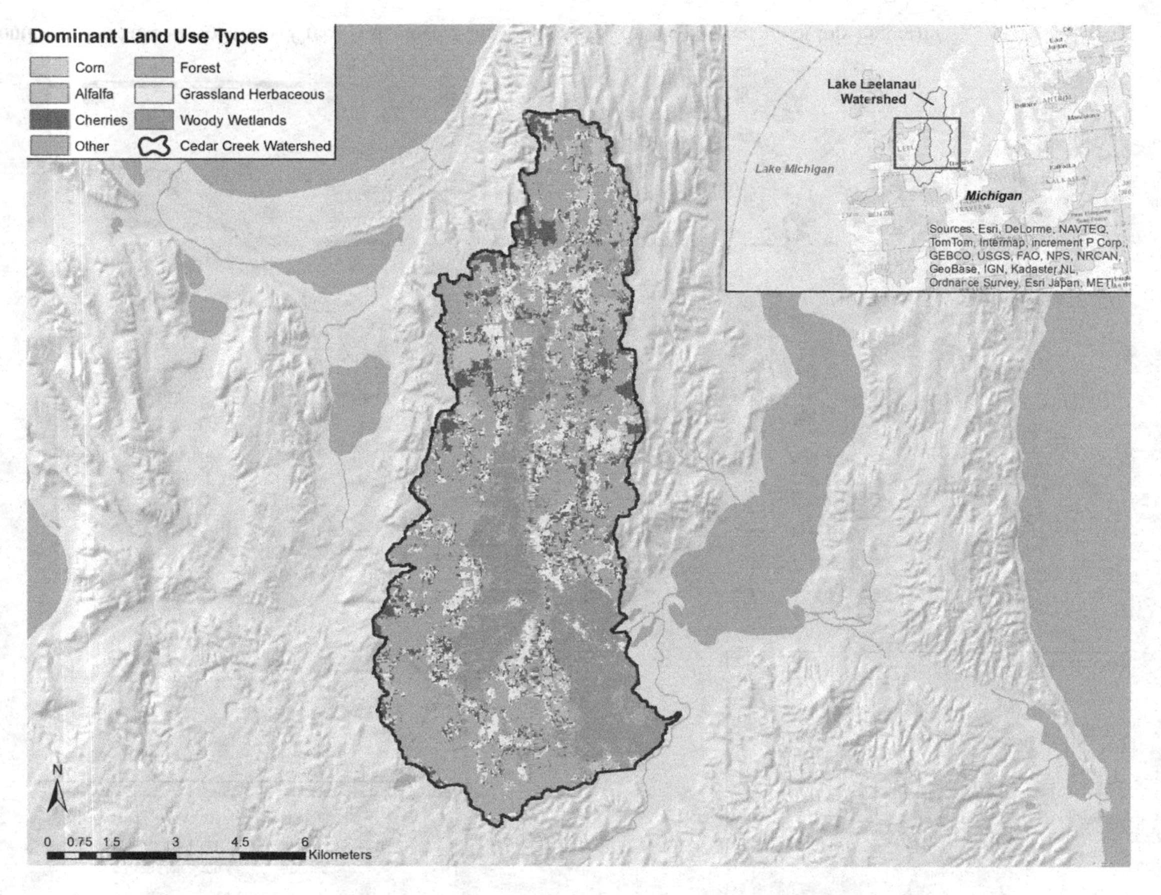

Fig. 11 Distribution of land use in Cedar Creek watershed, Michigan (see SI for a color version of this Figure)

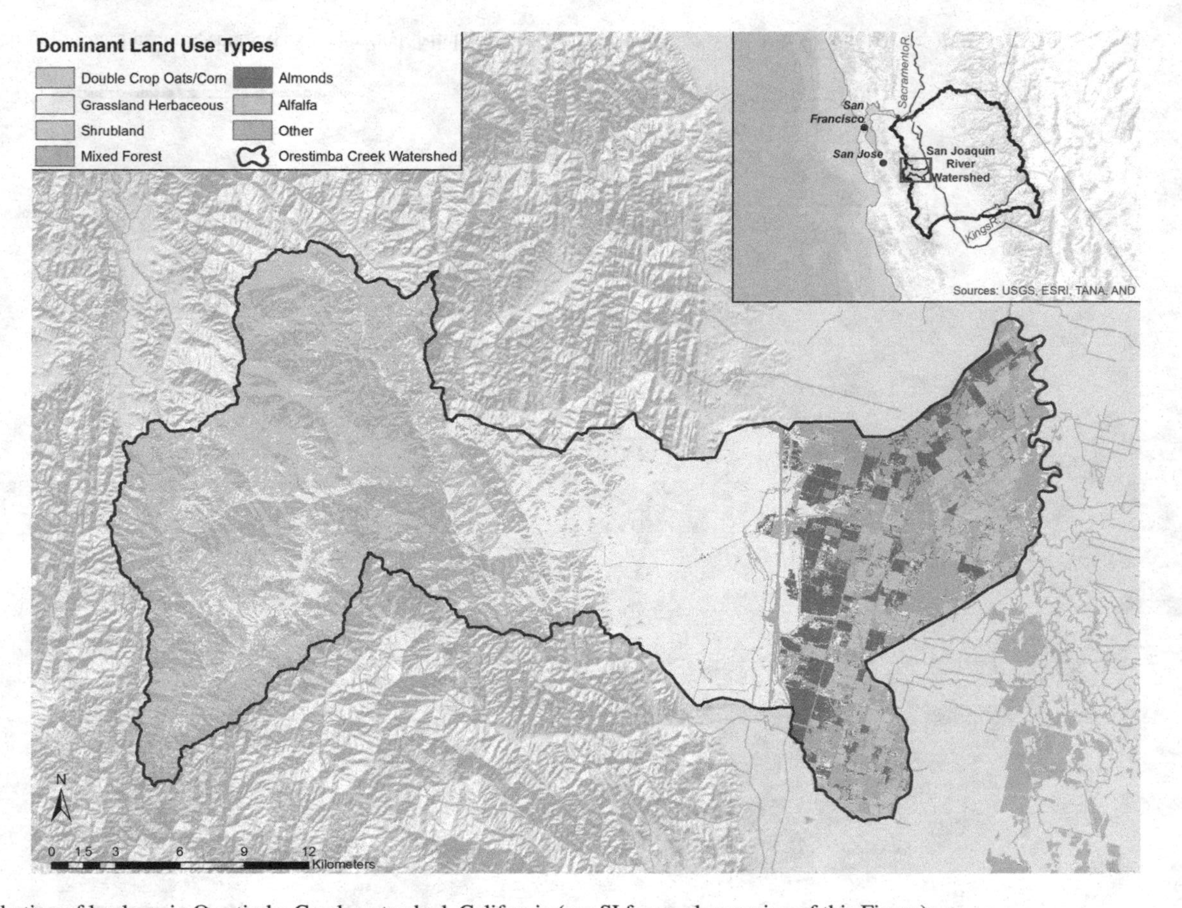

Fig. 12 Distribution of land use in Orestimba Creek watershed, California (see SI for a color version of this Figure)

rates, methods, and intervals between applications. Similarly, all eligible cropland in Cedar Creek was represented as the MI-cherries1 scenario. The applications were scheduled to occur over a 7-d application window to represent the reality that not all fields in a 7,000 12,000 ha watershed can receive applications on the same day.

An additional adjustment was made to the plant growth parameters for the Cedar Creek scenarios. In practice, applications of chlorpyrifos on cherries are directed to tree trunks and lower branches. With tart cherries, foliage may also be sprayed. The crop emergence and maturation dates in USEPA's scenario for MI-cherries resulted in an unrealistic 18, 47, and 0% of the CYP applied being depositing on the trees and 82, 53, and 100% on soil for each of the three applications per year, respectively. To simulate a more realistic proportion of deposition of chemical on the trees, the maximum canopy and maturity and harvest dates were modified to achieve deposition of 80% of the a.i. applied on the trees and 20% on the soil for all three applications.

Inputs of CPY for the California model were calculated slightly differently from the Georgia and Michigan watersheds. Applications of CPY were specific to the date, crop, and PLSS location in the California Pesticide Use Reporting database (CDPR 2012b). The California simulations were simulated for the more recent 9-yr period of 2000–2008 during which there were 293 actual CPY applications across the watershed.

Drift of CPY onto water surfaces was estimated with the AgDRIFT model to account for setback requirements on product labels. The ground spray was assumed to be applied with a 7.6 m (25-ft) spray buffer, high boom, and a fine to medium/coarse spray. Aerial spray was assumed to have a 45.7 m (150-ft) spray buffer and medium to coarse spray. Orchard applications were represented as an aerial spray (45.7 m spray buffer), because airblast (15.2 m spray buffer) is not specifically identified in the PUR database as an application method. Boom height and droplet sizes that result in the greatest drift allowed on CPY use labels were assumed. The proximity of treated fields to water was unknown, and therefore the area of surface water receiving spray drift was assumed to be in direct proportion of the watershed land area receiving applications.

Chemical properties used in the simulations of Dry Creek, Cedar Creek, and Orestimba Creek are summarized in Table 3. Two sets of simulations were conducted for each watershed using the 28-d and 96-d aerobic soil metabolism half-lives. Foliar degradation was not represented in the Orestimba Creek simulations because of occasional model crashes that prevented completion of the full set of simulations. Results are conservative for applications that occur to foliage for Orestimba Creek.

Daily estimates of masses of soluble and sediment-bound CPY were predicted at the outlet of each watershed with SWAT. Soluble mass values were converted to concentrations by dividing the mass of the soluble CPY (mg) by the volume of stream flow (L) for the day. Concentrations of sediment-bound CPY were calculated from the ratio of the daily bed load CPY mass (mg) to the mass of bed sediment in the outlet reach, assuming an active sediment layer of 3 cm (length of reach × width of reach × 0.03 m depth of sediment bed).

Table 8 Characteristics of modeled 96-h time-weighted mean chlorpyrifos concentrations in water draining the focus watersheds

Location of watershed		Concentration in water (µg L^{-1}) Half-life 28-d	96-d	Concentration in sediment (µg kg^{-1}) Half-life 28-d	96-d
California	Minimum	0.003	0.013	4.74	10.8
	Median	0.024	0.251	7.91	14.6
	90th centile	1.319	1.543	10.8	20.8
	Maximum	1.392	2.142	11.0	22.2
Georgia	Minimum	0.004	0.004	0.007	0.007
	Median	0.010	0.010	0.023	0.027
	90th centile	0.020	0.020	0.050	0.060
	Maximum	0.023	0.023	0.064	0.067
Michigan	Minimum	0.002	0.007	0.027	0.010
	Median	0.011	0.018	0.021	0.038
	90th centile	0.018	0.034	0.032	0.060
	Maximum	0.028	0.043	0.047	0.074

Values reflect the 90th centile of the annual maxima (1 in 10 yr maximum value)

Annual maximum concentrations. A probability analysis was conducted on the daily time-series predictions of CPY concentrations in the water and sediment for each watershed using the RADAR program (SI Appendix E). RADAR converts a daily time series to an annual maximum series for various exposure durations that are based on a rolling average calculation. The durations of exposure included the instantaneous maximum, 96-h, 21-d, 60-d, and 90-d average concentrations.

Maximum daily concentrations predicted for the California, Georgia, and Michigan watersheds were 3.2, 0.041, and 0.073 µg L^{-1}, respectively, with the 28-d aerobic soil metabolism half-life and 4.5, 0.042, and 0.122 µg L^{-1}, respectively, with the 96-d soil half-life. For sediments, the maximum daily concentrations predicted for the California, Georgia, and Michigan watersheds were 11.2, 0.077, and 0.058 µg kg^{-1}, respectively, with the 28-d half-life and 22.8, 0.080, and 0.087 µg kg^{-1}, respectively, with the 96-d soil half-life. The 96-h time weighted mean concentrations (TWMC), used in the risk assessment (Giddings et al. 2014) are summarized in Table 8. Concentrations associated with the minimum year, median year, 90th centile year, and maximum year are presented. Annual maxima concentrations for all years are provided in SI Appendix D.

The greatest annual concentrations of CPY in water and sediment were predicted in the California watershed, using the 96-d aerobic soil metabolism half-life. Predicted concentrations in water ranged from 0.003 to 1.39 µg L^{-1} for the 28-d half-life simulations, increasing slightly to a range of 0.013–2.14 µg L^{-1} for the 96-d half-life simulations. Similarly, CPY concentrations in sediment ranged from 4.74 to 11.0 µg kg^{-1} in simulations that used 28-d half-life, and 10.8–22.2 µg kg^{-1} in simulations that used a half-life of 96-d. Predicted annual concentrations in the Georgia and Michigan watersheds were less than those for California. The Georgia watershed had a maximum 96-h TWMC of 0.023 µg L^{-1} in water for the 28-d and

Table 9 Duration of exposures events in the focus watersheds for the 30-yr simulation period

Location of watershed	Georgia		Michigan		California	
Event threshold μg L^{-1}	0.1	1.0	0.1	1.0	0.1	1.0
Duration of events for 28-d half-life						
No. of events	0.0	0.0	0.0	0.0	10	4
Min. duration (d)	n/a	n/a	n/a	n/a	1	1
Max. duration (d)	n/a	n/a	n/a	n/a	11	2
Median duration (d)	n/a	n/a	n/a	n/a	1	1.5
Avg. duration (d)	n/a	n/a	n/a	n/a	2.9	1.5
Duration of events for 96-d half-life						
No. of events	0.0	0.0	3	0.0	35	10
Min. duration (d)	n/a	n/a	1	n/a	1	1
Max. duration (d)	n/a	n/a	1	n/a	15	6
Median duration (d)	n/a	n/a	1	n/a	1	1
Avg. duration (d)	n/a	n/a	1	n/a	2.3	1.9

96-d half-life simulations. The maximum 96-h TWMCs in sediment were 0.064 μg kg^{-1} and 0.067 μg kg^{-1} in the 28-d and 96-d half-life simulations, respectively. The Michigan watershed had a maximum 96-h TWMC of 0.028 μg L^{-1} and 0.043 μg L^{-1} in water for the 28-d and 96-d half-life simulations, respectively. Maximum 96-h TWMCs in sediment were 0.047 μg kg^{-1} and 0.074 μg kg^{-1} in the 28-d and 96-d half-life simulations, respectively.

Event duration analysis. Using RADAR, characterization of durations of exposure events was conducted on each time series record by using concentration thresholds of 0.1 and 1.0 μg L^{-1} for the three watersheds. In Table 9, we list each event that met the respective criteria for the model simulations using both half-life assumptions. Additional information on each event, including the maximum duration of the event, average concentration during the event, and the date when the event began are provided in SI Appendix E.

There were no events exceeding the concentration threshold for the Georgia watershed over the 30-yr simulation period with either the 28-d or 96-d half-life values. For the Michigan watershed, no events were predicted with the 28-d half-life and three events were predicted with the 96-d half-life. All three events were limited to 1 d. California was predicted to have the greatest number of exceedance events. At a 0.1 μg L^{-1} threshold, 10 events were predicted to occur with the 28-d half-life life, while 35 occurred with the 96-d half-life. For the 96-d half-life and the 0.1 μg L^{-1} threshold, the median event duration was 1 d and the longest duration was 15 d.

4 Discussion

CPY is not detected frequently or at large concentrations in surface waters of the U.S. Monitoring data reported in the USGS, CDPR, and WDOE databases indicate maximum water CPY concentrations ranging from 0.33 to 3.96 μg L^{-1}. These maxima represent approximately 17,700 sample analyses (911 locations) during 2002–2010 that were collected following the ban on residential uses and other label changes of CPY ca. 2001. However, detections are relatively infrequent. Characterization of surface water CPY concentrations in this 9-yr period from regional data sets that focused on use areas indicated that only 13–17% of collected samples contained detectable CPY concentrations. The 95th centile concentrations from these programs ranged from <0.01 to 0.3 μg L^{-1} across years.

Similarly, CPYO was not detected frequently or in large concentrations. The databases of the USGS, CDPR, and WDOE contained reports of only 25 detections of CPYO from 10,375 analyses (0.24% of samples) in surface water between 1999 and 2012. The concentrations reported in the 16 detections reported in the USGS databases (9,123) were all below the LOQ and neither the CDPR nor the WDOE databases contained any reports of detections of CPYO in surface waters.

One of the inherent limitations of monitoring data is that it may not be wholly representative of all locations. Because of logistical and resource issues, not all locations can be sampled with a frequency and duration that fully characterizes peaks in exposures, especially those of short duration. Thus, modeling was used to provide estimates of concentrations in surface waters across the country. From knowledge of rates of and frequency of application, environmental settings were identified as being more susceptible to runoff and CPY drift into surface water. These included areas of higher rainfall and heavier soils and those receiving the greatest single application rate or shortest intervals between multiple applications as identified from the sensitivity analysis of use-patterns.

Because it was not possible to model all locations in detail and all scenarios of CPY use in the USA, the analyses of sensitivity were used to identify three focus watersheds—Dry Creek in Georgia, Cedar Creek in Michigan, and Orestimba Creek in California. These watersheds were intended to provide realistic but reasonable worst-case predictions of concentrations of CPY in runoff water and sediment. The soil-water hydrologic model SWAT was combined with PRZM to predict these concentrations using a 30-y climatic record (1961–1990 for Dry Creek and Cedar Creek and 2000–2008 for Orestimba Creek) and local cropping and soils information. Two half-lives for aerobic soil metabolism of CPY in soil, 28 and 96 d, were selected for the purposes of modeling.

Estimated concentrations of CPY in water for the three watersheds were in general agreement with ambient monitoring data from 2002 to 2010 in the datasets of the USGS, CDPR, and WDOE. Maximum daily concentrations predicted for the California, Georgia, and Michigan watersheds were 3.2, 0.041, and 0.073 μg L^{-1}, respectively, with the 28-d aerobic soil metabolism half-life and 4.5, 0.042, and 0.122 μg L^{-1}, respectively, with the 96-d soil half-life. These compared favorably with maximum concentrations measured in surface water, which ranged from 0.33 to 3.96 μg L^{-1}. For sediments, the maximum daily concentrations predicted for the California, Georgia, and Michigan watersheds were 11.2, 0.077, and 0.058 μg kg^{-1}, respectively, with the 28-d half-life and 22.8, 0.080, and 0.087 μg kg^{-1}, respectively, with the 96-d soil half-life. Twelve detections out of 123 analyses were contained in the USGS, CDPR, and WDOE databases with concentrations reported from <2.0, up to 19 μg kg^{-1} with the exception of one value reported at 58.6 μg kg^{-1}. Again, the modeled values compared favorably with measured values. Duration and recovery intervals between CPY water concentrations that exceeded two different threshold values derived from toxicity data of Giddings et al. (2014) were also computed. Based on modeling with the half-life of 28 d, no exceedances were identified in the focus watersheds in Georgia or Michigan. Using the half-life of 96 d, only three exceedance events of 1-d duration each were identified in the Michigan focus-watershed. Frequency of exceedance was greater in the focus watershed from California. There were 10–35 exceedance events (depending on threshold level) during the 30-yr simulation period, or an average of less than one per year. Moreover, even in this worst-case focus-watershed, the median event exceedance duration was 1 d. The greater concentrations in Orestimba Creek are attributed to a higher frequency of applications, a higher frequency of runoff events (due to irrigation tail-water), and less stream-flow for dilution.

Several advantages and insights gained from the modeling of concentrations of CPY in surface waters were developed from range of modeling assumptions and field conditions simulated. Conservative assumptions were used in the modeling of concentrations in the focus-watersheds. Studies on environmental-fate of CPY have demonstrated a range of rates of degradation in soil, crops, and aquatic systems. Values from these data were selected to represent appropriate degradation and loss processes in the modeling. For example, the 90th centile confidence interval on the mean half-life was selected to err on the side of caution. Conservative assumptions were also used in configuring the Georgia and Michigan watersheds, in that all eligible crop acreage in each watershed was represented as if it were pecan or cherries, respectively; thereby, the soil properties and applications of CPY represented by the use-pattern selected produced the greatest estimates of exposure-concentrations. Model simulations for Orestimba Creek used reported applications of CPY, but field specific management practices to mitigate runoff and drift were not represented in the simulations. In addition, volatilization was not included in the California simulations. CPY drift for all three watersheds was assumed to occur with all treated crops having a proximity to water equal to the minimum setback requirements on the product label. The setback was used to simulate drift reductions to water, but reductions in pesticide loadings in runoff were not assumed to occur in the setback area.

Opportunities to verify the focus-watershed model results were limited for this study. No calibration of runoff water, soil erosion, or CPY properties was conducted. However, it was possible to model several field-specific runoff studies (Cryer and Dixon-White 1995; McCall et al. 1984; Poletika and Robb 1994; Racke 1993) by using the environmental fate properties employed in the modeling of the focus watersheds. Predicted volumes of runoff water, sediment, and CPY concentrations in runoff water and in eroded sediment were within an order of magnitude of the amounts measured in runoff studies, and they were neither consistently high nor consistently low, suggestive of a lack of model prediction bias.

The analyses used to characterize runoff of CPY at the national level incorporated a number of generalizations. For this reason, these analyses were only used to evaluate the relative potential for runoff of CPY as a guide in selecting the focus watersheds. The most significant generalization was representing all simulations as a generic crop (corn) with a unit application rate. Applications were set to occur relative to the typical date of corn emergence across the USA. Only two application scenarios were predicted—broadcast sprays to soil and to foliage. Volatilization was not simulated in the foliar application scenario. Granular applications and crop-specific label-type applications were not evaluated. The scenarios used in the sensitivity analysis on use-patterns are another relative indicator of the runoff potential of CPY because they represent a hypothetical environment—a 10-ha field draining into a 1-ha by 2-m deep pond. The pond remains at a constant volume that receives pesticide loads from drift and runoff, but not the corresponding influx of water that would occur during a runoff event.

The Orestimba Creek watershed in California was predicted to have the greatest exposure concentrations and the longest duration of exposures of the three watersheds. This is caused in part because of the relatively higher intensity of CPY's use, and because of the relatively small dilution and flushing that exists in the headwater channels. The watershed is under a water quality management plan administered by the Westside San Joaquin River Watershed Coalition to alter practices contributing to water quality issues associated with agricultural chemicals (CURES 2013; SJVDA 2010). The practices needing to be altered include education and outreach on pesticide application technologies and managing irrigation and storm-water runoff that were not taken into account in the model.

Characterization of CPY in the environment could be enhanced by including additional detail in all three watershed systems. These include heterogeneity in cropping and soils in Georgia and Michigan; management plans being introduced into Orestimba Creek, and weather specific drift estimates and application field proximity to water drainage-ways.

5 Summary

Concentrations of CPY in surface waters are an integral determinant of risk to aquatic organisms. CPY has been measured in surface waters of the U.S. in several environmental monitoring programs and these data were evaluated to characterize concentrations, in relation to major areas of use and changes to the label since 2001, particularly the removal of domestic uses. Frequencies of detection and 95th centile concentrations of CPY decreased more than fivefold between 1992 and 2010. Detections in 1992–2001 ranged from 10.2 to 53%, while 2002–2010 detections ranged from 7 to 11%. The 95th centile concentrations ranged from 0.007 to 0.056 μg L^{-1} in 1992–2001 and 0.006–0.008 μg L^{-1} in 2002–2010. The greatest frequency of detections occurred in samples from undeveloped and agricultural land-use classes. Samples from urban and mixed land-use classes had the smallest frequency of detections and 95th centile concentrations, consistent with the cessation of most homeowner uses in 2001. The active metabolite of CPY, CPYO, was not detected frequently or in large concentrations. In 10,375 analyses from several sampling programs conducted between 1999 and 2012, only 25 detections (0.24% of samples) of CPYO were reported and estimated concentrations were less than the LOQ.

Although the monitoring data on CPY provide relevant insight in quantifying the range of concentrations in surface waters, few monitoring programs have sampled at a frequency sufficient to quantify the time-series pattern of exposure. Therefore, numerical simulations were used to characterize concentrations of CPY in water and sediment for three representative high exposure environments in the U.S. The fate of CPY in the environment is dependent on a number of dissipation and degradation processes. In terms of surface waters, fate in soils is a major driver of the potential for runoff into surface waters and results from a number of dissipation studies in the laboratory were characterized. Aerobic degradation of CPY exhibits bi-phasic behavior in some soils; initial rates of degradation are greater than overall rates by factors of up to threefold. Along with fate in water, these data were considered in selecting parameters for the modeling concentrations in surface waters. An assessment of vulnerability to runoff was conducted to characterize the potential for CPY to be transported beyond a treated field in runoff water and eroded sediment across the conterminous U.S. A sensitivity analysis was performed on use practices of CPY to determine conditions that resulted in the highest potential runoff of CPY to aquatic systems to narrow the application practices and geographical areas of the country for selecting watersheds for detailed modeling. The selected focus-watersheds were Dry Creek in Georgia (production of pecans), Cedar Creek in Michigan (cherries), and Orestimba Creek in California (intensive agricultural uses). These watersheds provided realistic but reasonable worst-case predictions of concentrations of CPY in water and sediment.

Estimated concentrations of CPY in water for the three watersheds were in general agreement with ambient monitoring data from 2002 to 2010 in the datasets from US Geological Survey (USGS), California Department of Pesticide Regulation (CDPR), and Washington State Department of Ecology (WDOE). Maximum daily concentrations predicted for the watershed in California, Georgia, and Michigan were 3.2, 0.041, and 0.073 μg L^{-1}, respectively, with the 28-d aerobic soil metabolism half-life and 4.5, 0.042, and 0.122 μg L^{-1}, respectively, with the 96-d soil half-life. These estimated values compared favorably with maximum concentrations measured in surface water, which ranged from 0.33 to 3.96 μg L^{-1}. For sediments, the maximum daily concentrations predicted for the watersheds in California, Georgia, and Michigan were 11.2, 0.077, and 0.058 μg kg^{-1}, respectively, with the 28-d half-life and 22.8, 0.080, and 0.087 μg kg^{-1}, respectively, with the 96-d soil half-life. CYP was detected in 12 samples (10%) out of 123 sample analyses that existed in the USGS, CDPR, and WDOE databases. The concentrations reported in these detections were from <2.0, up to 19 μg kg^{-1}, with the exception of one value reported at 58.6 μg kg^{-1}. Again, the modeled values compared favorably with these measured values. Duration and recovery intervals between toxicity threshold concentrations of 0.1 and 1.0 μg L^{-1} were also computed. Based on modeling with the half-life of 28 d, no exceedance events were identified in the focus watersheds in Georgia or Michigan. Using the half-life of 96 d, only three events of 1-d duration only were identified in the Michigan focus-watershed. Frequency of exceedance was greater in the California focus watershed, though the median duration was only 1-d.

Acknowledgments The authors thank Jeff Wirtz of Compliance Services International; and J. Mark Cheplick, Dean A. Desmarteau, Gerco Hoogeweg, William J. Northcott, and Kendall S. Price, all of Waterborne Environmental for their contributions to this paper. We also acknowledge Yuzhou Luo, previously with the University of California, for providing the SWAT dataset for the Orestimba Creek watershed. We thank the anonymous reviewers of this paper for their suggestions and constructive criticism. Prof. Giesy was supported by the Canada Research Chair program, a Visiting Distinguished Professorship in the Department of Biology and Chemistry and State Key Laboratory in Marine Pollution, City University of Hong Kong, the 2012 "High Level Foreign Experts" (#GDW20123200120) program, funded by the State Administration of Foreign Experts Affairs, the P.R. China to Nanjing University and the Einstein Professor Program of the Chinese Academy of Sciences. Funding for this work was provided by Dow AgroSciences.

References

Banks KE, Hunter DH, Wachal DJ (2005) Chlorpyrifos in surface waters before and after a federally mandated ban. Environ Int 31:351–356

Batzer FR, Fontaine DD, White FH (1990) Aqueous photolysis of chlorpyrifos. DowElanco, Midland, MI, Unpublished Report

Budd R, O'Geen A, Goh KS, Bondarenko S, Gan J (2009) Efficacy of constructed wetlands in pesticide removal from tailwaters in the Central Valley, California. Environ Sci Technol 43: 2925–2930

Burns L (2004) Exposure analysis modeling system (EXAMS): user manual and system documentation. Version 2.98.04.06. Ecosystems Research Division U.S. Environmental Protection Agency, Athens, GA, EPA/600/R-00/081

Carousel RF, Imhoff JC, Hummel PR, Cheplick JM, Donigian AS, Jr, Suárez LA (2005) PRZM_3, A Model for Predicting Pesticide and Nitrogen Fate in the Crop Root and Unsaturated Soil Zones: Users Manual for Release 3.12.2. National Exposure Research Laboratory, Office of Research and Development, U.S. Environmental Protection Agency, Athens, GA

CDPR (2012a) Surface water database. California department of pesticide regulation. http://www.cdpr.ca.gov/docs/emon/surfwtr/surfcont.htm. Accessed 24 Feb 2012

CDPR (2012b) Pesticide use reporting data: user guide & documentation. California department of pesticide regulation. California Pesticide Information Portal (CalPIP). http://calpip.cdpr.ca.gov/. Accessed 28 Feb 2012

CEPA (2011a) State water resources control board. Surface water ambient monitoring program. California Environmental Protection Agency. http://www.waterboards.ca.gov/water_issues/programs/swamp/. Accessed Aug 2011

CEPA (2011b) Central valley regional water quality control board. Irrigated Lands Regulatory Program. California Environmental Protection Agency. http://www.swrcb.ca.gov/rwqcb5/water_issues/irrigated_lands/index.shtml/. Accessed Aug 2011

Cornell University Cooperative Extension (2012a) Crop and Pest Management Guidelines: 2012 New York and Pennsylvania Pest Management Guidelines for Grapes. http://ipmguidelines.org/Grapes/. Accessed Sept 2012

Cornell University Cooperative Extension (2012b) Crop and Pest Management Guidelines: 2012 Cornell Pest Management Guidelines for Commercial Turfgrass. http://ipmguidelines.org/Turfgrass/. Accessed Sept 2012

Cryer S, Dixon-White H (1995) A field runoff study of chlorpyrifos in southeast Iowa during the severe flooding of 1993. DowElanco, Indianapolis, IN, Unpublished Report

CURES (2013) Best Management Practices—Handbook & Information. Coalition for Urban/Rural Environmental Stewardship. available from http://www.curesworks.org/coalitions.asp, Davis, CA, USA pp >100

de Vette HQM, Schoonmade JA (2001) A study on the route and rate of aerobic degradation of ^{14}C-chlorpyrifos in four European soils. Dow AgroSciences, Indianapolis, IN, Unpublished Report

Domagalski JL, Munday C (2003) Evaluation of Diazinon and Chlorpyrifos Concentrations and Loads, and Other Pesticide Concentrations, at Selected Sites in the San Joaquin Valley, California, April to August, 2001. U.S. Geological Survey, Sacramento, CA. Water-Resources Investigations Report 03-4088

AgroSciences D (2009) Lorsban advanced insecticide supplemental labeling. Dow AgroSciences LLC, Indianapolis, IN

AgroSciences D (2010) Lorsban advanced insecticide label. Dow AgroSciences LLC, Indianapolis, IN

ECOFRAM (1999) ECOFRAM Aquatic Final Draft Reports. United States Environmental Protection Agency, Washington, DC. http://www.epa.gov/oppefed1/ecorisk/aquareport.pdf

Ensminger M, Bergin R, Spurlock F, Goh KS (2011) Pesticide concentrations in water and sediment and associated invertebrate toxicity in Del Puerto and Orestimba Creeks, California, 2007–2008. Environ Monit Assess 175:573–587

Fendinger NJ, Glotfelty DE (1990) Henry's law constants for selected pesticides, PAHs, and PCBs. Environ Toxicol Chem 9:731–736

FOCUS (2012) Overview of FOCUS. Forum for the Co-ordination of pesticide fate models and their Use. http://viso.ei.jrc.it/focus/. Accessed Nov 2012

Gesch DB (2007) The national elevation dataset digital elevation model technologies and applications: the DEM Users manual. In: Maune D (ed) Digital elevation model technologies and applications: the DEM Users manual, 2nd edn. American Society for Photogrammetry and Remote Sensing, Bethesda, MD, pp 99–118

Giddings JM, Williams WM, Solomon KR, Giesy JP (2014) Risks to aquatic organisms from the use of chlorpyrifos in the United States. Rev Environ Contam Toxicol 231:119–162

Gill SL, Spurlock FC, Goh KS, Ganapathy C (2008) Vegetated ditches as a management practice in irrigated alfalfa. Environ Monit Assess 144:261–267

Glotfelty DE, Seiber JN, Liljedahl LA (1987) Pesticides in fog. Nature Biotech 325:602–605

Gomez LE (2009) Use and benefits of chlorpyrifos in US agriculture. Dow AgroSciences, Indianapolis, IN, Unpublished Report

Hoogeweg CG, Williams WM, Breuer R, Denton D, Rook B, Watry C (2011) Spatial and Temporal Quantification of Pesticide Loadings to the Sacramento River, San Joaquin River, and Bay-Delta to Guide Risk Assessment for Sensitive Species. CALFED Science Grant #1055. CALFED Science Program, Sacramento,CA.

Hoogeweg CG, Denton DL, Breuer R, Williams WM, TenBrook P (2012) Development of a spatial-temporal co-occurrence index to evaluate relative pesticide risks to threatened and endangered species. In: Racke KD, McGaughey BD, Cowles JL, Hall AT, Jackson SH, Jenkins JJ, Johnston JJ (eds) Pesticide regulation and the endangered species Act, vol 1111. American Chemical Society, Washington, DC, pp 303–323

Johnson HM, Domagalski JL, Saleh DK (2011) Trends in pesticide concentrations in streams of the western United States, 1993-2005. J Am Water Res Assoc 47:265–286

Jones RL, Russell MH (eds) (2001) FIFRA environmental model validation task force final report. American Crop Protection Association, Washington, DC. http://femvtf.com/femvtf/Files/FEMVTFbody.pdf

Kennard LM (1996) Aerobic aquatic degradation of chlorpyrifos in a flow-through system. DowElanco, Indianapolis, IN, Unpublished Report

Long RF, Hanson BR, Fulton A (2010) Mitigation techniques reduce sediment in runoff from furrow-irrigated cropland. Calif Agric 64:135–140

Luo Y, Zhang X, Liu X, Ficklin D, Zhang M (2008) Dynamic modeling of organophosphate pesticide load in surface water in the northern San Joaquin Valley watershed of California. Environ Pollut 156:1171–1181

Luo Y, Zhang M (2010) Spatially distributed pesticide exposure assessment in the Central Valley, California, USA. Environ Pollut 158:1629–1637

MacDougall D, Crummett WB (1980) Guidelines for data acquisition and data quality evaluation in environmental chemistry. Anal Chem 52:2242–2249

Mackay D, Giesy JP, Solomon KR (2014) Fate in the environment and long-range atmospheric transport of the organophosphorus insecticide, chlorpyrifos and its oxon. Rev Environ Contam Toxicol 231:35–76

Martin JD, Eberle M (2009) Adjustment of pesticide concentrations for temporal changes in analytical recovery, 1992–2010. U.S. Geological Survey, Reson, VA

Martin JD, Eberle M, Nakagaki N (2011) Sources and preparation of data for assessing trends in concentrations of pesticides in streams of the United States, 1992–2010. U.S. Geological Survey, Reson, VA

McCall PJ, Oliver GR, McKellar RL (1984) Modeling the runoff potential and behavior of chlorpyrifos in a terrestrial-aquatic watershed. Dow Chemical, Midland, MI, Unpublished Report

NASQAN (2012) USGS National Stream Quality Accounting Network. United States Geological Survey. http://water.usgs.gov/nasqan/. Accessed 24 Feb 2012

NAWQA (2012) USGS National Water Quality Assessment Data Warehouse. United States Geological Survey. http://infotrek.er.usgs.gov/apex/f?p=NAWQA:HOME:0. Accessed 24 Feb 2012

NCWQR (2012) Tributary Monitoring Program. National Center for Water Quality Research. http://www.heidelberg.edu/wql/tributarydatadownload. Accessed Feb 2013

Neitsch SL, Arnold JG, Kiniry JR, Srinivasan R, Williams JR (2010) Soil and water assessment tool theoretical documentation, version 2009.TR-265. Grassland, Soil and Water Research Laboratory, Agricultural Research Service, Temple, TX

Orang MN, Snyder RL, Matyar JC (2008) Survey of irrigation methods in California in 2001. J Irrig Drain Eng 134:96–100

Parker R, Arnold JG, Barrett M, Burns L, Carrubba L, Neitsch SL, Snyder NJ, Srinivasan R (2007) Evaluation of three watershed-scale pesticide environmental transport and fate models. J Am Water Res Assoc 43:1424–1443

Poletika NN, Robb CK (1994) A field runoff study of chlorpyrifos in Mississippi delta cotton. DowElanco, Indianapolis, IN, Unpublished Report

Racke KD (1993) Environmental fate of chlorpyrifos. Rev Environ Contam Toxicol 131:1–151

Racke KD, Juberg DR, Cleveland CB, Poletika NN, Loy CA, Selman FB, Bums CJ, Bartels MJ, Price PS, Shaw MC, Tiu CC (2011) Dow AgroSciences' Response to EPA's Preliminary Human Health Assessment for Chlorpyrifos Registration Review. Dow AgroSciences, Indianapolis, IN. EPA-HQ-OPP-2008-0850-0025. Report lD: KDR09062011

Reeves GL, Mackie JA (1993) The aerobic degradation of ^{14}C-chlorpyrifos in natural waters and associated sediments. Dow Agrosciences, Indianapolis, IN, Unpublished Report

Rotondaro A, Havens PL (2012) Direct flux measurement of chlorpyrifos and chlorpyrifos-oxon emissions following applications of Lorsban advanced insecticide to alfalfa. Dow AgroScience, Indianapolis, IN, Unpublished Report

Ryberg KR, Vecchia AV, Martin JD, Gilliom RJ (2010) Trends in Pesticide Concentrations in Urban Streams in the United States, 1992–2008. U.S. Geological Survey, Reston, VA. U.S. Geological Survey Scientific Investigations Report 2010–5139

SJVDA (2010) Westside San Joaquin River Watershed Coalition. Focused Watershed Management Plan II. Westley Wasteway, Del Puerto Creek, Orestimba Creek. San Joaquin Valley Drainage Authority, available from http://www.waterboards.ca.gov/centralvalley/water_issues/irrigated_lands/management_plans_reviews/coalitions/westside/index.shtml, Los Banos, CA, USA pp 16

Snyder NJ, Williams WM, Denton DL, Bongard CJ (2011) Modeling the effectiveness of mitigation measures on the Diazinon labels. In: Goh KS, Bret BL, Potter TL, Gan J (eds) Pesticide mitigation strategies for surface water quality. DC pp, American Chemical Society, Washington, pp 227–257

Solomon KR, Williams WM, Mackay D, Purdy J, Giddings JM, Giesy JP (2014) Properties and uses of chlorpyrifos in the United States. Rev Environ Contam Toxicol 231:13–34

Starner K, Spurlock FC, Gill S, Feng H, Hsu J, Lee P, Tran D, White J (2005) Pesticide residues in surface water from irrigation-season monitoring in the San Joaquin Valley, California. Bull Environ Contam Toxicol 74:920–927

Sullivan DJ, Vecchia AV, Lorenz DL, Gilliom RJ, Martin JD (2009) Trends in Pesticide Concentrations in Corn-Belt Streams, 1996–2006. United States Geological Survey, Reston, VA. United States Geological Survey Scientific Investigations Report 2009–5132

Suntio LR, Shiu WY, Mackay D, Seiber JN, Glotfelty D (1987) A critical review of Henry's constants for pesticides. Rev Environ Contam Toxicol 103:1–59

Teske ME, Bird SL, Esterly DM, Ray SL, Perry SG (2002) A User's Guide for AgDRIFT® 2.0.05, Regulatory Version. CDI, Ewing, NJ. C.D.I. Report No. 01-02

USDA-ERS (2000) Farm Resource Regions. United States Department of Agriculture, Economic Research Service, Washington, DC. AIB-760 http://webarchives.cdlib.org/wayback.public/UERS_ag_1/20111128195215/http://www.ers.usda.gov/Briefing/ARMS/resourceregions/resourceregions.htm#older

USDA (2012) National Agricultural Statistics Service. United States Department of Agriculture. http://www.nass.usda.gov/Surveys/Guide_to_NASS_Surveys/Chemical_Use/index.asp. Accessed April, Aug, Sept 2012

USDHHS (1997) Toxicological profile for chlorpyrifos. United States Department of Health and Human Services, Washington, DC, TP 84

USEPA (1999) Reregistration eligibility science chapter for chlorpyrifos: fate and environmental risk assessment chapter. United States Environmental Protection Agency, Washington, DC

USEPA (2000) Drinking Water Screening Level Assessment Part A: Guidance for Use of the Index Reservoir in Drinking Water Exposure Assessments. United States Environmental Protection Agency. Pesticide Science Policy. Office of Pesticide Programs, Washington DC. http://www.epa.gov/pesticides/trac/science/reservoir.pdf

USEPA (2006) EPA Reach File references. United States Environmental Protection Agency, Washington, DC. http://www.epa.gov/waters/doc/rfindex.html

USEPA (2009) Guidance for Selecting Input Parameters in Modeling the Environmental Fate and Transport of Pesticides: Version 2.1. United States Environmental Protection Agency, Office of Pesticide Programs, Environmental Fate and Effects Division, Washington, DC. http://www.epa.gov/oppefed1/models/water/

USEPA (2011) Revised Chlorpyrifos Preliminary Registration Review Drinking Water Assessment. United States Environmental Protection Agency, Office of Chemical Safety and Pollution Prevention, Washington, DC, USA. PC Code 059101 http://www.epa.gov/oppsrrd1/registration_review/chlorpyrifos/EPA-HQ-OPP-2008-0850-DRAFT-0025%5B1%5D.pdf

USEPA (2012) Specified Ingredient Incidents for the Date Range 01/01/2002 to 06/15/2012. United States EPA Office of Pesticide Programs, Washington, DC

USGS (2013) NAWQA Database. United States Geological Survey. http://infotrek.er.usgs.gov/servlet/page?_pageid=543&_dad=portal30&_schema=PORTAL30. Accessed 2 Feb 2013

WDOE (2012) Environmental Information Management Database. Washington State Department of Ecology. http://www.ecy.wa.gov/eim/. Accessed 3 April 2012

Weston D, Holmes R, You J, Lydy M (2005) Aquatic toxicity due to residential use of pyrethroid insecticides. Environ Sci Technol 39:9778–9784

Zhang X, Starner K, Spurlock F (2012) Analysis of chlorpyrifos agricultural use in regions of frequent surface water detections in California, USA. Bull Environ Contam Toxicol 89: 978–984

Risks to Aquatic Organisms from Use of Chlorpyrifos in the United States

Jeffrey M. Giddings, W. Martin Williams, Keith R. Solomon, and John P. Giesy

1 Introduction

Effects of chlorpyrifos (CPY) in aquatic ecosystems are dependent on duration and magnitude of exposure and toxicity to individual species. This paper is focused on potential effects of CPY on aquatic organisms and ecosystems based on properties and current uses of CPY (Mackay et al. 2014; Solomon et al. 2014) and probabilities of exposure as determined by measurement in monitoring programs and predictions of simulation models (Williams et al. 2014). Exposures, toxicity, and risks to birds, other terrestrial wildlife, and pollinators are assessed in two additional companion papers (Cutler et al. 2014; Moore et al. 2014). This paper follows the framework for ecotoxicological risk assessment (ERA) developed by the US Environmental Protection Agency (USEPA 1992, 1998, 2004), and builds on a previous assessment of risks posed by CPY in surface waters of North America (Giesy et al. 1999).

Like many risk assessments, the previous assessment of the risk of CPY in surface waters (Giesy et al. 1999) was tiered. Lower tiers of risk assessments incorporate less data and therefore make conservative assumptions when characterizing hazards and

---

The online version of this chapter (doi:10.1007/978-3-319-03865-0_5) contains supplementary material, which is available to authorized users.

J.M. Giddings (✉)
Compliance Services International, 61 Cross Road, Rochester, MA 02770, USA
e-mail: jgiddings@ComplianceServices.com

W.M. Williams
Waterborne Environmental Inc., Leesburg, VA, USA

K.R. Solomon
Centre for Toxicology, School of Environmental Sciences, University of Guelph, Guelph, ON, Canada

J.P. Giesy
Department of Veterinary Biomedical Sciences and Toxicology Centre, University of Saskatchewan, 44 Campus Dr., Saskatoon, SK S7N 5B3, Canada

J.P. Giesy and K.R. Solomon (eds.), *Ecological Risk Assessment for Chlorpyrifos in Terrestrial and Aquatic Systems in the United States*, Reviews of Environmental Contamination and Toxicology 231, DOI 10.1007/978-3-319-03865-0_5, © The Author(s) 2014

risks. If the criteria for lower tiers do not indicate risk, further refinements of the risk assessment are not needed. This was not the case in the earlier assessment where lower-tier risk criteria were exceeded and potential hazards to aquatic organisms were identified (Giesy et al. 1999). However, these potential hazards were not consistent with the lack of incident reports, such as fish kills, attributable to use of chlorpyrifos in agriculture (Giesy et al. 1999). Refinement of the earlier ERA by the use of Species Sensitivity Distributions (SSDs) and measured concentrations of CPY in surface waters showed that, in almost all locations in the U.S., risks associated with use of CPY in agriculture were either negligible or *de minimis* (Giesy et al. 1999).

Since the 1999 ERA for CPY was written, there have been refinements in the process of risk assessment and additional data have become available for toxicity and exposures in surface waters. In addition, there have been changes in the labeled uses of CPY (Solomon et al. 2014); most notably, removal of termite control and residential uses of CPY in 2001. The former uses involved large rates of application (with attendant potential for large environmental exposures), and the changes in the labels significantly reduced exposures from relatively uncontrolled uses in urban environments (Banks et al. 2005; Phillips et al. 2007). Availability of additional data and changes in use patterns prompted the reassessment of risks to aquatic organisms from use of CPY in agriculture in the U.S., the results of which are presented here. Since lower-tier assessment had already indicated risk for CPY in surface waters (Giesy et al. 1999), the lower tiers were omitted from this ERA. The current assessment focused on a refined approach that employed SSDs and results of community-level studies in microcosms and mesocosms ("cosms") as points of departure for toxicity, and refined modeling of concentrations in surface waters (Williams et al. 2014, in this volume) to characterize exposures. Concentrations of CPY measured in surface waters were used as a check on the estimates of exposures predicted by use of simulation models (Williams et al. 2014) and as another line of evidence in the ERA.

2 Problem Formulation for Risk Assessment

Risk assessments, particularly ERAs, use a formal process of problem formulation (PF) to narrow the focus of the assessment to address key questions and, from these, develop risk hypotheses (USEPA 1998). Several components of the PF have been addressed in detail in companion papers and will only be summarized here.

2.1 Exposures to Chlorpyrifos

A conceptual model for exposures to CPY (Fig. 1) was constructed from environmental properties data that are presented in the companion papers (Mackay et al. 2014; Solomon et al. 2014; Williams et al. 2014). As several studies have noted, urban uses were a significant source of historical exposures to CPY in surface water

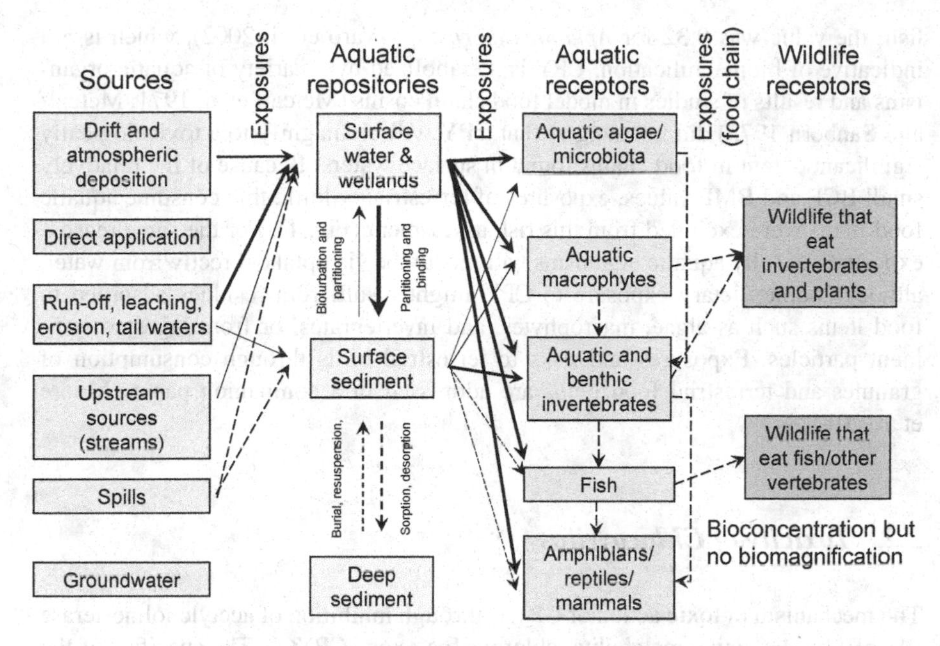

Fig. 1 The conceptual model for exposures of aquatic organisms to chlorpyrifos in surface waters. The weights of the *arrows* indicate importance of the pathway of exposure

(Banks et al. 2005; Phillips et al. 2007) and sediments (Ding et al. 2010) until label changes occurred in 2001. All current labeled uses of CPY are in agriculture, thus sources from urban uses were omitted from the conceptual model. As is discussed in more detail in Williams et al. (2014), CPY can enter surface waters via several routes (Fig. 1), although some are of lesser importance. CPY is not registered for direct application to surface waters and the relatively large K_{OC} (973–31,000 mL g^{-1} Solomon et al. 2014) mitigates against leaching and movement via groundwater. Spills might occur but, because they are episodic and cannot be predicted, the current labeled uses were the focus of this review. The major potential sources for exposures of aquatic organisms are run-off, erosion, and tail waters with lesser inputs via drift of sprays and deposition from the atmosphere (Williams et al. 2014).

CPY adsorbs to surficial sediments in water-bodies, and residues of CPY have been detected in surficial sediments in streams and creeks in areas of use. In a few cases, observed toxicity in sediments collected from agricultural drains, creeks, and rivers in California was linked to the presence of CPY (Phillips et al. 2012; Weston et al. 2012). However, in most locations CPY contributed less to toxicity of sediments than other pesticides such as pyrethroids (Amweg et al. 2006; Ding et al. 2010; Ensminger et al. 2011; Phillips et al. 2006).

CPY is bioconcentrated and/or bioaccumulated into aquatic organisms to a limited extent. Measures of bioconcentration factors (BCFs), bioaccumulation factors (BAFs), and biota-sediment accumulation factors (BSAFs) are relatively small (Mackay et al. 2014). One literature report gave a biomagnification factor (BMF) in fish; the value was 0.32 for *Aphanius iberus* sp. (Varo et al. 2002), which is not indicative of biomagnification. CPY is metabolized by a variety of aquatic organisms and results of studies in model food chain cosms (Metcalf et al. 1971; Metcalf and Sanborn 1975) did not suggest that CPY will biomagnify to a toxicologically significant extent in food chains found in surface waters. Because of the relatively small BCF and BMF values, exposures of terrestrial wildlife that consume aquatic food items were excluded from this risk assessment (Fig. 1). For the same reason, exposures of fully aquatic organisms will mostly be via uptake directly from water, although some dietary exposure to CPY might result from residues adsorbed to food items such as algae, macrophytes, and invertebrates, or from ingested sediment particles. Exposures and risks to terrestrial birds through consumption of granules and terrestrial food items are addressed in a companion paper (Moore et al. 2014).

2.2 *Toxicity of Chlorpyrifos*

The mechanism of toxic action of CPY is through inhibition of acetylcholinesterase (AChE) by the active metabolite, chlorpyrifos oxon (CPYO). The specifics of the mode of action are discussed in greater detail in the companion paper (Solomon et al. 2014). Inhibition of AChE by CPYO is reversible and, in the case of sublethal exposures, recovery of AChE can occur. AChE is a key enzyme in the nervous systems of most animals, and direct effects of CPY will occur at much smaller exposure concentrations than in organisms that lack the target enzyme, such as plants. Insects and crustaceans are generally more sensitive to CPY than are fish or amphibians (Giesy et al. 1999). The primary focus of this ERA is CPY and CPYO. Other metabolites and breakdown products, such as trichloropyridinol (TCP), are much less toxic (USEPA 2008) and are not addressed.

Effects of CPY on animals range from lethality to minor symptoms from which animals recover. Most testing of toxicity of CPY to aquatic animals has used lethality as the measurement endpoint, with results usually expressed as the LC50 or the no-observed-adverse-effect-concentration (NOAEC). Fish are an exception since a number of studies have characterized sublethal and behavioral responses.

2.2.1 Sublethal Effects on Aquatic Animals

Several studies have reported effects of CPY on behavior of arthropods and fish. Interpretation of these studies presents difficulties, because it is not always clear if the observed responses are alterations in normal behavior specifically induced by the pesticide or changes in behavior in response to general stress or symptomology of the toxicity. This distinction can be addressed in specifically designed tests such as have been used to assess aversion to ingestion by birds (Moore et al. 2014). In studies of the prawn *Macrobrachium rosenbergii*, effects on feeding were observed at concentrations less than half the 24-h LC_{50} of 0.7 μg L^{-1} (Satapornvanit et al. 2009). These effects persisted for at least 4 h after cessation of exposure to CPY, but it is not known if these were truly behavioral responses or symptoms of sublethal poisoning.

A number of studies of sublethal effects of CPY on fish have been conducted, some of which have focused on olfactory perception and others on behavior. Much of the research on effects of CPY and other pesticides on behavior has focused on migratory species of salmon because of their societal importance and the need for migratory species to be able to sense chemicals in water to successfully navigate to breeding waters. Exposure of the olfactory epithelium of Coho salmon (*Oncorhynchus kisutch*) to 0.7 μg CPY L^{-1} caused a 20% loss of sensory function as measured by neurophysiological response to salmonid bile salt and L-serine (Sandahl et al. 2004). Since these studies were conducted in the laboratory, the changes in sensory function were not evaluated at the level of the whole-animal. Several studies have linked effects of CPY on the sensory epithelium and behavior to inhibition of AChE. Working with juvenile Coho salmon (*Oncorhynchus kisutch*) exposed to CPY at concentrations between 0.6 and 2.5 μg L^{-1}, Sandahl et al. (2005) showed that spontaneous swimming rate and food strikes were correlated (r^2 0.58 and 0.53, respectively) with inhibition of AChE activity in the brain. Other studies of effects of CPY in the same species showed that thresholds for different behaviors were related to inhibition of AChE (Tierney et al. 2007). Thresholds for effects of CPY on swimming behavior ranged from 20 to 35% inhibition of AChE. Zebrafish (*Danio rerio*) exposed to 220 μg CPY L^{-1} for 24 h exhibited impaired swimming behavior ($p < 0.01$) and a concentration-response relationship was observed at concentrations greater than 35 μg L^{-1} (Tilton et al. 2011). Similarly, locomotory behavior of mosquito fish, *Gambusia affinis*, was affected by exposure to 60 μg CPY L^{-1} for 20 d (Rao et al. 2005). Although exposures to concentrations of CPY of 100 and 200 μg L^{-1} caused depression of whole-body AChE (≈60% of controls) in tadpoles of *Rana sphenocephala*, there were no effects on swim-speed or vulnerability to predation (Widder and Bidwell 2006).

Most of the reported behavioral responses of fish to CPY were related to inhibition of AChE. These observations are consistent with current understanding of functions of AChE in the nervous system. It is also not surprising that mixtures of carbamate and organophosphorus pesticides have the same effect as single compounds and that they act additively and sometimes synergistically (Laetz et al. 2009). However, when assessed in the context of actual exposures in the environment, risks are small as the pesticides must co-occur temporally and spatially to cause ecological effects. Even when total potencies of mixtures of insecticides are considered, exposures that inhibit AChE at concentrations greater than the threshold for effects on behavior rarely occur in key locations for valued species, such as salmon in the Pacific NW (Moore and Teed 2012).

Although effects on behavior due to inhibition of AChE can be observed in vertebrates, these have not been experimentally related to effects on survival, development, growth, and reproduction of individuals or ecosystem stability or function in a quantitative manner. Therefore, they cannot be incorporated into an ERA at this time.

Little information has been reported on the effects of CPY or other insecticides on behaviors of other aquatic vertebrates and, to our knowledge, there have been no robust extrapolations of effects on behavior to the endpoints of survival, development, growth and reproduction. For this reason, we excluded behavioral responses from this risk assessment, because it is still uncertain as to how one interprets the data, either for CPY or all other pesticides.

Few studies on aquatic organisms have reported effects on reproduction directly caused by CPY. Chronic exposure of the guppy (*Poecilia reticulata*) to CPY (commercial formulation) for 14 d at nominal concentrations of 0.002 and 2 μg L^{-1} resulted in concentration-related reductions in the frequency of reproductive behavior (gonopodial thrusts) in males (De Silva and Samayawardhena 2005). The number of young born per female over the 14-d period was reduced from an average of 27 in the controls to 24 in pairs exposed to 0.002 μg CPY L^{-1} and 8 in pairs exposed to 2 μg L^{-1}. Activity of AChE was not reported in this study, so it is difficult to relate these chronic effects to response of AChE in other studies on behavior or to shorter exposure durations in the field (Williams et al. 2014). Another study on tadpoles of *Rana dalmatina* (Bernabo et al. 2011) reported that exposures to concentrations of 25 or 50 μg CPY L^{-1} from Gosner stage 25 to 46 (57 d) increased the incidence of testicular ovarian follicles (TOFs). This observation was reported at environmentally unrealistic concentrations and is the only report of this response for CPY; no other reports of TOFs in fish or amphibians were found in the literature. There appeared to be no effects of these exposures on mortality or time to metamorphosis. No measurements of AChE activity were reported and the effects on reproduction were not characterized. Because of the paucity of data, we excluded the effects of CPY on reproduction from this assessment. However, we indirectly addressed the endpoints in some of the cosm studies, where significant changes in reproduction of invertebrates would likely be encompassed in responses at the population level.

2.2.2 Toxicity of CPY and Temporality of Exposures

Frequency, duration, and intervals between exposures to CPY will influence responses observed in receptor organisms. These differences in response will result from variations in toxicokinetics and toxicodynamics of CPY in the environment and in individual organisms. The hydraulics of surface waters and variability of inputs from uses and precipitation events shape the types of exposures to pesticides experienced by organisms in flowing waters (Bogen and Reiss 2012). These are further altered by the individual properties of pesticides, such as rates of degradation and/or tendency for partitioning into sediments. As illustrated elsewhere in this volume (Williams et al. 2014), most exposures to CPY in flowing waters are less than 2 d in duration and are followed by periods of lesser or no exposure. These episodic exposures are typical of what is observed in flowing waters for pesticides in general and are relevant to this risk assessment.

As was pointed out in an earlier risk assessment of CPY (Giesy et al. 1999), exposures via the matrix of the organism (water in this case) are driven by thermodynamic processes, such as partitioning into the organism as well as kinetic processes related to rates of diffusion, transport, etc. This means that the critical body burden associated with the threshold of toxicity is not reached until sometime after exposure is initiated. This has been demonstrated for CPY (Giesy et al. 1999) and other organophosphorus insecticides (Bogen and Reiss 2012). The relationship between time of exposure and toxicity is reciprocal, with shorter exposures at greater concentrations resulting in the same level of response as lesser concentrations for longer durations. This reciprocal relationship was demonstrated in studies of effects of CPY on *Daphnia magna* exposed to CPY for varying durations (Naddy et al. 2000). For example, continuous exposures to 0.25 μg L^{-1} CPY resulted in 100% mortality in 5 d, while exposures for 1 d followed by transfer to clean water resulted in only 17% mortality, and then only after 16 d (Naddy et al. 2000). Whether this is the result of a delayed (latent) response or other causes, including regeneration of AChE activity, is uncertain; however, given the recovery observed in other Crustacea (below), the latter is a more plausible explanation. Where multiple episodic exposures occur, recovery from toxic effects between exposures can affect responses of exposed organisms. This was demonstrated in the greater response of *D. magna* exposed to the same concentration of CPY for 1×12 h compared to animals exposed for 2×6 h, 3×4 h, or 4×3 h with a 24-h interval between pulses (Naddy and Klaine 2001). Here, the interval between exposures likely provided time for detoxification and excretion of CPY, recovery of the target enzyme AChE by dephosphorylation (k_3 in Fig. 4 in Solomon et al. 2014), and/or synthesis of new AChE (Naddy and Klaine 2001).

Recovery of AChE inhibited by CPY has been observed in arthropods and fish. After exposure of *D. magna* to the 24-h LC_{50} concentration, whole-body activity of AChE (50% of unexposed control at time of removal to uncontaminated water) recovered to control activity within 24 h when animals were moved to clean water (Barata et al. 2004). After exposures of larvae of the midge *Kiefferulus calligaster* to 0.38, 1.02, or 1.26 μg L^{-1} CPY for 3 d, concentration-dependent depression of activity of AChE as great as 90% was observed (Domingues et al. 2009). When transferred to fresh medium for a further 3 d, AChE activity returned to control values. Similar recovery of activity of AChE was observed in the shrimp (*Paratya australiensis*) exposed to CPY for 96 h at concentrations from 0.001 to 0.1 μg L^{-1} and then moved to clean medium for 48 h or 7 d (Kumar et al. 2010). Complete recovery was dependent on exposure concentration and recovery time. Recovery after a 7-d exposure to 0.025 μg L^{-1} occurred in 7 d, but after exposure to 0.1 μg L^{-1}, recovery was not complete within 7 d. Whether this recovery resulted from dephosphorylation of AChE or synthesis of new AChE is not known. However, it is clear that recovery occurs and that recovery times are of the order of 1 to ~7 d.

Studies of fish exposed to CPY suggest that recovery of AChE in fish takes longer than in arthropods. No recovery of brain- or muscle-AChE was observed within a 4-d period in mosquitofish (*Gambusia affinis*) exposed to 100 μg CPY L^{-1} for 24 h (Boone and Chambers 1996). This exposure resulted in 70% inhibition of these enzymes. In another study of the same species, exposure to 297 μg CPY L^{-1} for 96 h resulted in 80% inhibition of brain-AChE (Kavitha and Rao 2008), but activity had recovered to control levels after 20 d in clean water. AChE activity in the brain of Nile tilapia,

Oreochomis niloticus, exposed to 10 μg CPY L^{-1} for 24 h declined to 47% of control values but, after transfer to clean water, recovered to 55% after 7 d and 63% after 14 d (Chandrasekera and Pathiratne 2005). After 14-d exposures of the guppy (*Poecilia reticulata*) to 0.325 μg CPY L^{-1}, activity of whole-body AChE was 22% of that in control fish (van der Wel and Welling 1989). Following removal to clean water for a further 14 d, activity of AChE had recovered to 40% of that in control fish. Similar observations have been reported for other organophosphorus insecticides. For example, activity of brain-AChE in Atlantic salmon parr (*Salmo salar*) exposed to formulated fenitrothion (50% inhibition at initiation of recovery) and transferred to fresh water, recovered to 66% of control values in 7 d and 93% in 42 d (Morgan et al. 1990). It is not known if AChE recovery rates in fish differ among organophosphorus insecticides having *O*-methyl (fenitrothion) or *O*-ethyl (CPY) substituents, and whether there are models, with which recovery rates can be extrapolated.

If, as is generally suggested (Morgan et al. 1990), recovery of phosphorylated AChE in fish requires synthesis of new enzyme, rates of recovery would be slow and dependent on rates of metabolism and the physiological and biochemical characteristics of fishes, which appear to be unknown. In the absence of having a model for predicting recovery periods, empirical observations suggest that inter-exposure intervals of the order of 4–8 wk might be required for complete recovery of AChE in fish. This period was incorporated into the ERA (Sect. 4.2).

2.3 Protection Goals and Assessment Endpoints

Protection goals and assessment endpoints are strongly linked and do not change as higher tiers or refinements are applied in the ERA. The protection goals applied in this assessment were to protect populations and communities of most aquatic organisms most of the time and at most locations. Specifically, Species Sensitivity Distributions (SSDs) were used (Posthuma et al. 2002) for crustaceans, insects, and fish to calculate the 5th centile (also referred to as the HC5) as a community-focused endpoint. Because of functional redundancy and resiliency, some effects on a small proportion of species can be tolerated in an ecosystem and the 5th centile of these distributions has been shown to be generally protective of ecosystems and the services that they provide (Brock et al. 2006; Maltby et al. 2005). Furthermore, based on results of studies in the field and in cosms, exposures equivalent to the 5th centile appear to not cause adverse effects on populations or communities. This is due in part to reduced bioavailability compared to exposures of organisms under laboratory conditions, and to more rapid dissipation of CPY under field conditions.

2.4 Conceptual Models of Effects

Based on the likely effects of CPY on aquatic animals, a conceptual model for effects was constructed to serve as a guide for developing risk questions and

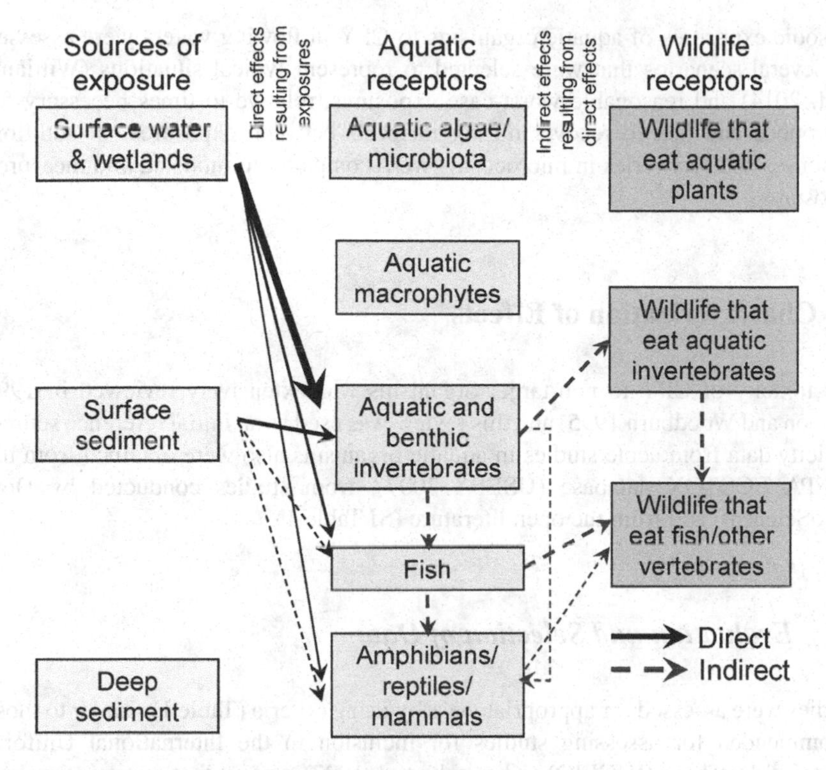

Fig. 2 Conceptual model for effects of chlorpyrifos on aquatic organisms in surface waters. The weights of the *arrows* indicate importance of the pathway of exposure

hypotheses (Fig. 2). Fish and amphibians are less sensitive to direct effects than crustaceans and insects but they could be affected indirectly via alterations of the food web (Fig. 2). As discussed above, exposure via the food chain was not considered in this ERA.

2.5 Analysis Plan

The previous assessment of risks of CPY in surface waters of the U.S. (Giesy et al. 1999) began with a lower-tier deterministic characterization of risk quotients (RQs, also referred to as hazard quotients, HQs) and then advanced through several tiers of refinement to a probabilistic assessment of risks based on comparisons of SSDs to distributions of measured concentrations of CPY in surface waters. Because lower tiers are designed to be conservative and to be applied when few data are available, they are not applicable to CPY, for which there is a wealth of data. Thus, the risk assessment was focused on the upper, more refined, tiers. SSDs were used to characterize acute effects, and these distributions were compared to concentrations predicted by simulation models and concentrations measured in surface waters.

Episodic exposures of aquatic organisms to CPY in flowing waters were assessed for several scenarios that were selected to represent typical situations (Williams et al. 2014) and reasonable worst-case exposures, relative to times necessary for arthropods and fish to recover during intervals between exposures. In addition, responses and recoveries in microcosms were compared to modeled and measured exposures.

3 Characterization of Effects

The toxicity of CPY to non-target organisms was extensively reviewed in 1995 (Barron and Woodburn 1995) and this review was used as an initial reference source. Toxicity data from acute studies in aquatic organisms also were obtained from the USEPA ECOTOX database (USEPA 2007), from studies conducted by Dow AgroSciences, and from the open literature (SI Table 1).

3.1 Evaluation and Selection of Data

Studies were assessed for appropriateness by using criteria (Table 1) similar to those recommended for assessing studies for inclusion in the International Uniform Chemical Database (IUCLID) (Klimisch et al. 1997), except that numerical values were assigned to the individual criteria. Scores used to characterize studies are described below, and these were mostly used to assess data from the open literature. Guideline studies conducted under Good Laboratory Practice (GLP) with full Quality Assurance/Quality Control (QA/QC) were given the maximum score unless they had <5 concentrations of exposures, the recommended number for guideline studies such as those of OECD.

Applied experimental procedures were scored from 1 to 5 (Table 1). A score of 1 was assigned if the design was inadequately described or incorrect and a 5 if it was complete, such as a guideline study conducted under GLP and with a clear protocol. Examples of factors considered when judging study quality were: incomplete description of the methods, inappropriate designs such as pseudoreplication and lack of appropriate controls, lack of information about test organisms, replicates, or number of test subjects per replicate, lack of an adequate description of the purity or form of the test substance, inappropriate statistical comparisons, lack of details about husbandry of organisms, lack of details on analytical methods, etc.

The use of QA/QC was scored from 1 to 5 (Table 1). If there was full QA/QC, the score was 5. Scores of 2–4 were assigned based on the amount of QC, such as, for example, measurements of exposure concentrations at the start of the study only (score = 2) or measurements of exposures and other parameters at regular intervals (score = 3).

Table 1 Criteria for assessing strength of the methods of toxicity studies on chlorpyrifos and its oxon

Strength of methods	1	2	3	4	5
Description of the experimental and procedures	Inadequately described or incorrect	Critical information on methods missing	Critical information on methods and procedures provided	Procedures described so that the study could be replicated or spirit of GLP	Guideline study or complete detailed protocol or full GLP
QA/QC	None	Some QC (e.g., exposures measured once)	More QC (e.g., exposures measured on a regular basis)	Robust QA/QC	Full QA/QC
Transparency of data	Critical data not provided	Summary data only	Summary data with variance characterized	Key raw data provided	All raw data provided
Concentrations tested	Only 1 conc. tested	2 concs. tested	3 concs. Tested or field study >3 sites	4 concs. tested	≥5 concs. tested
Descriptors from Klimisch et al. (1997)	Not assignable	Not reliable	With restriction	Without restrictions	

Transparency of data was scored from 1 to 5 (Table 1). If critical data were not provided, a score of 1 was assigned. If full raw data were provided, the score was 5. If data were provided in tables and graphs, intermediate scores were assigned depending on the clarity of the data and the description of variance, etc.

Most concentration-response study guidelines require that five concentrations be used in a toxicity test; therefore, five test concentrations were used to define a maximum score (5). It was recognized that, in some circumstances, the use of fewer concentrations could still provide useful data, particularly if the concentration-response curve was steep. These studies were, however, assigned lesser scores.

The overall evaluation of the strength of the methods was obtained from the computed average of the scores. Toxicity data for inclusion in the risk assessment were selected based on the overall score. Guideline and GLP studies with full QA and QC, and studies with scores of ≥4, were preferred. Studies with scores of <4 ≥3 were included as qualified values and those with scores <3 were not included.

Additional information was also used to select data for use when multiple results were available for the same species. For example, flow-through exposures were selected over renewal and renewal over static. However, if only data from static exposures were available, they were used with caution. Where several stages of the same species were tested, the more sensitive stage (based on the EC or LC value) was used. For example, data from larval amphibians were selected over embryos, which are generally less sensitive (Richards and Kendall 2002). The details of toxicity data included and excluded from SSDs are indicated in SI Table 1.

Data for saltwater and freshwater, Palearctic and Nearctic, tropical and temperate organisms were not separated, as differences in sensitivity between these groups have been shown to be minimal, and their 5th centile concentrations (HC5s) are not significantly different (Maltby et al. 2005). Data from studies that used the formulated product were included as were those with the active ingredient; however, if data on both formulated and active ingredient were available, only data for the active ingredient were used. If toxicity values for more than one study were available for a species and they were of equal quality, the geometric mean of these values was used to construct the SSD.

For aquatic organisms, the most frequently reported toxicity data were effect concentrations (ECs) that cause some magnitude of effect. For instance, the EC_{50} is the concentration that causes a 50% change in a measurement endpoint, such as growth or reproduction. When the effect is mortality, it is expressed as the lethal concentration that causes 50% mortality in a specified duration of exposure, i.e., 96-h LC_{50}. The HC5, based on acute LC_{50} values, has been found to be protective of responses to CPY at the ecosystem level (Maltby et al. 2005).

All durations of exposure from 2 to 5 d were included; durations >5 d were excluded from the SSDs. Analysis of the exposure profiles (Williams et al. 2014) showed that concentrations greater than toxicity values were of short duration (median = 1 d) and that acute toxicity data were the most appropriate for the assessment. When toxicity values were reported for different periods of exposure, data for the longest period of exposure up to 5 d were included in the SSDs. Toxicity values excluded from the data set were LOEL, LOEC, NOEL, NOEC, MATC, unspecified measures of effect, responses such as behavior, which are difficult to link to survival, development, and reproduction, and data from strains of insects that had been selected for resistance to chlorpyrifos. LC values such as LC_5, LC_{10}, LC_{90}, and LC_{99} were infrequently reported, cannot be combined with LC_{50s}, and were excluded from the assessment.

3.2 *Species Sensitivity Distributions*

As outlined in the Analysis Plan (Sect. 2.5), SSDs were used to characterize the toxicity of CPY to aquatic organisms. Data for different taxa were separated to better assess responses in relation to protection goals which might differ between taxa, for example, invertebrates and fish. Using SSDs to characterize toxicity of CPY is different from using SSDs to develop guidelines and criteria (CCME 2007, 2008).

SSDs were constructed and 5th centiles and their confidence intervals calculated with the aid of the SSD Master Version 3.0 software (CCME 2013). This software is a series of macro statements that are executed in Microsoft Excel. Raw data are entered into a spreadsheet and a cumulative frequency distribution (CFD) is fitted using the plotting positions calculated from the Hazen equation and log-transformed toxicity values to produce an SSD. The data are then fitted to several models (normal, logistic, Extreme Value, and Gumbel) and information on goodness of fit, HC5 and confidence interval for each model is provided. Graphical displays of the SSDs are provided and can be inspected to select those that provide best fit of the data in the region of interest, such as the lower tail of the SSD.

3.3 *Ecotoxicological Profiles by Taxon*

Data were evaluated by taxonomic groups, based on the mode of action and likely sensitivity. These groupings included: plants, crustaceans, insects, fish, amphibians, and other invertebrates. Where insufficient data were available to construct a SSD ($\geq$8 species), such as for algae, amphibians, and invertebrates (other than crustaceans and insects), the data are presented in narrative. Results for each taxon are discussed in the following sections.

Plants. The only data on toxicity of CPY to plants were those generated with algae. All of the four species assessed (Table 2) were saltwater algae. Since there were <8 data points and only one study met the criterion for inclusion in the analysis, an SSD was not derived. The range of EC_{50s} for algae was from 138 to 769 μg CPY L^{-1}, which indicated that algae are relatively tolerant of exposure to chlorpyrifos. Given the mode of action and the lack of a critical mechanism of action or appropriate target site in plants, this is not surprising. Because of this lack of sensitivity, algae were not considered further in the ERA. It is very unlikely that plants

Table 2 Toxicity values for chlorpyrifos used in this assessment (complete data and codes for the test item, exposure and medium are shown in SI Table 1)

Species	Exposure duration (d)	Endpoint	GM (µg L^{-1})[a]	n	Test item	Exposure type	Medium
Algae							
Isochrysis galbana	4	EC50 (growth)	138	1	F	S	SW
Thalassiosira pseudonana	4	EC50 (growth)	148	1	F	S	SW
Skeletonema costatum	3	EC50 (growth)	298	5	F	S	SW
Dunaliella tertiolecta	4	EC50 (growth)	769	1	A	S	SW
Amphibia							
Xenopus laevis	4	LC50	134	2	A	R	FW
Lithobates clamitans clamitans	4	LC50	236	1	A	R	FW
Rana dalmatina	4	LC50	5174	1	A	S	FW
Crustacea							
Daphnia ambigua	2	LC50	0.035	1	A	S	FW
Ceriodaphnia dubia	4	LC50	0.054	2	A	S	FW
Gammarus pulex	4	LC50	0.07	1	A	F	FW
Hyalella azteca	2	LC50	0.10	1	A	S	FW
Moina australiensis	2	LC50	0.10	1	A	S	FW
Gammarus lacustris	4	LC50	0.11	1	F	S	FW
Daphnia pulex	3	LC50	0.12	1	A	R	FW
Palaemonetes pugio	4	LC50	0.15	2	A	R	FW
Neomysis integer	4	LC50	0.16	2	F	F	SW
Gammarus palustris	4	LC50	0.19	1	A	R	SW
Daphnia carinata	2	LC50	0.19	3	A	S	FW
Macrobrachium rosenbergii	2	LC50	0.30	1	F	S	FW
Daphnia longispina	4	LC50	0.30	1	A	R	FW
Paratya australiensis	4	LC50	0.33	1	A	R	FW
Simocephalus vetulus	4	LC50	0.50	1	A	R	FW
Daphnia magna	4	LC50	0.82	1	A	S	FW
Amphiascus tenuiremis	4	LC50	1.47	2	A	S	SW
Procambarus sp.	4	LC50	1.55	1	A	S	FW
Gammarus fossarum	4	LC50	2.90	1	A	R	FW
Orconectes immunis	4	LC50	6.00	1	A	F	FW
Asellus aquaticus	4	LC50	8.58	1	A	S	FW
Eriocheir sinensis	4	LC50	30.5	4	F	R	SW
Neocaridina denticulata	4	LC50	457	1	A	S	FW
Fish							
Menidia menidia	4	LC50	0.53	3	F	F	SW
Leuresthes tenuis	4	LC50	1.1	11	F	F	SW
Menidia peninsulae	4	LC50	1.3	1	A	F	SW
Menidia beryllina	4	LC50	4.2	1	A	F	SW
Fundulus heteroclitus	4	LC50	4.65	1	F	S	SW
Pungitius pungitius	4	LC50	4.70	1	A	F	FW
Atherinops affinis	4	LC50	4.97	2	F	S	SW

(continued)

Table 2 (continued)

Species	Exposure duration (d)	Endpoint	GM (μg L⁻¹)[a]	n	Test item	Exposure type	Medium
Poecilia reticulata	4	LC50	7.2	1	A	R	FW
Cyprinus carpio	4	LC50	8.00	1	F	S	FW
Oncorhynchus mykiss	4	LC50	8.49	2	A	F	FW
Gasterosteus aculeatus	4	LC50	8.5	1	A	F	FW
Lepomis macrochirus	4	LC50	10	1	A	F	FW
Leuciscus idus	4	LC50	10	1	A	R	FW
Oncorhynchus clarki	4	LC50	11	3	F	S	FW
Aphanius iberus	3	LC50	18	1	F	R	SW
Sander vitreus	2	LC50	18	4	A	S	FW
Melanotaenia fluviatilis	4	LC50	122	1	F	R	FW
Pimephales promelas	4	LC50	207	4	A	F	FW
Salvelinus namaycush	4	LC50	244	1	F	F	FW
Oryzias latipes	2	LC50	250	1	A	R	FW
Rutilus rutilus	4	LC50	250	1	A	R	FW
Gambusia affinis	4	LC50	298	1	A	R	FW
Opsanus beta	4	LC50	520	1	A	R	SW
Ictalurus punctatus	4	LC50	806	1	A	F	FW
Carassius auratus	4	LC50	>806	1	A	F	FW
Insects							
Deleatidium sp.	2	LC50	0.05	1	F	S	FW
Chironomus riparius	4	LC50	0.17	2	F	NR	FW
Atalophlebia australis	4	LC50	0.24	1	A	R	FW
Simulium vittatum	2	LC50	0.28	1	A	S	FW
Cloeon dipterum	4	LC50	0.3	1	A	F	FW
Chironomus dilutus	4	LC50	0.62	2	A	S	FW
Anax imperator	4	LC50	1.98	1	A	S	FW
Plea minutissima	4	LC50	1.98	1	A	S	FW
Corixa punctata	4	LC50	2.00	1	A	F	FW
Sigara arguta	2	LC50	2.16	1	F	S	FW
Ranatra linearis	4	LC50	4.48	1	A	S	FW
Chaoborus obscuripes	4	LC50	6.6	1	A	F	FW
Notonecta maculata	4	LC50	7.97	1	A	S	FW
Xanthocnemis zealandica	2	LC50	8.44	1	F	S	FW
Paraponyx stratiotata	4	LC50	27.2	1	A	S	FW
Molanna angustata	2	LC50	>34	1	A	S	FW
Sialis lutariah	4	LC50	>300	1	A	S	FW
Rotifers							
Brachionus calyciflorus	2	LC50	12,000	1	F	S	FW
Mollusks							
Mytilus galloprovincialis	2	EC50 (development)	154	1	A	S	SW
Lampsilis siliquoidea	4	LC50	250	1	A	S	FW
Aplexa hypnorum	4	LC50	>806	1	A	F	FW

[a]GM = geometric mean where n > 1

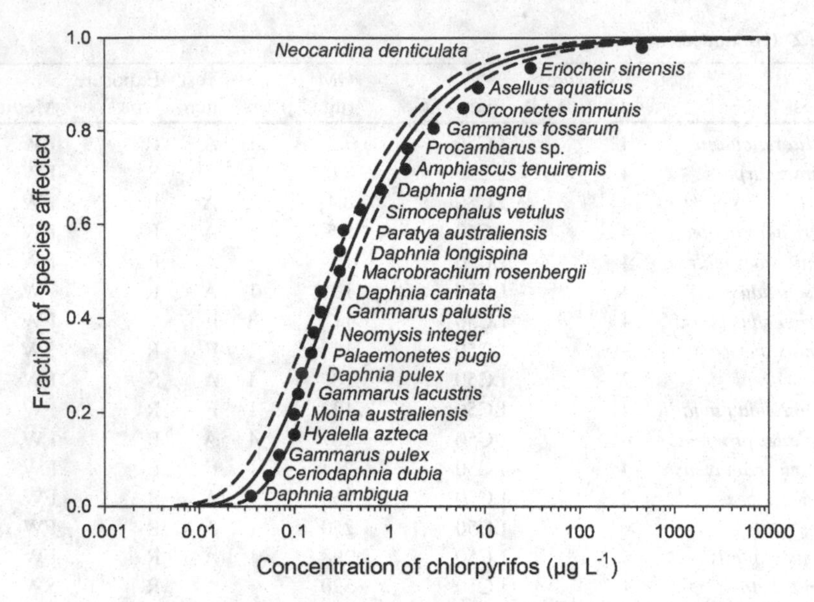

Fig. 3 Species Sensitivity Distribution (SSD) for chlorpyrifos in crustaceans. *Solid line* is the fitted Gumbel model. *Dashed lines* represent 95% confidence interval

would be affected by environmentally-relevant concentrations of CPY and that indirect effects would occur higher in the food web. Thus, protection of these other components of the food web would also be protective of plants in general and phytoplankton in particular.

Crustacea. Data for toxicity of CPY from 23 species of crustaceans met the criteria for inclusion in the analysis (Table 2). The range of LC_{50}s was from 0.035 to 457 µg CPY L^{-1}. The model for the cumulative frequency distribution (CFD) with the best fit was the Gumbel (SI Table 4, SI Fig. 1) and the SSD is shown in Fig. 3. The HC5 (95%CI) was 0.034 (0.022–0.051) µg CPY L^{-1} (SI Table 4).

Insects. Toxicity data for CPY from 17 species of aquatic insects met the criteria for inclusion in the analysis (Table 2). The range of LC_{50}s was from 0.05 to >300 µg CPY L^{-1}. Two values were reported as "greater than" and were included in the calculations of the ranks for constructing the SSD (Fig. 4). The Extreme Value model gave the best fit for the CFD (SI Table 4), but visual inspection of the plots of the various models showed that the fit in the lower tail was better for the Gumbel model (SI Fig. 1). The lower tail is where exceedences are more likely and, for this reason and for consistency with the other taxa, this model was used. The HC5 (95%CI) was 0.087 (0.057–0.133) µg CPY L^{-1} (SI Table 4).

Fish. Data for toxicity of CPY from 25 species of fish met the criteria for inclusion in the analysis (Table 2). One value reported as "greater than" was included in the calculations of the ranks for constructing the SSD (Fig. 5). The model that exhibited the best CFD fit was Gumbel (SI Table 4 and SI Fig. 1). The range of LC_{50}s was

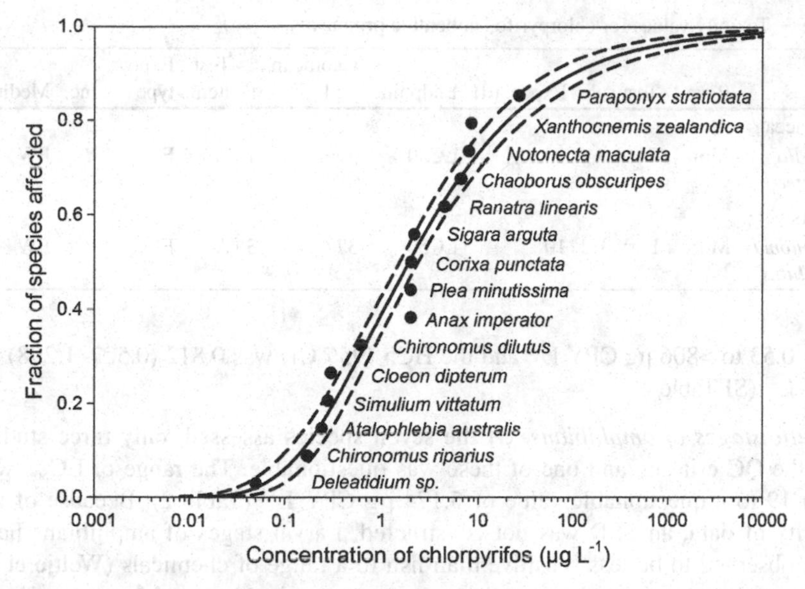

Fig. 4 Species Sensitivity Distribution (SSD) for chlorpyrifos in aquatic insects. *Solid line* is the fitted Gumbel model. *Dashed lines* represent 95% confidence interval. Two species (*Molanna angustata* and *Sialis lutariah*) with "greater than" LC_{50} values were included in the species rankings but not included in the model

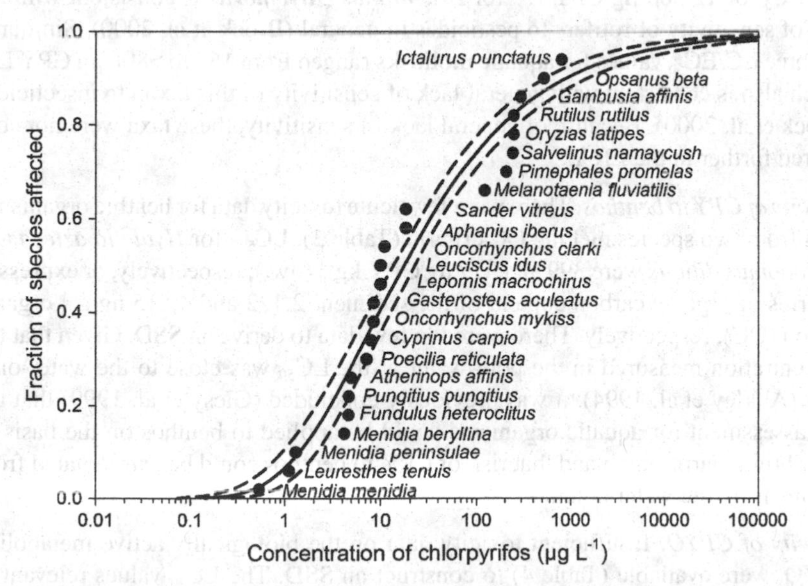

Fig. 5 Species Sensitivity Distribution (SSD) for chlorpyrifos in fish. *Solid line* is the fitted Gumbel model. *Dashed lines* represent 95% confidence interval. One species (*Carassius auratus*) with a "greater than" LC_{50} value was included in the species rankings but not included in the model

Table 3 Toxicity values for chlorpyrifos in benthic organisms

Species	Resp.[a]	Test sub.	Expos. (d)	Endpoint	Geomean µg kg^{-1}	n	Test item	Expos. type	Inc.	Medium
Crustacea										
Hyalella azteca	Mort	7–14 d	10	LC50	399	1	A	F	Y	FW
Insects										
Chironomus dilutus	Mort	Larv 3rd	10	LC50	377	3	A	F	Y	FW

from 0.53 to >806 µg CPY L^{-1} and the HC5 (95%CI) was 0.812 (0.507–1.298) µg CPY L^{-1} (SI Table 4).

Aquatic stages of amphibians. Of the seven species assessed, only three studies met the QC criteria, and one of these was questionable. The range of LC_{50}s was from 19 to a questionable value of 5,174 µg CPY L^{-1} (Table 2). Because of the paucity of data, an SSD was not constructed. Larval stages of amphibians have been observed to be less sensitive than fish to a range of chemicals (Weltje et al. 2013), and the toxicity data for fish can be extrapolated to and be protective of amphibians. Therefore, aquatic stages of amphibians were not considered further in the ERA.

Other invertebrates. Toxicity data for four other invertebrates were found (Table 2). The LC_{50} of 12,000 µg CPY L^{-1} for *Brachionus calyciflorus* is consistent with the lack of sensitivity of rotifers to pesticides in general (Brock et al. 2000). Similarly, the three LC/EC_{50} values for aquatic mollusks ranged from 154 to >806 µg CPY L^{-1}, which also is consistent with general lack of sensitivity of this taxon to insecticides (Brock et al. 2000). Because of general lack of sensitivity, these taxa were not considered further in the ERA.

Toxicity of CPY to benthos. There were few acute toxicity data for benthic organisms. Data from two species met the QC criteria (Table 3). LC_{50}s for *Hyalella azteca* and *Chironomus dilutus* were 399 and 377 µg CPY kg^{-1} (dwt), respectively, or expressed in terms of organic carbon content of the sediment 2,122 and 4,815 ng g^{-1} organic carbon (OC), respectively. There were too few data to derive an SSD. Given that the concentration measured in the pore-water at the LC_{50} was close to the water-only LC_{50} (Ankley et al. 1994), it was previously concluded (Giesy et al. 1999) that the risk assessment for aquatic organisms could be applied to benthos on the basis of equilibrium partitioning and that risk of CPY to benthos could be extrapolated from organisms in the water column.

Toxicity of CPYO. Insufficient toxicity data on the biologically active metabolite, CPYO, were available (Table 4) to construct an SSD. The LC_{50} values relevant to risks to surface-water organisms were a 48-h LC_{50} for *D. magna* of 1.9 µg CPYO L^{-1} and 96-h LC_{50} of 1.1 µg CPYO L^{-1} in the bluegill sunfish *L. macrochirus*. The only toxicity value reported for amphibians was in larval *Rana boylii*, but the reported LC_{50} value of >5 µg CPYO L^{-1} was from a study that did not meet the criteria for inclusion in the ERA. The LC_{50} value for CPYO in *D. magna* was larger than CPY

Table 4 Toxicity values for chlorpyrifos oxon in aquatic organisms

Species	Resp.[a]	Test sub.	Expos. (d)	Endpoint	μg L^{-1}	Test item	Expos. type	Inc.	Medium
Coral									
Acropora millepora	Fert	Embryo	0.125	EC50	>30	A	S	Q	SW
Acropora millepora	Meta.	Larva	0.75	EC50	0.39	A	S	Q	SW
Crustacea									
Daphnia magna	Mort.	<24-h old	2	LC50	1.9	A	F	Y	FW
Fish									
Lepomis macrochirus	Mort.	46 mm	4	LC50	1.1	A	F	Y	FW
Amphibians									
Rana boylii	Mort.	G32-44	4	LC50	>5	A	S	N	FW

(0.82 μg L^{-1}, Table 2), while that for *L. macrochirus* was less than that for CPY (4.2 μg L^{-1}, Table 2). Although CPYO is much more potent than CPY at the target (AChE), the molecule is very labile in aqueous solution (discussed below).

3.4 Evidence from Microcosms and Mesocosms

Chlorpyrifos has been the subject of a large number of studies in microcosms and mesocosms (Table 5). Because the distinction between microcosms and mesocosms is primarily semantic, they are referred to jointly as "cosms" in the following discussion. Data from cosms add several types of realism to assessment of the potential effects of chemicals on aquatic organisms (Graney et al. 1995). They allow for more realistic exposure scenarios because factors such as photolysis, microbial degradation and adsorption to aquatic plants and sediments (Giesy and Odum 1980; Graney et al. 1989) are included. Cosms also include dynamic interactions between and among species so that potential "ecosystem-level" effects can be evaluated, including predator-prey interactions in the larger systems (Giesy and Geiger 1980). Early cosm studies of CPY were reviewed by Leeuwangh (1994), Barron and Woodburn (1995), and Giesy et al. (1999). Since the publication of the Giesy et al. review, results of several cosm studies of CPY have featured prominently in comparisons of cosm studies, single-species laboratory toxicity tests, and regulatory benchmarks across classes of insecticides (Brock et al. 2000, 2006; Maltby et al. 2005; van Wijngaarden et al. 2005b). Several newer cosm studies have broadened the scope of conclusions about chlorpyrifos effects on aquatic communities to a wider range of locations and environmental conditions (Daam et al. 2008a, b; López-Mancisidor et al. 2008a, b; van Wijngaarden et al. 2005a; Zafar et al. 2011). The body of evidence from cosm studies is consistent in supporting the conclusion that concentrations of 0.1 μg CPY L^{-1} or less cause no ecologically significant effects on aquatic communities.

Table 5 Summary of microcosm and mesocosm studies performed with chlorpyrifos

Ref	Cosm	Exposure	Zooplankton			Macroinvertebrates			$NOAEC_{eco}$
			Taxa	Community	Recovery	Taxa	Community	Recovery	
Biever et al. (1994), Giddings (1993a, b); Giddings et al. (1997); Giddings (2011)	Outdoor 11 m^3, Kansas	Lorsban® 4E; 0.03, 0.1, 0.3, 1, 3 $\mu g\ L^{-1}$; single spray or 3× slurry @ 2-wk intervals	Copepods and cladocerans NOAEC = 0.1 $\mu g\ L^{-1}$; rotifers increased	PRC NOAEC = 3 $\mu g\ L^{-1}$ (single spray), 0.1 $\mu g\ L^{-1}$ (3× slurry)	All parameters recovered by end of study (single app), up to 0.3 $\mu g\ L^{-1}$ (3× slurry)	Most benthic taxa NOAEC = 0.1 $\mu g\ L^{-1}$ (single spray), 0.3 $\mu g\ L^{-1}$ (3× slurry)	PRC NOAEC = 0.3 $\mu g\ L^{-1}$	All taxa recovered at all treatment spray levels except Chironomidae at 1 $\mu g\ L^{-1}$; many taxa and PRC did not recover at 3 $\mu g\ L^{-1}$ slurry treatment	0.1 $\mu g\ L^{-1}$ (single spray or 3× slurry)
Brazner and Kline (1990)	Outdoor littoral enclosures, 55 m^3, Minnesota	Dursban®: 0.5, 5.0, and 20.0 $\mu g\ L^{-1}$ applied once	Cladocerans and ostracods decreased; rotifers decreased at low conc. but unaffected or increased at higher conc.	Not reported	Cladocerans, copepods, and rotifers recovered at 0.5 $\mu g\ L^{-1}$ but not at higher concentrations	Chironomids decreased	Not reported	Chironomids recovered at 0.5 $\mu g\ L^{-1}$ within 16 d	<0.5 $\mu g\ L^{-1}$
Brock et al. (1992, 1993, 1995)	Indoor 847 L, Netherlands, *Elodea nuttallii*-dominated (western waterweed)	Dursban® 4E 48%: 5 and 35 $\mu g\ L^{-1}$ applied once	Cladocerans, copepods, amphipods, isopods, and insects decreased; no effect on rotifers	Secondary effects observed: increase in periphyton, gastropods, oligochaetes	Some recovery by amphipods, complete or nearly complete recovery by isopods, cladocerans, and copepods	Insects reduced at 5 $\mu g\ L^{-1}$, eliminated at 35 $\mu g\ L^{-1}$	Not reported	At 5 $\mu g\ L^{-1}$, slight recovery, then disappearance again at end of study; no recovery at 35 $\mu g\ L^{-1}$	<5 $\mu g\ L^{-1}$
Brock et al. (1992, 1993)	Indoor 847 L, Netherlands, devoid of macrophytes	Dursban® 4E 48%; 35 $\mu g\ L^{-1}$ applied once	Cladocerans, copepods, isopods, insects decreased; amphipods eliminated; no effect on rotifers	Secondary effects observed: increase in phytoplankton, rotifers, bivalves, oligochaetes	Some recovery by cladocerans and copepods; full recovery by isopods; no recovery by amphipods	Insects eliminated, species not identified	Diversity associated with POM higher in open-water systems	Insects absent until final wk	<35 $\mu g\ L^{-1}$

Brock et al. (1995), Cuppen et al. (1995), Van Donk et al. (1995)	Indoor 847 L, Netherlands, *Elodea nuttallii*-dominated (western waterweed), nutrient-enriched system	Dursban® 4E 48%: 5 and 35 µg L^{-1} applied once	Copepods decreased, cladocerans eliminated., rotifers increased	Not reported	Cladocera reappeared after 10 wk, then increased; copepods surpassed pre-dose levels after 5 wk; rotifer populations decreased after their initial increase	Isopods decreased; *Gammarus pulex*, *Chaoborus obscuripes*, and *Plea minutissima* eliminated	Recoveries by *Proasellus meridianus* and *Lumbriculus variegatus* restored decomposition rates	Rapid recovery by *Proasellus meridianus* and some recovery by *Lumbricula variegatus*. Eliminated species did not reappear	<5 µg L^{-1}
Daam et al. (2008a)	Outdoor 1 m^3 Thailand	Dursban® 40 EC; 0.1, 1, 10, 100 µg L^{-1} applied once	Cladocerans (*Moina micrura*) most sensitive, NOAEC = 0.1 µg L^{-1}; other cladocerans (*Ceriodaphnia cornuta*), adult copepods, copepod nauplii NOAEC = 1 µg L^{-1} on wk-1, then recovered; some rotifers (*Keratella*) decreased at 100 µg L^{-1}, other rotifers increased	PRC affected at 1, 10, 100 µg L^{-1}; NOAEC = 0.1 µg L^{-1}	PRC recovered at 1 µg L^{-1} by d-14, at 10 µg L^{-1} by d-35, and at 100 µg L^{-1} by d-70; all taxa recovered by wk-10 at 100 µg L^{-1}	Pebble baskets; Conchostraca (clam shrimp) eliminated at three highest concentrations; ostracods eliminated at 10 µg L^{-1}; Corixidae decreased at 10 µg L^{-1}; snails, flatworms, mollusks increased; NOAEC = 0.1 µg L^{-1}	PRC NOAEC = 0.1 µg L^{-1}	PRC recovered by wk-10 at all concentrations; Corixidae recovered by wk-8; all taxa recovered by wk-10	0.1 µg L^{-1}
Daam et al. (2008b)	Outdoor 250 L, Thailand	Dursban® 40% AI; 1 µg L^{-1} applied once or 2× @ 2-wk interval	Cladocerans decreased; copepods, rotifers, ostracods increased	PRC affected in both treatments	PRC recovered by d-32 in both treatments	Not included in study	Not included in study	Not included in study	<1 µg L^{-1}

(continued)

Table 5 (continued)

			Zooplankton			Macroinvertebrates			
Ref	Cosm	Exposure	Taxa	Community	Recovery	Taxa	Community	Recovery	$NOAEC_{eco}$
Hurlburt et al. (1970)	Outdoor, 2.9–3.8 m^3, water level maintained twice weekly, California	Dursban® 40 EC, 0.01, 0.05, 0.10, 1.0 lb/A (2 $\mu g\ L^{-1}$ to 200 $\mu g\ L^{-1}$) applied four times	Dominant zooplankton (*Cyclops vernalis* and *Moina micrura*) increased while *Diaptomus pallidus* and rotifers decreased	Not reported	*M. micrura* and *C. vernalis* did not recover after first treatment of high doses. Lower doses: no recovery after second treatment. *D. pallidus* showed no consistent evidence of effect at low doses, did not develop populations at high doses	*Corisella decolor* was the only major insect Significant reduction at 10 $\mu g\ L^{-1}$ after first treatment. Low populations for all doses after third treatment	Not included in study	Gradual recovery for all but the highest dose. At 200 $\mu g\ L^{-1}$ recovery was not observed	<2 $\mu g\ L^{-1}$
Hurlburt et al. (1972)	Outdoor, 33 m^3, water level maintained twice weekly, California	Dursban® 40 EC, 7.2, 72 $\mu g\ L^{-1}$ applied three times	Dominant zooplankton (*Cyclops vernalis* and *Moina micrura*) reduced while *Diaptomus pallidus* (copepod) and rotifers increased	Strong negative correlation between cladoceran and rotifer populations	*C. vernalis* and *M. micrura* recovered in 1–3 wk at 7.2 $\mu g\ L^{-1}$ and in 3–6 wk at 72 $\mu g\ L^{-1}$	Notonectidae highly reduced as of 5 wk post-final treatment, except nymphs at 7.2 $\mu g\ L^{-1}$. Corixids declined in controls and adults were more numerous at 72 $\mu g\ L^{-1}$. Mayflies decimated at both doses, first treatment. *Tropisternus* and *Helophorus* also significantly more numerous at higher doses	Predaceous insects were 45% and 9% of control levels 5 wk after low and high dose treatments, respectively	Predaceous insects were affected and recovered slowly, while herbivorous insects were less affected and recovered more quickly	<7.2 $\mu g\ L^{-1}$

López-Mancisidor et al. (2008b)	Outdoor 11 m^3 Spain, plankton-dominated	Chas® 48 EC, 0.033, 0.1, 0.33 and 1 μg L^{-1}, sprayed 4× @ weekly intervals	Cyclopoid copepod and cladoceran density decreased at 1 μg L^{-1}; calanoid copepods increased; some rotifers increased; NOAEC = 0.33 μg L^{-1}	PRC affected at 0.33 μg L^{-1}; NOAEC = 0.1 μg L^{-1}	Number of arthropod taxa recovered by d-39; PRC recovered by d-81; total copepods and nauplii still less than controls on d-130; all taxa recovered within 12 wk below 1 μg L^{-1}	Not included in study	Not included in study	Not included in study	0.1 μg L^{-1} (nominal) = 0.169 μg L^{-1} (measured max) = 0.158 μg L^{-1} (max 7-d TWA)
López-Mancisidor et al. (2008a)	Outdoor 11 m^3 Spain, plankton-dominated	Chas® 48 EC, 0.1 and 1 μg L^{-1}, sprayed once	Cladocerans, copepods, some rotifers (*Keratella*) decreased at 1 μg L^{-1}; some rotifers (*Brachionus*) increased	PRC NOAEC = 0.1 μg L^{-1}	*Daphnia galeata* decreased to near zero, recovered quickly; *Keratella* still reduced on d-99	Not included in study	Not included in study	Not included in study	0.1 μg L^{-1}
Pusey et al. (1994)	Outdoor stream (45 m × 0.4 m), flow through conditions, Australia	Chemspray, 0.1 and 5.0 μg L^{-1} applied in 6 h pulses	Not included in study	Not included in study	Not included in study	Chironomids: NOAEC = 0.1 μg L^{-1}, decreased at 5 μg L^{-1}	NOAEC = 0.1 μg L^{-1}	Chironomid recovered by end of study (d-80)	0.1 μg L^{-1}
Van den Brink et al. (1995)	Indoor 847 L, Netherlands, freshwater	Dursban® 4E: 0.1 μg L^{-1}, continuous dosing, 7 wk	Cladocerans (*Daphnia galeata*), copepods, and ostracods generally decreased; rotifers (*Keratella quadrata*) increased	Not reported	No recovery; rotifer populations decreased after their initial increase	Isopods decreased. *Gammarus pulex* eliminated	Not reported	Very slight (not significant) recovery by *Proasellus meridianus*. No community recovery	<0.1 ng L^{-1}, continuous for 7 wk
Van Wijngaarden et al. (2005b)	Indoor 18 L; temperate and Mediterranean conditions	Dursban® 480, 0.01-10 μg AI L^{-1}	Cladocerans, copepod nauplii most sensitive; rotifers increased	PRC NOAEC = 0.1 μg AI L^{-1}	21 d under temperate conditions, >28 d under Mediterranean conditions	Not included in study	Not included in study	Not included in study	0.1 μg L^{-1}

(continued)

Table 5 (continued)

Ref	Cosm	Exposure	Zooplankton			Macroinvertebrates			NOAEC$_{eco}$
			Taxa	Community	Recovery	Taxa	Community	Recovery	
Ward et al. (1995)	Outdoor stream (45 m × 0.4 m), flow through conditions, Australia	Chemspray, 0.1 and 5.0 μg L^{-1} applied continuously over 21 d	Cladocerans, copepods, and ostracods decreased at both doses, more severely at 5 μg L^{-1}	Total invertebrate abundance and diversity reduced at both doses	Ostracods recovered at 0.1 μg L^{-1}; cladocerans and copepods did not recover, but low abundances pre-trial render this uncertain	13/19 chironomids and 8/36 nonchironomids were significantly reduced and many showed no recovery by 70 d	Notable sensitivity observed for all insects, even predators	Abundance and diversity slightly reduced at 0.1 μg L^{-1} and more severely reduced at 5 μg L^{-1}. Abundances of invertebrates at 5 μg L^{-1} recovered between 42 and 70 d after the first treatment	NA
Zafar et al. (2011)	Outdoor 1.3 m^3 (Netherlands)	480 g AI L^{-1} EC; three exposure regimes with similar TWA: 0.9 μg L^{-1} (single app), 0.3 μg L^{-1} (three apps @ 7 d), 0.1 μg L^{-1} (continuous 21-d)	Cladocerans most sensitive; also copepod nauplii; rotifers increased	Immediate PRC effect after single app; progressive effects in other treatment	PRC still significant in single-app treatment at end of study	*Cloeon dipterum* and *Chaoroborus* most sensitive in 0.9 μg L^{-1} but not affected in 0.1 μg L^{-1} continuous; also *Gammarus pulex*, Trichoptera	Immediate effect after single app; progressive effects in other treatments	PRC still significant at end of study; *Chaoborus* recovered after single app; *G. pulex* still reduced in 0.1 μg L^{-1} continuous at end of study	<0.9 μg L^{-1} (single app) <0.3 μg L^{-1} (three apps) <0.1 μg L^{-1} (continuous 21 d)

California ponds. Two studies on fates and effects of CPY were conducted in outdoor experimental ponds at Riverside, California (Hurlburt et al. 1970; 1972). In the first study, ponds were sprayed four times at 2-wk intervals at initial concentrations from 2 to 200 μg L^{-1}. In the second study, ponds were sprayed three times at 2-wk intervals to produce concentrations of 7.2 μg CPY L^{-1} and 72 μg L^{-1}. In both studies, the dominant zooplankton species, *Cyclops vernalis* and *Moina micrura*, were reduced, while *Diaptomus pallidus* and rotifers (especially *Asplanchna brightwelli*) increased. The increases in *D. pallidus* and rotifers were attributed to reduced predation and competition. *C. vernalis* and *M. micrura* recovered in 1–3 wk at 7.2 μg L^{-1} and in 3–6 wk at 72 μg L^{-1}. Predaceous insects (notonectids and corixids) were affected and recovered slowly, while herbivorous insects were less affected and recovered more quickly. No effects were observed on the mosquitofish (*Gambusia affinis*).

Minnesota littoral enclosures. In situ enclosures in the littoral region of a pond in Duluth, Minnesota were sprayed once with CPY to produce initial concentrations of 0.5, 5, and 20 μg L^{-1} (Brazner and Kline 1990; Siefert et al. 1989). Cladocerans (five species) and ostracods (*Cyclocypris*) decreased 4 d after treatment. Densities of copepods were slightly reduced in treated enclosures but were not significantly less than controls. Rotifers were reduced at 0.5 μg L^{-1}, but were unaffected (or increased) at the greater concentrations. Chironomids (the dominant insect group) were reduced 4 d after treatment; they recovered within 16 d at 0.5 μg L^{-1}, but remained less abundant than controls after 32 d at the greater concentrations. Other insects and the amphipod, *Hyalella azteca*, were reduced. Snails, planaria, and protozoa were unaffected or increased. Survival of bluegill sunfish (*L. macrochirus*) decreased at 5 and 20 μg L^{-1}. Survival of fathead minnows (*Pimephales promelas*) was unaffected, but the study authors reported that *P. promelas* growth was reduced due to a reduction in invertebrate abundance. As discussed by Giesy et al. (1999), the data from the study do not support this interpretation, and the effect on growth of *P. promelas*, if real, remains unexplained.

Kansas outdoor cosms. Outdoor pond cosms in Kansas were treated with CPY with initial concentrations of 0.03, 0.1, 0.3, 1, and 3 μg L^{-1} (Biever et al. 1994; Giddings 1993a, b; Giddings et al. 1997; Giddings 2011). In separate series of cosms, applications were made as a single surface spray, three CPY-treated slurry applications at 2-wk intervals, and a combination of the two. Results were similar in all three series, and only the single spray treatment will be summarized here.

The total abundance of copepods was reduced for 3 d at 0.3 μg CPY L^{-1} (recovery by d-15), for 29 d at 1 μg L^{-1} (recovery by d-43), and for 22 d at 3 μg L^{-1} (recovery by d-29). No effects were observed at 0.1 μg L^{-1}. The calanoid copepod *D. pallidus* was a notable exception to the general sensitivity of the copepods: at the highest treatment level (3 μg L^{-1}) *D. pallidus* increased soon after the chlorpyrifos application. An increase in *D. pallidus* after chlorpyrifos treatment was also observed by Hurlbert et al. (1972), who noted that the increase took place only after numbers of *C. vernalis* had decreased. The numbers of cyclopoids, as well as most calanoids other than *D. pallidus*, were also reduced in the Kansas cosms.

This pattern, observed in studies conducted 20 yr and 2,000 km apart, implies that (a) *D. pallidus* is less sensitive than cyclopoids to CPY, and (b) *D. pallidus* competes with cyclopoids for food, and therefore benefits from reductions in cyclopoid abundance. It has been reported that "the survival and reproduction of *D. pallidus* were substantially enhanced by the addition of rotifers to a threshold algal diet" (Williamson and Butler 1986).The increase in *D. pallidus* at 3 μg L^{-1} may therefore have been partly a result of the increase in abundance of rotifers (see below).

The cladocerans were slightly less abundant at 0.3 μg L^{-1} than in controls, but only on d-43. Pronounced effects occurred at 1 μg L^{-1} (recovery by d-57) and 3 μg L^{-1} (recovery by d-43). No effects on total numbers of cladocerans were observed at 0.1 μg L^{-1}. The most abundant cladocerans were *Chydorus sphaericus* and *Alona* sp. *C. sphaericus* was most sensitive to CPY, while *Alona* was less sensitive and appeared to benefit from changes that occurred at the greater concentrations.

There were no significant differences in abundance of rotifers among CPY treatment levels on any sample event. The total numbers of the two major rotifer groups, Ploima and Flosculariaceae, were also unaffected by treatment with CPY. Total numbers of zooplankton were not significantly reduced at any concentration on any sample event. The observed reductions in copepods and sensitive cladocerans were offset by increases in rotifers and more tolerant cladocerans.

Benthic insect communities in the cosms were dominated by Diptera and Ephemeroptera. Abundance of Diptera was significantly reduced at 0.3 and 1 μg L^{-1} on d-15 only, and at 3 μg L^{-1} from d-1 through d-29 (recovery by d-42). Treatment-related reductions in numbers of Ephemeroptera were found at 0.3 μg L^{-1} (d-1 only), 1 and 3 μg L^{-1} (d-1 and -15). There were no significant differences after d-15. Significant differences in total numbers of benthic insects occurred on d-1 and -15 at 0.3 and 1 μg L^{-1}, and from d-1 through d-29 at 3 μg L^{-1} (recovery by d-42). No effects on Diptera, Ephemeroptera, or total benthic insects were observed at 0.1 μg L^{-1}. In terms of invertebrate community structure (based on Principal Response Curve (PRC) analysis) (Giddings 2011) and abundance of sensitive populations within the community, no ecologically relevant effects occurred at 0.1 μg L^{-1}.

Australian stream cosms. Two studies in Australia reported effects of CPY on invertebrate communities in large outdoor experimental streams (Pusey et al. 1994; Ward et al. 1995). In the first study, 6-h pulses of CPY were applied at 0.1 and 5 μg L^{-1} and invertebrate community responses were monitored for 80 d. There were no effects at 0.1 μg L^{-1}. The abundance of chironomids, but not other invertebrate groups, was reduced at 5 μg L^{-1}. Invertebrate abundance recovered by the end of the study. In the second study, the same concentrations of CPY were applied continuously for 21 d. Abundance and diversity of invertebrates were slightly reduced at 0.1 μg L^{-1} and more severely reduced at 5 μg L^{-1}. Abundances of invertebrates at 5 μg L^{-1} recovered between 42 and 70 d after the first treatment. Snails became more abundant in the treated streams.

Dutch ditch cosms. CPY has been the subject of several studies in indoor and outdoor cosms representing Dutch ditches at the Winand Staring Center, The Netherlands. The indoor cosms (Brock et al. 1992; 1993; 1995; Cuppen et al. 1995; van den Brink et al. 1995; Van Donk et al. 1995) were sprayed once, with an initial concentration of 35 μg CPY L^{-1}. Direct effects were observed on cladocerans, copepods, amphipods, isopods, and insects. An algal bloom occurred, as had also been observed in the California ponds (Hurlburt et al. 1970; 1972) and in the Minnesota enclosures (Siefert et al. 1989). The researchers documented the recovery of the cosm-invertebrates as concentrations of CPY declined. Copepods and some cladoceran populations recovered when concentrations of CPY reached 0.2 μg L^{-1}; other cladocerans recovered when concentrations fell to 0.1 μg L^{-1}. Taxa with no recolonization sources (such as insects, amphipods, and isopods) did not recover, but cage studies showed that *Asellus aquaticus* could survive when concentrations decreased to 1.3 μg L^{-1}; and *Chaoborus obscuripes*, *Cloeon dipterum*, *Gammarus pulex* could survive when concentrations reached 0.2 μg CPY L^{-1}.

The outdoor ditch enclosures were sprayed once with CPY concentrations of 0.1, 0.9, 6, and 44 μg L^{-1} (van den Brink et al. 1996; van Wijngaarden et al. 1996). No effects were observed at 0.1 μg L^{-1}. At greater concentrations, numbers of macroinvertebrates were reduced and shifts were observed in the relative abundance of different functional groups (reductions in the proportion of gatherers, increases in the proportions of filter feeders and shredders). Most taxa, other than *G. pulex*, recovered rapidly at all concentrations. *G. pulex* could not recover because there was no source of recolonization.

In further studies, macrophyte-dominated outdoor Dutch ditch cosms were treated with CPY under three different exposure regimes: a single application of 0.9 μg L^{-1}, three applications of 0.3 μg L^{-1} at 7-d intervals, and continuous application of 0.1 μg L^{-1} for 21 d using a pump (Zafar et al. 2011). The three exposure regimes were designed to produce similar 21-d time-weighted averages of 0.1 μg L^{-1}. Under all exposure regimes, cladocerans and copepod nauplii were the most sensitive taxa of zooplankton, while numbers of rotifers increased. *C. dipterum* and *Chaoborus* sp. were the most sensitive taxa of macroinvertebrates. In both the zooplankton and insect communities, effects were observed immediately after the single application of 0.9 μg L^{-1} but occurred more slowly in the other treatments. Overall, effects on both the zooplankton and macroinvertebrate communities were more or less the same under all exposure regimes.

Indoor cosms simulating Mediterranean environments. Indoor, plankton-dominated microcosms were used to compare the responses of aquatic communities to CPY under conditions pertaining to Mediterranean regions (higher temperature and greater amounts of nutrients) with conditions representing cool temperate regions (van Wijngaarden et al. 2005a). CPY was applied once to give concentrations from 0.01 to 10 μg L^{-1} in water of microcosms. CPY dissipated more rapidly under Mediterranean than temperate conditions. As in previous studies, cladocerans and copepod nauplii were among the most sensitive taxa of zooplankton, while numbers of rotifers and adult copepods generally increased as the other groups declined.

The NOAEC for the most sensitive zooplankton populations and for the zooplankton community was 0.1 μg L^{-1} under both temperate and Mediterranean conditions. The phytoplankton community was altered and phytoplankton chlorophyll increased at 1 μg L^{-1}, but only under Mediterranean conditions. Overall, the study supported a community-level NOAEC value of 0.1 μg L^{-1}.

Spanish outdoor cosms. Outdoor plankton-dominated 11-m^3 cosms were treated with single CPY applications of 0.1 and 1 μg L^{-1} (López-Mancisidor et al. 2008b). Cladocerans, copepods, and some rotifers (*Keratella* sp.) decreased at 1 μg L^{-1}; other rotifers (*Brachionus* sp.) increased. *Daphnia galeata*, which had been severely reduced, recovered rapidly; *Keratella* sp. was still reduced on d-99. There were no effects at 0.1 μg L^{-1}. In a subsequent study, López-Mancisidor et al. (2008a) sprayed the cosms four times at weekly intervals to produce concentrations of 0.033, 0.1, 0.33, and 1 μg L^{-1}. Population densities of cyclopoid copepods and cladocerans decreased at 1 μg L^{-1}, while calanoid copepods and some rotifers increased. Community analysis using PRC indicated significant effects at 0.33 μg L^{-1}. All taxa recovered within 12 wk (9 wk after the final CPY application) except at the highest treatment level (1 μg L^{-1}). These studies indicated that CPY caused no effects on zooplankton after single or multiple exposures to CPY at 0.1 μg L^{-1}.

Thailand outdoor cosms. In a study in 1-m^3 outdoor cosms in Thailand treated once 0.1, 1, 10, and 100 μg CPY L^{-1}, cladocerans (*Moina micrura*) were the most sensitive zooplankton taxa, with significant reductions at 1, 10, and 100 μg L^{-1} (Daam et al. 2008a). Other cladocerans (*Ceriodaphnia cornuta*), adult copepods, and copepod nauplii were reduced for 1 wk at 1 μg L^{-1} and then recovered. Some rotifers (including *Keratella*) decreased at 100 μg L^{-1}, while other rotifer species increased. Among macroinvertebrates, Conchostraca (clam shrimp) were eliminated at the three greatest concentrations. Ostracods and corixids were reduced or eliminated at 10 μg CPY L^{-1}. Snails, flatworms, and mollusks increased in abundance. PRC analysis showed that both zooplankton and macroinvertebrate communities were affected at the three greatest concentrations of CPY. Zooplankton communities recovered by d-14 at 1 μg L^{-1}, d-35 at 10 μg L^{-1}, and d-70 at 100 μg L^{-1}; all taxa of macroinvertebrates recovered by d-70 at all concentrations. Similar studies on the effects of CPY on zooplankton were conducted in 250-L outdoor cosms in Thailand, treated either once or twice (7-d interval) with 1 μg L^{-1} CPY (Daam et al. 2008b). Cladocerans decreased, while copepods, rotifers, and ostracods increased; PRC indicated community recovery by d-32 in both treatments. Overall, the Thai cosm studies indicated a NOAEC of 0.1 μg L^{-1}.

Conclusions from cosm studies with CPY. Results of cosm studies of CPY summarized above (except the recent studies) were included in analyses by Brock et al. (2000), van Wijngaarden et al. (2005b), Maltby et al. (2005), and Brock et al. (2006), in which the cosm results were compared with single-species toxicity data and regulatory benchmarks for various groups of insecticides. The summarized conclusions of these reviews are consistent with the earlier Giesy et al. (1999) ecological risk assessment, and are:

- Sensitivity of species in cosms is similar to the sensitivity of the same or related taxa in laboratory toxicity tests for single species. For CPY and other acetylcho-

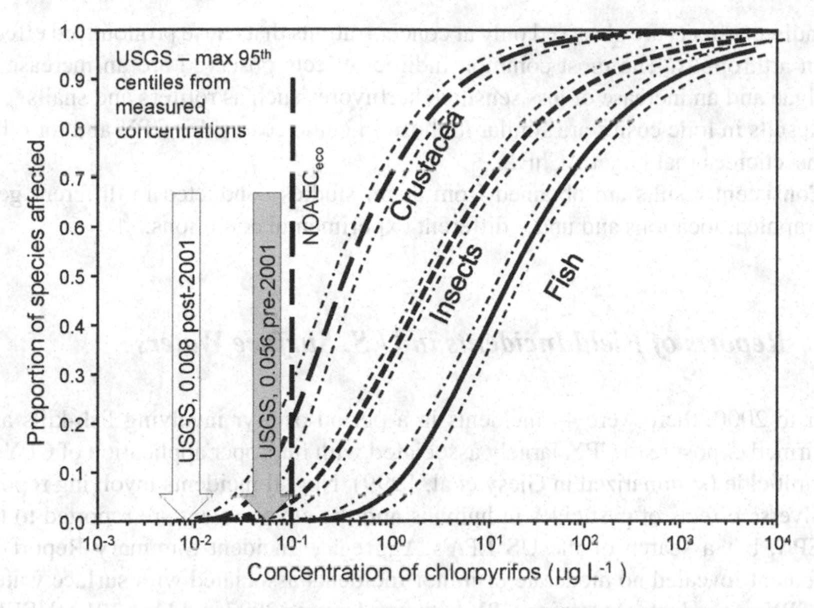

Fig. 6 Comparison of SSDs for 96-h toxicity values for chlorpyrifos and the $NOAEC_{eco}$ from cosm studies to greatest annual 95th centiles of concentrations reported by the US Geological Survey from surface waters samples collected before and after 2001. For a more detailed description of the data, see Williams et al. (2014)

linesterase (AChE) inhibitors, Amphipoda, Cladocera, Copepoda, Trichoptera, Ephemeroptera, and Diptera are the most sensitive taxa in cosms, with effects observed at 0.1- to 1-times the LC_{50} of the most sensitive standard test species. Effects on mollusks, annelids, and plants are observed at concentrations 10–100 times greater than the LC_{50} of the most sensitive species. Most rotifers are unaffected at even greater concentrations.

- The ecosystem-level NOAEC ($NOAEC_{eco}$) is the concentration in which "no, or hardly any, effects on the structure and functioning of the studied (model) ecosystem are observed" (Fig. 6, van Wijngaarden et al. 2005b). For CPY, the $NOAEC_{eco}$ for a single exposure is 0.1 µg L^{-1}. Effects are generally more severe with repeated or chronic exposure, but such exposure patterns are not typical for CPY (Williams et al. 2014). The $NOAEC_{eco}$ falls at the 23rd centile of the SSD for crustaceans, the most sensitive taxon for which data are available (Fig. 6). This implies that use of the HC5 in the risk characterization errs on the side of protection.
- For inhibitors of AChE and other insecticides, sensitive crustaceans and insects in static systems usually recover within 8 wk of a single pulsed exposure below the LC_{50} of the most sensitive species. For multiple applications, recovery occurs within 8 wk of the last application less than 0.1× the LC_{50} (van Wijngaarden et al. 2005b). The extent and rate of recovery in cosms is determined by exposure concentration, life cycle, and ecological factors such as the degree of isolation of the test system from sources of recolonization.

- Indirect effects are observed only at concentrations that cause pronounced effects on arthropods. The most common indirect effects observed are an increase of algae and an increase of less sensitive herbivores such as rotifers and snails.
- Results in lotic cosms are similar to those in lentic cosms, for CPY and for other insecticides that target AChE.
- Consistent results are obtained from cosm studies conducted in different geographical locations and under different experimental conditions.

3.5 Reports of Field Incidents in U.S. Surface Waters

Prior to 2000, there were 44 incidents in a period of 3 yr involving fish-kills and confirmed exposure to CPY, largely associated with improper application of CPY as a termiticide (summarized in Giesy et al. 1999). Not all incidents involving reports of adverse effects of pesticides in humans and the environment are reported to the USEPA, but a search of the US EPA's Aggregate Incident Summary Report by Ingredient revealed no moderate or minor incidents associated with surface waters and CPY or products containing CPY between January 2002 and June 2012 (USEPA 2012a). A total of 1,548 incidents were included in this dataset. However, some incidents were reported in the US EPA's database of Specified Ingredient Incidents (USEPA 2012b). The database contained 666 data records from the U.S. and other locations, and 4 were associated with verified exposure to CPY and kills of fish and/or invertebrates. All four incidents appeared to be related to misuse and included improper use of CPY as a termiticide in Alabama in June 2002, incorrect aerial application of a mixture of CPY and cyfluthrin in Lavender Canal in California, Feb 2003, and a similar incident on the Boone River, Iowa with a mixture of CPY and pyraclostrobin in Aug 2009. One fish-kill incident was due to a spill or deliberate release of several pesticides, including CPY, in Grape and Core Creeks in North Carolina in May 2003 (Incident # I014123). Several thousand fish were killed, and CPY was detected at 1.33 and 5.1 μg L^{-1} in Core Creek (along with pebulate and fenamiphos). One sample in Grape Creek contained CPY at 24 g L^{-1} (described as an emulsion), clearly from a major spill or deliberate release. This sample also contained sulfotep (0.51 g L^{-1}), diazinon (0.74 g L^{-1}), malathion (9.5 g L^{-1}), and fenamiphos (1.6 g L^{-1}). All incidents were linked to misuse, and there was no indication that normal use of CPY in agriculture has resulted in fish kills.

4 Characterization of Risks

Where sufficient data were available, such as for surface waters, risks posed by CPY to aquatic organisms were characterized by comparison of measured and predicted concentrations to SSDs of acute toxicity values. For sediments, where fewer toxicity data were available, simple quotients of exposure concentrations to single toxicity values (risk quotients, RQs) were utilized.

Table 6 Risk quotients for maximum 95th centile measured concentrations of CPY in surface waters of the U.S. before and after 2001

Taxon	Crustacea	Insects	Fish
HC5 (μg L^{-1})	0.034	0.091	0.820
RQ for greatest annual 95th centile pre-2001 (0.056 μg L^{-1})[a]	1.65	0.64	0.07
RQ for greatest annual 95th centile post-2001 (0.008 μg L^{-1})[a]	0.24	0.09	0.01

[a]See Sect. 2.1 and Fig. 3 in Williams et al. (2014) for the derivation of the greatest annual 95th centile concentrations

4.1 Risks from Measured Exposures

Risks for CPY in surface waters. There is a relatively large database of measured concentrations of CPY in surface waters (see detailed characterization in Sect. 3.1 in Williams et al. 2014). In almost all of these data sets, frequency of sampling was too small to allow exposures to be characterized as 96-h time-weighted-mean concentrations for direct comparison to 96-h toxicity values. However, comparisons in relation to changes in the use of CPY were possible. The greatest annual 95th centiles of concentrations measured in surfaces waters by the US Geological Survey before and after the introduction of new labels in 2001 clearly shows the reductions in exposures and risks that resulted from the changed use pattern (Fig. 6 and Table 6). Based on 95th centiles and the HC5, risks for fish in either period were small. An extensive review of the toxicity screening data from 2004 to 2009 in samples of surface waters of the Central Valley of California (Hall and Anderson 2012) confirmed that the reductions in concentrations of CPY after 2001 (see Sect. 4 in Williams et al. 2014) were reflected in reductions in the frequency of detection of toxicity mediated by CPY.

Risks for CPY in sediments. The toxicity of CPY in sediments in areas of intensive use, has infrequently been reported in studies conducted recently (Sect. 2.1). Comparison of the 10-d LC_{50} toxicity values for *H. azteca* and *C. dilutus* (Table 3) to the greatest concentration (58.6 μg kg^{-1}) measured in sediments (Sect. 3.2 in Williams et al. 2014) gave RQs of 0.15 and 0.16 for the two species. These RQs are only slightly above the Level of Concern for non-endangered species (USEPA 2004) and are consistent with toxicity testing of sediments from areas of intensive use since 2000 (Sect. 2.1).

Risks from CPYO. As discussed in a companion paper (Solomon et al. 2014), CPYO is formed from CPY in the environment and *in vivo* but has seldom been detectable in surface waters (see Sect. 4 in Williams et al. 2014). The National Water Quality Assessment (NAWQA) database included results of 7,098 analyses for CPYO in surface water samples between 1999 and 2012 (NAWQA 2012). CPYO was detected in 16 samples (detection rate of 0.23%), and the greatest estimated concentration (i.e., >LOD but <LOQ) was 0.0543 μg CPYO L^{-1}. Similar results were found in the National Stream Quality Accounting Network (NASQAN) database (NASQAN 2012),

where the rate of detection for 2,025 analyses of surface water between 2001 and 2012 was 0.44%, and the greatest estimated concentration was 0.0356 μg CPYO L^{-1}. Databases of pesticide concentrations in surface waters of California (288 analyses, CDPR 2012) and Washington State (964 analyses, WDOE 2012) contained no detections for CPYO. In a study of pesticides in surfaces waters at various elevations above the Central Valley of California (LeNoir et al. 1999), CPYO was detected at concentrations ranging from 0.024 to 0.037 μg CPYO L^{-1} as compared to CPY which ranged from 0.089 to 0.124 μg CPY L^{-1} at the same locations. Thus, the frequency of detection was small and the concentrations, when measurable, also were small.

Risks from measured concentrations of CPYO were all small. The RQ for the greatest measured concentration of CPYO in surface waters (0.054 μg L^{-1}) and the LC_{50} of 1.1 μg L^{-1} for the most sensitive freshwater (FW) organism tested (*L. macrochirus*) was 0.049, which is below the level of concern (LOC) for highly valued species (USEPA 2004).

The small estimated risks from CPYO are supported by several lines of evidence. CPYO is formed from CPY in the atmosphere and is detected in air near sites of application and at more distant locations (see discussion in Mackay et al. 2014). Because CPYO is more polar than CPY (log KOW of 2.89 vs. 5, Tables 5 and 6 in Mackay et al. 2014) it would be expected to partition into precipitation and accumulate to a greater extent than CPY in surface waters. However, this does not occur; for example, LeNoir et al. (1999) showed that CPYO was detected in water at smaller concentrations than CPY, the opposite of those in air sampled at the same locations. The most likely reason for this is the greater rate of hydrolysis of CPYO compared to CPY with half-lives of 13 d vs. 30–50 d, respectively (Tables 7 and 6 in Mackay et al. 2014). Since CPYO is more polar than CPY (Mackay et al. 2014), it would not be expected to be taken up into and accumulate in organisms as much as CPY. Finally, because CPYO is the active toxic form of CPY and is transformed *in vivo*, the toxicity of CPYO would be implicitly included in toxicity testing in the laboratory and cosms where animals are exposed to CPY. For all these reasons, environmental risks from CPYO were smaller than those for CPY. Therefore, a separate and detailed risk assessment was not required.

4.2 Risks from Modeled Exposures to CPY

Probabilistic analysis of risks. The higher-tier modeling of CPY concentrations in surface waters for three scenarios of intensive use and vulnerability to runoff and contamination of surface waters (Sect. 6.2 in Williams et al. 2014) provided frequency distributions of annual maximum 96-h time-weighted mean concentrations. These values could then be compared to distributions of 48- to 96-h toxicity values from the SSDs (Sect. 3.2) using probabilistic approaches. To characterize the risks graphically, these values were used to construct joint probability curves (JPCs, ECOFRAM 1999; Giesy et al. 1999). Reference lines proposed for interpretation of JPCs (Moore et al. 2010) were added to the graphs.

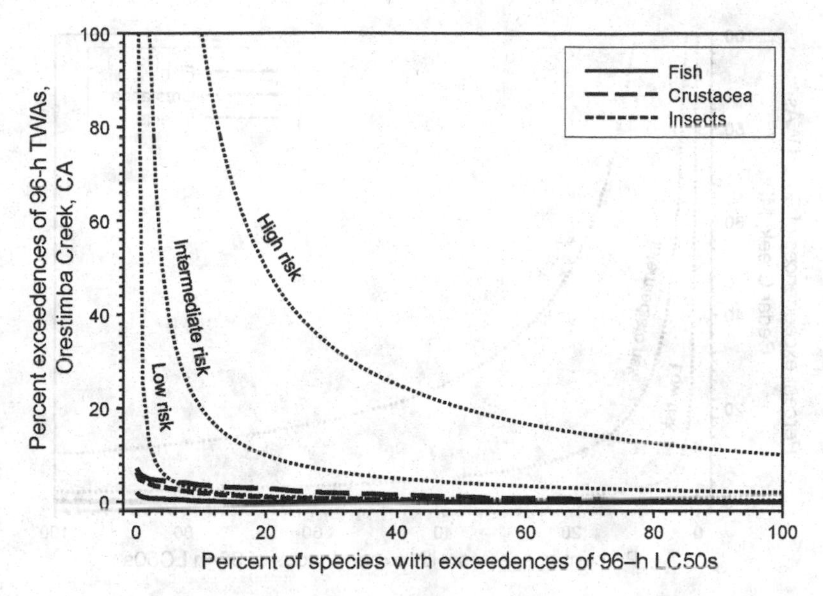

Fig. 7 Joint probability curves for estimated 96-h time-weighted average (TWA) concentrations of chlorpyrifos in Orestimba Creek, CA modeled from Jan 1, 2000 to Dec 31, 2009 and species sensitivity distributions for fish, Crustacea, and insects

The JPCs for the concentrations modeled in Orestimba Creek, CA (Fig. 7) showed that fish and insects were at *de minimis* risk. The line for crustaceans was slightly above the reference line for low risk, indicating that some species of crustaceans are at low (but not *de minimis*) risk of adverse effects in this use scenario.

The JPC for concentrations modeled in another focus-scenario, Cedar Creek, MI (Fig. 8) indicated *de minimis* risk for crustaceans, insects, and fish. The modeled concentrations in the other focus-scenario, Dry Creek, GA were smaller than those in Cedar Creek, MI (Table 8 in Williams et al. 2014), hence, these risks also were *de minimis* (JPC not shown). Overall, the probabilistic analyses of these data suggest that risks from direct effects of CPY on fish are *de minimis* in all areas of use. In most areas of use, as exemplified by the modeling of concentrations in Cedar Creek, MI and Dry Creek, GA, risks to insects and crustaceans will be *de minimis* as well. Low risk is also predicted for crustaceans in Orestimba Creek, CA, an intensive-use scenario that reasonably exemplifies the worst-case.

There were insufficient toxicity data for CPY in sediment to conduct a probabilistic assessment of risk. However, comparison of the 10-d LC_{50} toxicity values for *H. azteca* and *C. dilutus* (Table 3) to the maximum modeled values of 22.2, 0.067, and 0.074 μg kg^{-1} resulted in RQs of 0.06, <0.001, and <0.001 for Orestimba Creek, Cedar Creek, and Dry Creek, respectively. The RQs for Cedar Creek and Dry Creek are well below the Level of Concern for all species. The RQ for Orestimba Creek is slightly greater than the Level of Concern (0.05; EPA 2004) for endangered and threatened (listed) invertebrates, but below the Level of Concern for other (non-listed) species.

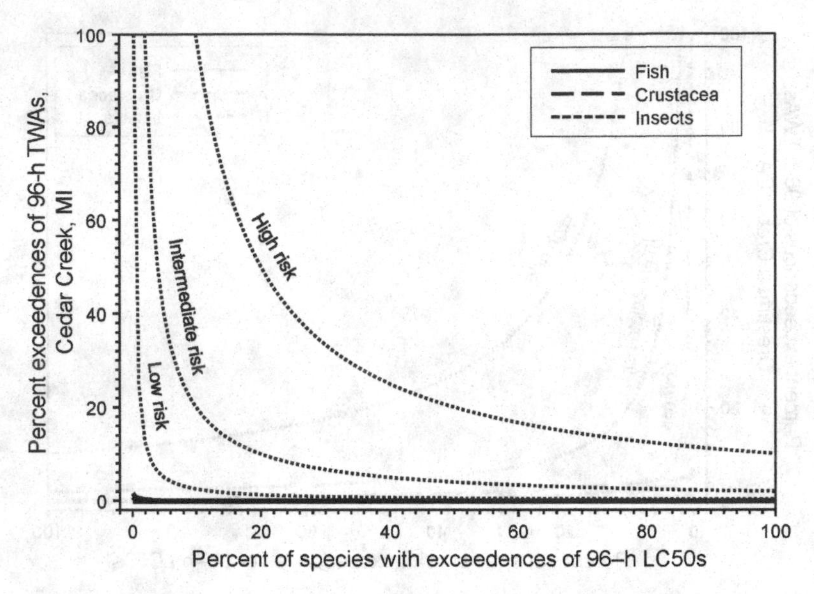

Fig. 8 A joint probability curve for estimated 96-h time-weighted mean concentrations of chlorpyrifos in Cedar Creek, MI modeled from Jan 1, 1961 to Dec 31, 1990. For details of the modeling of exposures, see Williams et al. (2014)

Risks from repeated exposures to CPY. The risks of CPY were further evaluated by considering the duration of exposure and the time between exposures that were predicted by SWAT for watersheds in Michigan, Georgia, and California. The analysis was conducted using the RADAR program (ECOFRAM 1999; Williams et al. 2014). RADAR analyzes the daily time-series of exposure estimates to identify events in which concentrations exceed a pre-defined threshold, calculate the duration of each event, and determine the time between events (recovery time). The NOAECeco from the cosm studies, 0.1 μg L^{-1}, was used as the threshold in this analysis. The full results are presented in SI Appendix E of Williams et al. (2014). No events occurred in the Georgia watershed, and none in the Michigan watershed when the 28-d half-life was used. With a 96-d half-life, there were three events in the Michigan watershed, all of 1-d duration and with at least 1,240 d between events. The short event durations and long intervals between events imply that no exposures in the Michigan watershed would result in ecologically significant effects.

Over the 10-yr simulation in the California watershed, there were 10 events (28-d half-life) or 35 events (96-d half-life), in which concentrations exceeded 0.1 μg L^{-1}. The minimum and median event durations in the California watershed were 1 d for both half-lives, and the maximum event durations were 11 d and 15 d using the 28-d and 96-d half-lives, respectively. Recovery times ranged from 1 to >1,892 d. Using either half-life, recovery times in half of the events were greater than 14 d, long enough for toxicodynamic recovery from AChE inhibition in crustaceans and insects (Sect. 2.2). About one-third of these events had recovery times greater than 56 d, long enough for toxicodynamic recovery in fish (Sect. 2.2) and for ecological recovery in cosms (Sect. 3.4). Only three or four events (for the 96-d and 28-d half-lives, respectively) had durations of 4 d or greater and recovery times less than 56 d. These results suggest that ecologically significant single and repeated exposure events were rare, even in the high-exposure California scenario.

5 Conclusions

This ecological risk assessment of CPY and its oxon CPYO built upon a previous assessment (Giesy et al. 1999) and was refined to address changes in the labeled uses, different use patterns, and new toxicity data. Exposure data were taken from Williams et al. (2014), which characterizes measured and modeled concentrations of CPY in surface waters of the U.S.

The major pathway for exposures to CPY in surface waters is direct accumulation from water, rather than through diet or from sediments. CPY adsorbs strongly to sediments, and this mitigates exposures to benthic invertebrates via sediment. CPY's sediment behavior is consistent with the fact that toxicity is less frequently observed to occur via sediment than water under field testing conditions. The focus of the ERA was thus directed mostly to surface waters and water-column organisms.

Because exposures to CPY in flowing surface waters are episodic with durations usually less than 2 d (Williams et al. 2014), recovery of organisms between pulses can reduce overall risks, but frequent pulses with short recovery periods could result in cumulative damage and cumulative risks. The few studies that have characterized recovery of the target enzyme (AChE) from CPY suggest that invertebrates recover more rapidly than fish. These recovery periods were from 1 to ~7 d for invertebrates, and periods of the order of 4–8 wk might be required for complete recovery of AChE in fish. These periods were considered in the risk assessment. In situations where there is potential for multiple pulsed exposures, a more complex model could be developed that includes accumulation, time to effects and species-specific rates of recovery of AChE. In this assessment of risk, to be conservative, it was assumed that recovery in all organisms would be at the upper bound of observed times (2–8 wk). This assumption likely results in an overestimate of risk.

Characterization of acute toxicity of CPY showed that crustaceans were most sensitive to CPY (HC5 = 0.034 μg CPY L^{-1}), closely followed by insects (HC5 = 0.087 μg CPY L^{-1}). Fish were less sensitive (HC5 = 0.812 μg CPY L^{-1}). The little data available for aquatic stages of amphibians suggested that they were less sensitive to CPY than fish. Thus fish were protective of amphibians, and amphibians would only need to be considered in an ERA if fish were affected. This was not the case for CPY.

Assessment of the results of a large number of studies of the effects of CPY in cosms suggested that the no observed adverse effect concentration in these systems ($NOAEC_{eco}$) was 0.1 μg L^{-1}. These data were derived from single and multiple exposures to CPY and support the conclusion that the HC5s for insects and crustaceans from acute toxicity studies are predictive and protective of toxicity under conditions more relevant to the field. Results for cosms thus provided another line of evidence for characterization of the risks of CPY under conditions that are more representative of conditions in the field.

Risks to aquatic organisms from measured exposures were assessed by comparing the 95th centile concentrations to the HC5s for the SSDs. These data may not fully capture peak exposures but suggested that there were *de minimis* risks for all aquatic organisms from exposures measured after use patterns were changed in 2001. The analyses also showed that risks had decreased from those prior to 2001, which leads to the conclusion that the changes made in 2001 and 2005 to the labeled use patterns, and possibly other changes in general pesticide stewardship, mitigated CPY exposures and reduced risks.

Estimated exposures from models for three focus-scenarios, representing greater vulnerability to exposures than other use scenarios (Williams et al. 2014), allowed the assessment of risks based on 96-h time-weighted-mean concentrations that were matched to the 48–96 h toxicity data. Based on the joint probabilities of distributions of data for exposure and toxicity, we concluded that risks for fish and aquatic stages of insects were *de minimis* in all three regions. However, in the intensive-use scenario of Orestimba Creek, in California, risk to crustaceans was greater and deemed to be not *de minimis*. Further analysis of risks from repeated exposures to CPY, in these three focus-scenarios, confirmed the *de minimis* risks to crustaceans, insects, and fish in the focus-scenarios in GA and MI. Repeated exposures in Orestimba Creek, CA suggested small risks to fish, insects, and crustaceans. We concluded that repeated exposures to insects and crustaceans would not be ecologically relevant because of the potential for rapid recovery in these taxa. Risks for fish may be somewhat greater because there is more uncertainty regarding recovery of the target enzyme AChE and because of their longer reproductive cycles. The lack of fish-kills since 2002 in the U.S. that were associated with confirmed exposure to CPY is consistent with the small risks to fish and the smaller exposures in surface waters since the change in the labeled uses.

Too few data on toxicity of CPYO were available to conduct a probabilistic risk assessment but, on the basis of the available data and the large margins of exposure, we concluded that risks of CPYO to aquatic organisms were *de minimis*. CPYO is the active metabolite of CPY, and its toxicity is subsumed by the parent CPY. It is thus not surprising that CPYO's toxicity is similar to that of CPY. CPYO is more rapidly hydrolyzed in water and is more polar than CPY and is less likely to be taken up into aquatic organisms. Detections of CPYO in surface waters were infrequent, and the concentrations were all less than toxicologically significant values for the one fish and one invertebrate for which data were available.

This ERA was supported by several strong data-sets. There is a good database of toxicity values for CPY, and many of these tests are of high quality. They are certainly sufficient to characterize acute toxicity to insects, crustaceans, and fish. There are also large sets of data for measured values in surface waters in a number of locations, including areas of intensive use, where greater exposures would be expected. Several studies conducted in cosms, some of excellent quality, are available to provide points of reference for the SSDs and information on recovery of invertebrates from exposures to CPY. These strengths have helped reduce uncertainty in the ERA since these cosms included a number of taxa for which there were few toxicity data from laboratory studies. These cosms provide data on responses of aquatic organisms to CPY under realistic conditions.

Just as there were strengths in the ERA there were several areas of uncertainty, some more relevant than others. There were few usable toxicity data from amphibians, but evaluations of the relative sensitivity of fish and amphibians to several classes of toxicants (Weltje et al. 2013) suggest that toxicity data from fish can provide equivalency for amphibians. There were few data on recovery of AChE, the target enzyme for CPY, in aquatic organisms, and this is an uncertainty in the analysis of the relevance of the duration between exposure-events. Because of this, longer and more conservative durations were used in the assessment. In addition, this is an uncertainty that is relevant to all organophosphorus insecticides, as they share the same target enzyme and toxicodynamics of recovery.

There was uncertainty with regard to the demonstrated effects of CPY on behavior and the relevance of these to survival, growth, development, and reproduction (SGDR). Pesticides that target the nervous system are expected to cause effects on behavior, but it is difficult to determine the relevance of these responses to SGDR. For invertebrates in cosms, all responses, including those mediated by behavioral effects, are subsumed into the responses and recovery of exposed populations and communities and are reflected in the $NOAEC_{eco}$ of 0.1 μg L^{-1}. Data to extrapolate behavioral responses to SGDR for fish and other vertebrates are not available for CPY or, for that matter, all other pesticides that target the nervous system. This is a general uncertainty that has still to be addressed in the science of ERA.

When this ERA was initiated, there was uncertainty about the relevance of the formation of CPYO from the parent, CPY, and how this might influence risks. While still somewhat uncertain, this issue is judged to be of lesser relevance than that of CPY itself. There are several lines of evidence to support this conclusion. The oxon is an integral component of the toxicodynamics of CPY and is formed *in vivo*. Toxicity of the oxon in aquatic organisms is not vastly or consistently different from that of the parent CPY, and, to some degree is included in the toxicity studies with CPY. The oxon is more rapidly hydrolyzed in the environment, partitions more into water, and is less likely to bioconcentrate into organisms than CPY (see discussions in the companion paper, Mackay et al. 2014).

6 Summary

The risk of chlorpyrifos (CPY) to aquatic organisms in surface water of North America was assessed using measured concentrations in surface waters and modeling of exposures to provide daily concentrations that better characterize peak exposures. Ecological effects were compared with results of standard laboratory toxicity tests with single species as well as microcosm and mesocosm studies comprised of complex aquatic communities. The upper 90th centile 96-h concentrations (annual maxima) of chlorpyrifos in small streams in agricultural watersheds in Michigan and Georgia were estimated to be ≤0.02 μg L^{-1}; in a reasonable worst-case California watershed, the 90th centile 96-h annual maximum concentrations ranged from 1.32 to 1.54 μg L^{-1}. Measured concentrations of chlorpyrifos are less than estimates from simulation models. The 95th centile for more than 10,000 records compiled by the US Geological Survey was 0.008 μg L^{-1}. Acute toxicity endpoints for 23 species of crustaceans ranged from 0.035 to 457 μg L^{-1}; for 18 species of aquatic insects, from 0.05 to 27 μg L^{-1}; and for 25 species of fish, from 0.53 to >806 μg L^{-1}. The No Observed Adverse Effect Concentration ($NOAEC_{eco}$) in more than a dozen microcosm and mesocosm studies conducted in a variety of climatic zones, was consistently 0.1 μg L^{-1}. These results indicated that concentrations of CPY in surface waters are rarely great enough to cause acute toxicity to even the most sensitive aquatic species. This conclusion is consistent with the lack of fish-kills reported for CPY's normal use in agriculture in the U.S.

Analysis of measured exposures showed that concentrations in surface waters declined after labeled use-patterns changed in 2001, and resulted in decreased risks for crustaceans, aquatic stages of insects, and fish. Probabilistic analysis of 96-h time-weighted mean concentrations, predicted by use of model simulation for three focus-scenarios selected for regions of more intense use of CPY and vulnerability to runoff, showed that risks from individual and repeated exposures to CPY in the Georgia and Michigan watersheds were *de minimis*. Risks from individual exposures in the intense-use scenario from California were *de minimis* for fish and insects and low for crustaceans. Risks from repeated exposures in the California intense-use scenario were judged not to be ecologically relevant for insects and fish, but there were some risks to crustaceans. Limited data show that chlorpyrifos oxon (CPYO), the active metabolite of CPY is of similar toxicity to the parent compound. Concentrations of CPYO in surface waters are smaller than those of CPY and less frequently detected. Risks for CPYO in aquatic organisms were judged to be *de minimis*.

Several uncertainties common to all AChE inhibitors were identified. Insufficient data were available to allow interpretation of the relevance of effects of CPY (and other pesticides that also target AChE) on behavior to assessment endpoints such as survival, growth, development, and reproduction. Data on the recovery of AChE from inhibition by CPY in fish are limited. Such data are relevant to the characterization of risks from repeated exposures, and represent an uncertainty in the assessment of risks for CPY and other pesticides that share the same target and toxicodynamics. More intensive monitoring of areas of greater use and more comprehensive models of cumulative effects that include rates of accumulation, metabolism and recovery of AChE in the more sensitive species would be useful in reducing this uncertainty.

Acknowledgements We wish to thank Julie Anderson for assistance with review and collation of the toxicity data. We thank the anonymous reviewers of this paper for their suggestions and constructive criticism. Prof. Giesy was supported by the Canada Research Chair program, a Visiting Distinguished Professorship in the Department of Biology and Chemistry and State Key Laboratory in Marine Pollution, City University of Hong Kong, the 2012 "High Level Foreign Experts" (#GDW20123200120) program, funded by the State Administration of Foreign Experts Affairs, the P.R. China to Nanjing University and the Einstein Professor Program of the Chinese Academy of Sciences. Funding for this project was provided by Dow AgroSciences.

References

Amweg EL, Weston DP, You J, Lydy MJ (2006) Pyrethroid insecticides and sediment toxicity in urban creeks from California and Tennessee. Environ Sci Technol 40:1700–1706

Ankley GT, Call DJ, Cox JS, Kahl MD, Hoke RA, Kosian PA (1994) Organic carbon partitioning as a basis for predicting the toxicity of chlorpyrifos in sediments. Environ Toxicol Chem 13:621–626

Banks KE, Hunter DH, Wachal DJ (2005) Chlorpyrifos in surface waters before and after a federally mandated ban. Environ Int 31:351–356

Barata C, Solayan A, Porte C (2004) Role of B-esterases in assessing toxicity of organophosphorus (chlorpyrifos, malathion) and carbamate (carbofuran) pesticides to *Daphnia magna*. Aquat Toxicol 66:125–139

Barron MG, Woodburn KB (1995) Ecotoxicology of chlorpyrifos. Rev Environ Contam Toxicol 144:1–93

Bernabo I, Gallo L, Sperone E, Tripepi S, Brunelli E (2011) Survival, development, and gonadal differentiation in *Rana dalmatina* chronically exposed to chlorpyrifos. J Exp Zool A Ecol Genet Physiol 315:314–327

Biever RC, Giddings JM, Kiamos M, Annunziato MF, Meyerhoff R, Racke K (1994) Effects of chlorpyrifos on aquatic microcosms over a range of off-target spray drift exposure levels. In: BCPC (ed.) Proceedings, Brighton Crop Protection Conference on Pests and Diseases. Vol. 3, BCPC, London, UK. pp 1367–1372

Bogen KT, Reiss R (2012) Generalized Haber's law for exponential concentration decline, with application to riparian-aquatic pesticide ecotoxicity. Risk Anal 32:250–258

Boone JS, Chambers JE (1996) Time course of inhibition of cholinesterase and aliesterase activities, and nonprotein sulfhydryl levels following exposure to organophosphorus insecticides in mosquitofish (*Gambusia affinis*). Fundam Appl Toxicol 29:202–207

Brazner JC, Kline ER (1990) Effects of chlorpyrifos on the diet and growth of larval fathead minnows, *Pimephales promelas* in littoral enclosures. Can J Fish Aquat Sci 47:1157–1165

Brock TCM, Arts GHP, Maltby L, van den Brink PJ (2006) Aquatic risks of pesticides, ecological protection goals and common aims in EU legislation. Integr Environ Assess Manag 2:e20–e46

Brock TCM, Bogaert M, Bos AR, Breukelen SWF, Reiche R, Terwoert J, Suykerbuyk REM, Roijackers RMM (1992) Fate and effects of the insecticide Dursban® 4E in indoor *Elodea* dominated and macrophyte-free freshwater model ecosystems: II. Secondary effects on community structure. Arch Environ Contam Toxicol 23:391–409

Brock TCM, Roijackers RMM, Rollon R, Bransen F, van der Heyden L (1995) Effects of nutrient loading and insecticide application on the ecology of *Elodea*-dominated freshwater microcosms: II. Responses of macrophytes, periphyton and macroinvertebrate grazers. Arch Hydrobiol 134:53–74

Brock TCM, Van Wijngaarden RPA, Van Geest GJ (2000) Ecological risks of pesticides in freshwater ecosystems. Part 2: Insecticides. Alterra, Wageningen, The Netherlands, 089

Brock TCM, Vet JJRM, Kerkhofs MJJ, Lijzen J, Zuilekom WJ, Gijlstra R (1993) Fate and effects of the insecticide Dursban® 4E in indoor *Elodea* dominated and macrophyte-free freshwater model ecosystems: III. Aspects of ecosystem functioning. Arch Environ Contam Toxicol 25: 160–169

CCME (2008) Canadian water quality guidelines for the protection of aquatic life—chlorpyrifos. Canadian Council of Ministers of the Environment, Ottawa, ON, Canada. http://ceqg-rcqe.ccme.ca/download/en/164

CCME (2007) Canadian water quality guidelines for the protection of aquatic life. A protocol for the derivation of water quality guidelines for the protection of aquatic life 2007. Canadian Council of Ministers of the Environment, Ottawa, ON, Canada. http://ceqg-rcqe.ccme.ca/

CCME (2013) Determination of hazardous concentrations with species sensitivity distributions, SSD Master. Canadian Council of Ministers of the Environment, Ottawa, ON, Canada

CDPR (2012) Surface water database. California Department of Pesticide Regulation. http://www.cdpr.ca.gov/docs/emon/surfwtr/surfcont.htm. Accessed 24 Feb 2012

Chandrasekera LK, Pathiratne A (2005) Response of brain and liver cholinesterases of Nile tilapia, *Oreochromis niloticus*, to single and multiple exposures of chlorpyrifos and carbosulfan. Bull Environ Contam Toxicol 75:1228–1233

Cuppen JGM, Gylstra R, Sv B, Budde BJ, Brock TCM (1995) Effects of nutrient loading and insecticide application on the ecology of *Elodea*-dominated freshwater microcosms: responses of macroinvertebrate detritivores, breakdown of plant litter, and final conclusions. Arch Hydrobiol 134:157–177

Cutler GC, Purdy J, Giesy JP, Solomon KR (2014) Risk to pollinators from the use of chlorpyrifos in the United States. Rev Environ Contam Toxicol 231:219–265

Daam MA, Crum SJ, Van den Brink PJ, Nogueira AJ (2008a) Fate and effects of the insecticide chlorpyrifos in outdoor plankton-dominated microcosms in Thailand. Environ Toxicol Chem 27:2530–2538

Daam MA, Van den Brink PJ, Nogueira AJ (2008b) Impact of single and repeated applications of the insecticide chlorpyrifos on tropical freshwater plankton communities. Ecotoxicology 17:756–771

De Silva PM, Samayawardhena LA (2005) Effects of chlorpyrifos on reproductive performances of guppy (*Poecilia reticulata*). Chemosphere 58:1293–1299

Ding Y, Harwood AD, Foslund HM, Lydy MJ (2010) Distribution and toxicity of sediment-associated pesticides in urban and agricultural waterways from Illinois, USA. Environ Toxicol Chem 29:149–157

Domingues I, Guilhermino L, Soares AM, Nogueira AJ, Monaghan KA (2009) Influence of exposure scenario on pesticide toxicity in the midge *Kiefferulus calligaster* (Kieffer). Ecotoxicol Environ Saf 72:450–457

ECOFRAM (1999) ECOFRAM aquatic final draft reports. United States Environmental Protection Agency, Washington, DC. http://www.epa.gov/oppefed1/ecorisk/aquareport.pdf

Ensminger M, Bergin R, Spurlock F, Goh KS (2011) Pesticide concentrations in water and sediment and associated invertebrate toxicity in Del Puerto and Orestimba Creeks, California, 2007–2008. Environ Monit Assess 175:573–587

Giddings JM (1993a) Chlorpyrifos (Lorsban 4E): outdoor aquatic microcosm test for environmental fate and ecological effects. Springborn Laboratories for Dow Chemical, Wareham, MA (unpublished report)

Giddings JM (1993b) Chlorpyrifos (Lorsban 4E): outdoor aquatic microcosm test for environmental fate and ecological effects of combinations of spray and slurry treatments. Springborn Laboratories for Dow Chemical, Wareham, MA (unpublished report)

Giddings JM (2011) Invertebrate communities in outdoor microcosms treated with chlorpyrifos: reanalysis of data reported in Giddings 1992. Dow AgroSciences, Indianapolis, IN (unpublished report)

Giddings JM, Biever RC, Racke KD (1997) Fate of chlorpyrifos in outdoor pond microcosms and effects on growth and survival of bluegill sunfish. Environ Toxicol Chem 16:2353–2362

Giesy JP, Geiger RA (1980) Large scale microcosms for assessing fates and effects of trace contaminants. In: Giesy JP (ed) Microcosms in ecological research. Department of Energy, Technical Information Center, Oak Ridge, TN, pp 304–318

Giesy JP, Odum EP (1980) Microcosmology: the theoretical basis. In: Giesy JP (ed) Microcosms in ecological research. Department of Energy Technical Information Center, Oak Ridge, TN, pp 1–13

Giesy JP, Solomon KR, Coates JR, Dixon KR, Giddings JM, Kenaga EE (1999) Chlorpyrifos: ecological risk assessment in North American aquatic environments. Rev Environ Contam Toxicol 160:1–129

Graney RA, Giesy JP, DiToro D (1989) Mesocosm experimental design strategies: advantages and disadvantages in ecological risk assessment. In: Voshell JR Jr (ed) Using mesocosms to assess the aquatic ecological risk of pesticides: theory and practice. Entomological Society of America, Lanham, MD, pp 74–88

Graney RL, Giesy JP, Clark JR (1995) Field studies. In: Rand GM (ed) Fundamentals of aquatic toxicology: effects, environmental fate and risk assessment. Taylor & Francis, Bristol, PA, pp 257–305

Hall L Jr, Anderson RD (2012) Historical trends analysis of 2004 to 2009 toxicity and pesticide data for California's Central Valley. J Environ Sci Health A 47:801–811

Hurlburt SH, Mulla MS, Keith JO, Westlake WE, Dhsch ME (1970) Biological effects and persistence of Dursban in freshwater ponds. J Econ Entomol 63:43–52

Hurlburt SH, Mulla MS, Willson HR (1972) Effects of an organophosphorus insecticide on the phytoplankton, zooplankton and insect populations of fresh-water ponds. Ecol Monogr 42:269–299

Kavitha P, Rao JV (2008) Toxic effects of chlorpyrifos on antioxidant enzymes and target enzyme acetylcholinesterase interaction in mosquito fish, *Gambusia affinis*. Environ Toxicol Pharmacol 26:192–198

Klimisch H-J, Andreae M, Tillmann U (1997) A systematic approach for evaluating the quality of experimental toxicological and ecotoxicological data. Regul Toxicol Pharmacol 25:1–5

Kumar A, Doan H, Barnes M, Chapman JC, Kookana RS (2010) Response and recovery of acetylcholinesterase activity in freshwater shrimp, *Paratya australiensis* (Decapoda: Atyidae) exposed to selected anti-cholinesterase insecticides. Ecotoxicol Environ Saf 73:1503–1510

Laetz CA, Baldwin DH, Collier TK, Hebert V, Stark JD, Scholz NL (2009) The synergistic toxicity of pesticide mixtures: implications for risk assessment and the conservation of endangered Pacific salmon. Environ Health Perspect 117:348–353

Leeuwangh P (1994) Comparison of chlorpyrifos fate and effects in outdoor aquatic micro-and mesocosms of various scale and construction. In: Hill IR, Heimbach F, Leeuwangh P, Matthiesen P (eds) Freshwater field tests for hazard assessment of chemicals. Lewis Publishers, Boca Raton, FL, pp 217–248

LeNoir JS, McConnell LL, Fellers GM, Cahill TM, Seiber JN (1999) Summertime transport of current-use pesticides from California's Central Valley to the Sierra Nevada mountain range, USA. Environ Toxicol Chem 18:2715–2722

López-Mancisidor P, Carbonell G, Fernandez C, Tarazona JV (2008a) Ecological impact of repeated applications of chlorpyrifos on zooplankton community in mesocosms under Mediterranean conditions. Ecotoxicology 17:811–825

López-Mancisidor P, Carbonell G, Marina A, Fernandez C, Tarazona JV (2008b) Zooplankton community responses to chlorpyrifos in mesocosms under Mediterranean conditions. Ecotoxicol Environ Saf 71:16–25

Mackay D, Giesy JP, Solomon KR (2014) Fate in the environment and long-range atmospheric transport of the organophosphorus insecticide, chlorpyrifos and its oxon. Rev Environ Contam Toxicol 231:35–76

Maltby L, Blake NN, Brock TCM, van den Brink PJ (2005) Insecticide species sensitivity distributions: the importance of test species selection and relevance to aquatic ecosystems. Environ Toxicol Chem 24:379–388

Metcalf RL, Sanborn JR (1975) Pesticides and environmental quality in Illinois. Ill Nat Hist Surv Bull 31:381–436

Metcalf RL, Sanga GK, Kapoor IP (1971) Model ecosystems for the evaluation of pesticide biodegradability and ecological magnification. Environ Sci Technol 5:709–713

Moore DRJ, Teed RS, Greer C, Solomon KR, Giesy JP (2014) Refined avian risk assessment for chlorpyrifos in the United States. Rev Environ Contam Toxicol 231:163–217

Moore DRJ, Fischer DL, Teed RS, Rodney SI (2010) Probabilistic risk-assessment model for birds exposed to granular pesticides. Integr Environ Assess Manag 6:260–272

Moore DRJ, Teed RS (2012) Risks of carbamate and organophosphate pesticide mixtures to salmon in the Pacific Northwest. Integr Environ Assess Manag 9(1):70–78

Morgan MJ, Fancey LL, Kiceniuk JW (1990) Response and recovery of brain acetylcholinesterase activity in Atlantic salmon *Salmo salar* exposed to fenitrothion. Can J Fish Aquat Sci 47: 1652–1654

Naddy RB, Johnson KA, Klaine SJ (2000) Response of *Daphnia magna* to pulsed exposures of chlorpyrifos. Environ Toxicol Chem 19:423–431

Naddy RB, Klaine SJ (2001) Effect of pulse frequency and interval on the toxicity of chlorpyrifos to *Daphnia magna*. Chemosphere 45:497–506

NASQAN (2012) USGS national stream quality accounting network. United States Geological Survey. http://water.usgs.gov/nasqan/. Accessed 24 Feb 2012

NAWQA (2012) USGS national water quality assessment data warehouse. United States Geological Survey. http://infotrek.er.usgs.gov/apex/f?p=NAWQA:HOME:0. Accessed 24 Feb 2012

Phillips BM, Anderson BS, Hunt JW, Huntley SA, Tjeerdema RS, Kapellas N, Worcester K (2006) Solid-phase sediment toxicity identification evaluation in an agricultural stream. Environ Toxicol Chem 25:1671–1676

Phillips BM, Anderson BS, Hunt JW, Siegler K, Voorhees JP, Tjeerdema RS, McNeill K (2012) Pyrethroid and organophosphate pesticide-associated toxicity in two coastal watersheds (California, USA). Environ Toxicol Chem 31:1595–1603

Phillips PJ, Ator SW, Nystrom EA (2007) Temporal changes in surface-water insecticide concentrations after the phase out of diazinon and chlorpyrifos. Environ Sci Technol 41:4246–4251

Posthuma L, Suter GW, Traas T (2002) Species sensitivity distributions in risk assessment, species sensitivity distributions in ecotoxicology. CRC, Boca Raton, FL, 564

Pusey BJ, Arthrington A, McLean J (1994) The effects of a pulsed application of chlorpyrifos on macroinvertebrate communities in an outdoor artificial stream system. Ecotoxicol Environ Saf 27:221–250

Rao JV, Begum G, Pallela R, Usman PK, Rao RN (2005) Changes in behavior and brain acetylcholinesterase activity in mosquito fish, *Gambusia affinis* in response to the sub-lethal exposure to chlorpyrifos. Int J Environ Res Pub Health 2:478–483

Richards SM, Kendall RJ (2002) Biochemical effects of chlorpyrifos on two developmental stages of *Xenopus laevis*. Environ Toxicol Chem 21:1826–1835

Sandahl JF, Baldwin DH, Jenkins JJ, Scholz NL (2005) Comparative thresholds for acetylcholinesterase inhibition and behavioral impairment in coho salmon exposed to chlorpyrifos. Environ Toxicol Chem 24:136–145

Sandahl JF, Baldwin DH, Jenkins JJ, Scholz NL (2004) Odor-evoked field potentials as indicators of sublethal neurotoxicity in juvenile coho salmon (*Oncorhynchus kisutch*) exposed to copper, chlorpyrifos, or esfenvalerate. Can J Fish Aquat Sci 61:404–413

Satapornvanit K, Baird DJ, Little DC (2009) Laboratory toxicity test and post-exposure feeding inhibition using the giant freshwater prawn *Macrobrachium rosenbergii*. Chemosphere 74: 1209–1215

Siefert RE, Lozano SJ, Brazner JC, Knuth ML (1989) Littoral enclosures for aquatic field testing of pesticides: effects of chlorpyrifos on a natural system. In: Voshell JR Jr (ed) Using mesocosms to assess the aquatic ecological risk of pesticides: theory and practice. Entomological Society of America, Lanham, MD, pp 57–73

Solomon KR, Williams WM, Mackay D, Purdy J, Giddings JM, Giesy JP (2014) Properties and uses of chlorpyrifos in the United States. Rev Environ Contam Toxicol 231:13–34

Tierney K, Casselman M, Takeda S, Farrell T, Kennedy C (2007) The relationship between cholinesterase inhibition and two types of swimming performance in chlorpyrifos-exposed coho salmon (*Oncorhynchus kisutch*). Environ Toxicol Chem 26:998–1004

Tilton FA, Bammler TK, Gallagher EP (2011) Swimming impairment and acetylcholinesterase inhibition in zebrafish exposed to copper or chlorpyrifos separately, or as mixtures. Comp Biochem Physiol C 153:9–16
USEPA (2012a) Aggregate incident summary report by ingredient for the date range 01/01/2002 to 06/15/2012. United States Environmental Protection Agency Office of Pesticide Programs, Washington, DC
USEPA (2007) ECOTOXicology Database System. Version 4.0. United States Environmental Protection Agency, Office of Pesticide Programs, Environmental Fate and Effects Division, United States EPA, Washington, D. http://www.epa.gov/ecotox/. Accessed March 2012
USEPA (1992) Framework for ecological risk assessment. United States Environmental Protection Agency, Washington, DC, EPA/630/R-92/001
USEPA (1998) Guidelines for ecological risk assessment. United States Environmental Protection Agency, Washington, DC
USEPA (2004) Overview of the ecological risk assessment process in the office of pesticide programs: endangered and threatened species effects determinations. United States Environmental Protection Agency, Office of Prevention, Pesticides, and Toxic Substances, Office of Pesticide Programs, Washington, DC
USEPA (2008) Registration review—preliminary problem formulation for ecological risk and environmental fate, endangered species and drinking water assessments for chlorpyrifos. United States Environmental Protection Agency, Office of Pesticide Programs, Washington, DC
USEPA (2012b) Specified ingredient incidents for the date range 01/01/2002 to 06/15/2012. United States Environmental Protection Agency, Office of Pesticide Programs, Washington, DC
van den Brink PJ, Ev D, Gylstra R, Crum SJH, Brock TCM (1995) Effects of chronic low concentrations of the pesticides chlorpyrifos and atrazine in indoor freshwater microcosms. Chemosphere 31:3181–3200
van den Brink PJ, Van Wijngaarden RPA, Lucassen WGH, Brock TCM, Leeuwangh P (1996) Effects of the insecticide Dursban® 4E (active ingredient chlorpyrifos) in outdoor experimental ditches: II. Invertebrate community responses and recovery. Environ Toxicol Chem 15:1143–1153
van der Wel H, Welling W (1989) Inhibition of acetylcholinesterase in guppies (*Poecilia reticulata*) by chlorpyrifos at sublethal concentrations: methodological aspects. Ecotoxicol Environ Saf 17:205–215
Van Donk E, Prins H, Voogd HM, Crum SJH, Brock TCM (1995) Effects of nutrient loading and insecticide application on the ecology of *Elodea*-dominated freshwater microcosms: I. Responses of plankton and zooplanktivorous insects. Arch Hydrobiol 133:417–439
van Wijngaarden RP, Brock TC, Douglas MT (2005a) Effects of chlorpyrifos in freshwater model ecosystems: the influence of experimental conditions on ecotoxicological thresholds. Pest Manag Sci 61:923–935
van Wijngaarden RPA, Brink PJ, Crum SJH, Oude Voshaar JH, Brock TCM, Leeuwangh P (1996) Effects of the insecticide Dursban® 4E (active ingredient chlorpyrifos) in outdoor experimental ditches: II. Invertebrate community responses and recovery. Environ Toxicol Chem 15: 1133–1142
van Wijngaarden RPA, Brock TCM, van den Brink PJ (2005b) Threshold levels of insecticides in freshwater ecosystems, a review. Ecotoxicology 14:353–378
Varo I, Serrano R, Pitarch E, Amat F, Lopez FJ, Navarro JC (2002) Bioaccumulation of chlorpyrifos through an experimental food chain: study of protein HSP70 as biomarker of sublethal stress in fish. Arch Environ Contam Toxicol 42:229–235
Ward S, Arthington AH, Pusey BJ (1995) The effects of a chronic application of chlorpyrifos on the macroinvertebrate fauna in an outdoor artificial stream system: species responses. Ecotoxicol Environ Saf 30:2–23
WDOE (2012) Environmental information management database. Washington State Department of Ecology. http://www.ecy.wa.gov/eim/. Accessed 3 Apr 2012
Weltje L, Simpson P, Gross M, Crane M, Wheeler JR (2013) Comparative acute and chronic sensitivity of fish and amphibians: a critical review of data. Environ Toxicol Chem 32(5):984–994

Weston DP, Ding Y, Zhang M, Lydy MJ (2012) Identifying the cause of sediment toxicity in agricultural sediments: the role of pyrethroids and nine seldom-measured hydrophobic pesticides. Chemosphere 90:958–964

Widder PD, Bidwell JR (2006) Cholinesterase activity and behavior in chlorpyrifos-exposed *Rana sphenocephala* tadpoles. Environ Toxicol Chem 25:2446–2454

Williams WM, Giddings JM, Purdy J, Solomon KR, Giesy JP (2014) Exposures of aquatic organisms resulting from the use of chlorpyrifos in the United States. Rev Environ Contam Toxicol 231:77–118

Williamson CE, Butler NM (1986) Predation on rotifers by the suspension-feeding calanoid copepod *Diaptomus pallidus*. Limnol Oceanogr 31:393–402

Zafar MI, Van Wijngaarden RP, Roessink I, Van den Brink PJ (2011) Effects of time-variable exposure regimes of the insecticide chlorpyrifos on freshwater invertebrate communities in microcosms. Environ Toxicol Chem 30:1383–1394

Refined Avian Risk Assessment for Chlorpyrifos in the United States

Dwayne R.J. Moore, R. Scott Teed, Colleen D. Greer, Keith R. Solomon, and John P. Giesy

1 Introduction

Chlorpyrifos (O,O-diethyl O-(3,5,6-trichloro-2-pyridinyl) phosphorothioate; CPY) is a widely used, organophosphorus insecticide that was first registered in the United States in 1965. It is available in flowable and granular formulations under the trademark Lorsban® and is registered in many countries for control of pests in soil or on foliage. Birds are potentially at risk following application of CPY because: (1) they forage in areas that could be treated with the pesticide, and (2) CPY has been shown to be toxic to birds under laboratory conditions when they were exposed to ecologically relevant concentrations in the diet. Here we present a refined assessment of risk to birds from application of granular or flowable formulations of CPY to crops in the United States at rates and frequencies of use approved on the current product labels. This assessment focused on bird species that are known to frequently forage in crop fields treated with CPY.

The online version of this chapter (doi:10.1007/978-3-319-03865-0_6) contains supplementary material, which is available to authorized users.

D.R.J. Moore (✉) • C.D. Greer
Intrinsik Environmental Sciences (US), Inc., New Gloucester, ME, USA
e-mail: dmoore@intrinsikscience.com

R.S. Teed
Intrinsik Environmental Sciences, Inc., Carleton University Technology and Training Center – Suite 3600, Ottawa, ON, Canada

K.R. Solomon
Centre for Toxicology, School of Environmental Sciences, University of Guelph, Guelph, ON, Canada

J.P. Giesy
Department of Veterinary Biomedical Sciences and Toxicology Centre, University of Saskatchewan, 44 Campus Dr., Saskatoon, SK S7N 5B3, Canada

J.P. Giesy and K.R. Solomon (eds.), *Ecological Risk Assessment for Chlorpyrifos in Terrestrial and Aquatic Systems in the United States*, Reviews of Environmental Contamination and Toxicology 231, DOI 10.1007/978-3-319-03865-0_6, © The Author(s) 2014

Mammals are far less sensitive than birds to acute exposures of chlorpyrifos (see review in Solomon et al. 2001). In addition, mammals are less exposed to granular CPY because they do not consume grit to aid digestion as do birds (Solomon et al. 2001). Therefore, risks due to exposure to CPY are likely to be greater for birds than for mammals. We did not conduct a refined risk assessment for mammals because any mitigations stemming from the avian risk assessment should also be protective of mammals foraging in treated fields.

This assessment builds upon past assessments of CPY, including the most recent EPA re-registration assessment (USEPA 1999) and a refined risk assessment to birds by Solomon et al. (2001). Using a conservative, screening-level risk assessment approach, the USEPA (1999) concluded that single and multiple applications of CPY potentially pose risks to birds. However, a more refined assessment of exposure based on simulations and analyses of field studies and incident reports demonstrated that the risks of exposure to flowable and granular CPY were small (Solomon et al. 2001). Solomon et al. (2001) used a probabilistic individual-based model to predict mortality for eight focal species exposed to flowable and granular CPY in corn. The model predicted that the eight focal species would not experience any mortality. Since the completion of the assessments by USEPA (1999) and Solomon et al. (2001), the labels have been amended to require buffer zones, reduce single and seasonal application rates, reduce the number of applications per season, and increase the minimum re-treatment intervals (USEPA 2009). In addition, EPA is preparing a new assessment of risk of flowable and granular formulations of CPY to birds that will make use of their standard screening-level risk assessment approach (USEPA 2008a). An initial draft of the reassessment by EPA was to be released for public comment in the latter part of 2013 (USEPA 2009).

A refined assessment of risks posed by labeled uses of CPY in agriculture in the USA was conducted to reflect changes made to the label and the availability of new information and methods for conducting exposure assessments of birds to insecticides. We initiated the preparation of this assessment by carefully formulating the problem to be addressed. Such problem formulation establishes the scope of the assessment, including defining the routes of exposure to be considered, focal species of birds and patterns of use. The problem formulation concludes with an analysis plan. Subsequent sections describe the methods and assumptions for assessing exposure and effects and characterization of risks. This paper is part of a series that describes the properties and environmental chemodynamics (Solomon et al. 2014), long-range atmospheric transport (Mackay et al. 2014), concentrations in aquatic environments (Williams et al. 2014), risks to aquatic organisms (Giddings et al. 2013), and risks to pollinators (Cutler et al. 2014) of CPY.

2 Problem Formulation

The goal of problem formulation is to develop a plan for the analysis that will guide the assessment of risks to terrestrial birds. To accomplish this task, the following topics are briefly reviewed: (1) patterns and amounts of CPY used; (2) formulations;

(3) transformation products in the environment; (4) routes of exposure; and (5) mode of toxic action and thresholds for effects. The information cogent to these topics was used to create a conceptual model and identify focal avian species for the assessment. The problem formulation concludes with a list of exposure scenarios that were included in the refined avian risk assessment and an overview of the analysis plan.

2.1 Patterns of Use

Chlorpyrifos is an organophosphate insecticide that provides broad spectrum control of insects in cereal, oil, forage, nut, and vegetable crops (Solomon et al. 2014; USEPA 2011). The focus of this assessment is on representative current use flowable and granular formulations of Lorsban, i.e., Lorsban® Advanced and Lorsban® 15G, respectively.

Chlorpyrifos is registered for use on a wide variety of crops including Brassica vegetables, corn, onion, peanut, sugar beet, sunflower, and tobacco for the granular formulation (Lorsban 15G), as well as alfalfa, Brassica vegetables, citrus, corn, cotton, grape, mint, onion, peanut, pome and stone fruits, soybean, sugar beet, sunflower, sweet potato, tree nuts, and wheat for the flowable formulations (e.g., Lorsban Advanced) (Gomez 2009; Solomon et al. 2014). The greatest amounts of CPY used in 2007 were applied to soybean, corn, almond, apple, alfalfa, wheat, and pecan (Gomez 2009; Solomon et al. 2014). Chlorpyrifos is widely used in the Midwest and Plains regions, California, Florida and Georgia because these are the primary growing areas for many row crops, citrus and tree nuts. For additional information on CPY use patterns, see Sect. 2.8 and Solomon et al. (2014).

2.2 Formulations

The focus of this risk assessment is on two formulations, the granular and the flowable. Lorsban 15G (Dow AgroSciences 2008) is a clay-based (e.g., montmorillonite, bentonite) granular formulation containing 15% active ingredient. Lorsban Advanced (Dow AgroSciences 2009) is a flowable formulation, specifically an emulsion in water that contains 40.2% active ingredient.

2.3 Metabolites of CPY in the Environment

The fate and transport of CPY in the environment is reviewed in Mackay et al. (2014). In this section, we focus on degradates of CPY in the environment that could be relevant to the avian risk assessment. The major transformation product of hydrolysis of CPY in alkaline soil is 3,5,6-trichloro-2-pyridinol (TCP) (Solomon et al. 2014). TCP is non-toxic to birds at concentrations greater than what would be encountered in the environment (acute LD_{50} >1,000 mg ai kg^{-1} bwt, chronic LC_{50}s of 500–5,600 mg ai kg^{-1} in the diet) (Campbell et al. 1990; Long et al. 1990; Miyazaki and Hodgson 1972). Transformation products of TCP are also not toxic to birds at concentrations observed in the environment (Racke 1993). The oxon of CPY (CYPO; *O*-ethyl *O*-(3,5,6-trichloro-2-pyridinol) phosphorothionate) is formed *in vivo* in birds by oxidative desulfuration (Testai et al. 2010). This metabolite is shorter lived in the environment than CPY and is rapidly degraded via hydrolysis to TCP and diethylphosphate (Mackay et al. 2014). The oxon of CPY is toxic to non-target organisms, including birds, but poses little risk because it is formed in very small quantities in the terrestrial environment and is rapidly degraded (Bidlack 1979; Chapman and Harris 1980; de Vette and Schoonmade 2001). The Pesticide Management Regulatory Agency of Canada (PMRA) (2007) did not consider CPYO to pose a risk to birds. Another transformation product, 3,5,6-trichloro-2-methoxypyridine (TMP) has also been reported in aerobic biodegradation studies (Bidlack 1979; Racke 1993). The half-life of TMP is similar to that of TCP and is not toxic to birds at concentrations observed in the environment (Racke 1993; Reeves 2008). For the reasons cited above, TCP, CPYO, and TMP were not considered in this refined assessment of risks of CPY to birds.

2.4 Routes of Exposure for Birds in Terrestrial Environments

Based on the physical and chemical properties of CPY (Solomon et al. 2014), bioaccumulation of CPY could occur. It is not likely, however, to be a significant pathway of exposure for birds because CPY is rapidly metabolized with a half-life of approximately 1-d (Barron and Woodburn 1995; Mackay et al. 2014; Racke 1993; Smith et al. 1967). Because CPY has a half-life of 2–5-d on foliage under field conditions (Williams et al. 2014), acute exposure is the primary concern. Although not persistent in the field, there is the potential for chronic exposure because flowable CPY may be applied up to four times per season with intervals as short as 10-d. Therefore, for flowable CPY, both acute and chronic risks to birds were estimated. Because granular CPY can only be applied once per season (Solomon et al. 2014), only acute risks were estimated for this formulation.

Whether applied as a granular or flowable formulation, wind and rain cause penetration of CPY into soil (Solomon et al. 2001), and volatilization from moist soil surfaces is rapid (HSDB 2013; Mackay et al. 2014). Once below the soil surface, CPY is much less available to birds. The most likely routes of exposure of birds to CPY following application of the flowable formulation are through the ingestion of residues on plants and prey and in drinking water. Exposure to CPY through inhalation, dermal contact, and preening are also routes of exposure for birds following application of flowable CPY. However, the results of several studies conducted with turkeys (*Meleagris gallopavo*) (Kunz and Radeleff 1972; McGregor and Swart 1968, 1969) indicate that uptake from dermal exposure directly from soil and vegetation sprayed at maximum allowable rates on the Lorsban Advanced label, and any subsequent preening would not cause adverse effects (Solomon et al. 2001). By performing an analysis with the USEPA (2010) Screening Tool for Inhalation

Risk (STIR), version 1.0, we determined that inhalation of CPY would not be a significant exposure pathway for birds. In that analysis, airblast application at the largest permitted, single application rate on the Lorsban Advanced label (i.e., 6.23 kg ha^{-1} (5.6 lb ai A^{-1} for oranges)), and a vapor pressure of 0.0000202 mm Hg (0.00269 Pa) (Solomon et al. 2014; USEPA 2009) were assumed. When the predicted exposure was compared to the lowest oral LD_{50} for birds, i.e., 5.62 mg ai kg^{-1} bwt for common grackle (Schafer and Brunton 1979), the results of STIR predicted that exposure via inhalation "is not likely significant" for birds exposed to CPY on treated fields immediately after application at the maximum permitted rate on the label for Lorsban Advanced. Thus, the most important routes of exposure for birds following application of flowable CPY are ingestion of residues on food items and ingestion of water from on-field puddles and other drinking water sources (e.g., dew). These routes of exposure were the focus of the avian risk assessment for flowable CPY.

Following application of the flowable formulation, CPY can reach offsite soil, water bodies, terrestrial vegetation, and insects from spray drift, runoff and erosion (Williams et al. 2014). Some CPY will dissipate into the air, either becoming airborne during application or volatilizing from treated surfaces (Mackay et al. 2014). Chlorpyrifos in air may be transported by wind and deposited offsite, but amounts will be small (Mackay et al. 2014). Direct application of CPY to streams, lakes, and ponds is not permitted by product labels. Because exposure will be greatest on treated fields, assessment of risks of flowable CPY to birds was estimated for birds foraging on-field. Risks to birds foraging off-field would be much less.

Following application of granular CPY, unincorporated and intact granules might be directly ingested by birds while they are foraging for grit (Luttik and de Snoo 2004; Moore et al. 2010b, c). When water has collected on the soil surface where granules have been applied, birds might ingest dissolved CPY from pooled water. However, farmers do not normally apply granular pesticides when soil is saturated with water or when significant precipitation is expected within a day or two of application. Exposure via dermal contact is expected to be minimal because the CPY that exists in granules is unlikely to be available for transport across feathers and the bird epidermis. This assumption is supported by the results of a study involving penned turkeys (*M. gallopavo*) that were exposed to 5% CPY in granules applied to soil at a rate of 3.36 kg ha^{-1} (3 lb ai A^{-1}) (Price et al. 1972). The only labeled crop having a higher maximum application rate for granular CPY (Lorsban 15G) is peanuts (i.e., 4.48 kg ha^{-1}). No toxicity occurred during the 4-wk study, indicating that turkeys did not accumulate significant amounts of CPY through their feet or feathers. Thus, the focus of the assessment of for the granular formulation was on birds exposed to CPY granules while foraging for grit to aid digestion.

2.5 Toxicity and Mode of Action

As with other organophosphorus pesticides, CPY is rapidly absorbed following ingestion in food and water. It then undergoes oxidative metabolism to form CPYO, which is the chemical primarily responsible for toxicity (Solomon et al. 2014). The oxon of CPY binds to the enzyme that hydrolyzes the neurotransmitter acetylcholine, i.e., acetylcholinesterase (AChE). The resulting accumulation of acetylcholine causes overstimulation of cholinergic synapses (Testai et al. 2010). Exposure to CPY in birds can be detected biochemically as reduced activity of AChE in blood plasma or brain (Parsons et al. 2000; Testai et al. 2010). Other symptoms of toxicity include loss of mass, ruffled appearance, loss of coordination, reduced reaction to sound and movement, wing droop, prostrate posture, weakness of lower limbs, lethargy, gaping, salivation, muscle fasciculation, convulsions, and death (Gallagher et al. 1996).

Acute dietary studies have been conducted to determine toxicity in birds. However, birds tend to avoid treated food with high CPY concentrations (see SI Appendix 3, Sect. 1.2), which limits the usefulness of studies in which CPY is fed in the diet to derive dose-response relationships. Five-d GLP (Good Laboratory Practice) dietary studies yielded LC_{50} values of 2,772 mg ai kg^{-1} diet for northern bobwhite (*Colinus virginianus*) (Beavers et al. 2007), and 1,083 mg ai kg^{-1} diet for mallards (*Anas platyrhynchos*) (Long et al. 1991). Based on the results of acute, oral gavage studies, LD_{50}s ranged from 5.62 mg ai kg^{-1} bwt for adult common grackles (*Q. quiscula*) (Schafer and Brunton 1979) to 112 mg ai kg^{-1} bwt for mallard ducklings (*A. platyrhynchos*) (Hudson et al. 1984). In an 8-wk study, in which adult mallards were fed CPY, a NOEC (no-observed effects concentration) of 100 mg ai kg^{-1} diet was observed (Fink 1977). Reduced consumption of food and production of fewer eggs, and overt signs of toxicity, such as ataxia, ruffled appearance, weakness of lower limbs, and lethargy, were observed at the lowest observed effect concentration (LOEC) of 215 mg ai kg^{-1} diet or greater concentrations. A reproductive study in mallard of CPY effects, in which adults were exposed to treated diet for 9-wk prior to egg laying and for 8-wk during egg laying, reported a NOEC of 25 mg ai kg^{-1} diet and a LOEC of 125 mg ai kg^{-1} diet (Fink 1978a). A similar study of CPY effects on reproduction of the northern bobwhite reported a NOEC at the greatest concentration tested, 125 mg ai kg^{-1} diet (Fink 1978b).

2.6 Conceptual Model

A conceptual model provides a written and visual description of possible exposure routes between ecological receptors and a stressor. The model includes hypotheses for how a stressor might come into contact with and affect receptors. These hypotheses are derived by use of professional judgment and information available on sources of exposure, characteristics of the stressor (e.g., chemistry, fate, and transport), ecosystems at risk, and anticipated effects to birds. The conceptual model for evaluating potential risks to birds from the application of CPY as a flowable (Fig. 1) product illustrates that the most likely routes of exposure of birds are ingestion of foliage, seeds, fruits, insects, and drinking water from pools or foliage in the treated area. For granular CPY (Fig. 2), exposure is most likely to be the result of direct consumption of granules mistaken for grit.

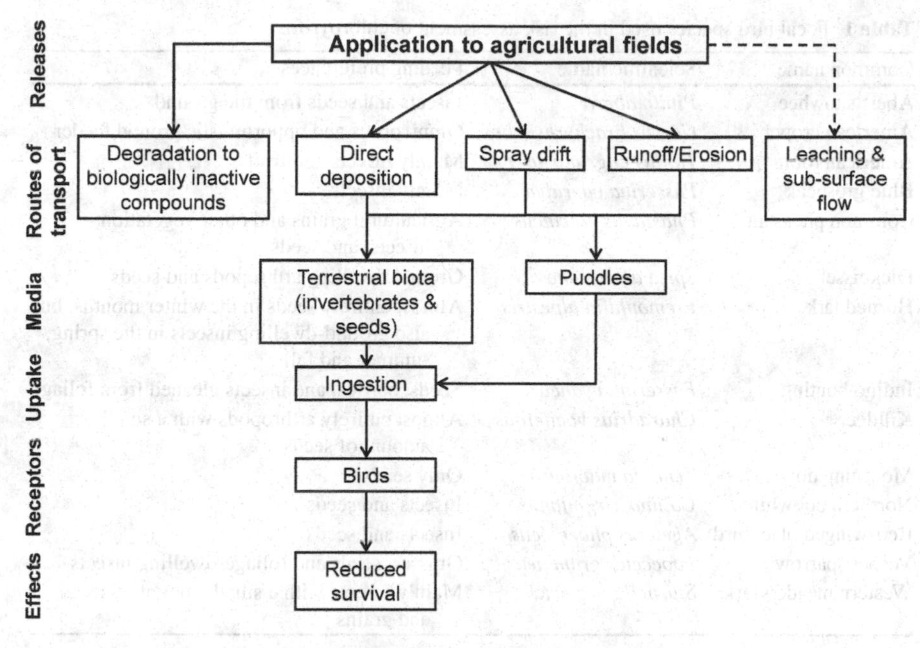

Fig. 1 Conceptual model for exposure of birds to flowable chlorpyrifos

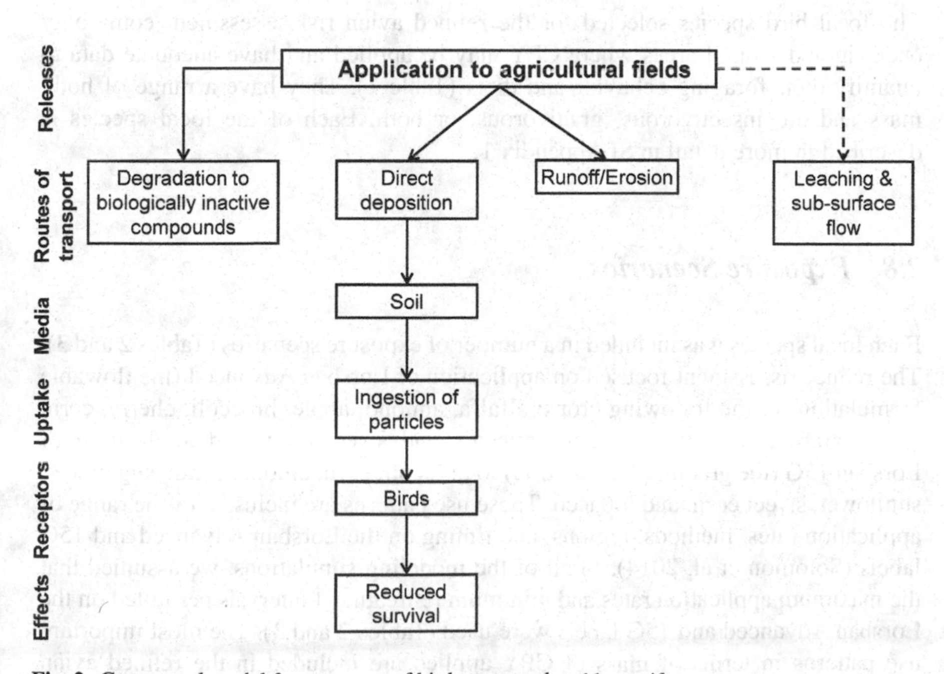

Fig. 2 Conceptual model for exposure of birds to granular chlorpyrifos

Table 1 Focal bird species used in the risk assessment of chlorpyrifos

Common name	Scientific name	Feeding preferences
Abert's towhee	*Pipilo aberti*	Insects and seeds from the ground
American crow	*Corvus brachyrhynchos*	Omnivorous and opportunistic ground feeder
American robin	*Turdus migratorius*	Mainly insects and fruit
Blue grosbeak	*Passerina caerulea*	Mainly insects
Common pheasant	*Phasianus colchicus*	Agricultural grains and other vegetation, insects, and seeds
Dickcissel	*Spiza americana*	Ground-dwelling arthropods and seeds
Horned lark	*Eremophilia alpestris*	Almost entirely seeds in the winter months, but also ground-dwelling insects in the spring, summer and fall
Indigo bunting	*Passerina cyanea*	Seeds, berries, and insects gleaned from foliage
Killdeer	*Charadrius vociferous*	Almost entirely arthropods with a small amount of seeds
Mourning dove	*Zenaida macroura*	Only seeds
Northern bobwhite	*Colinus virginianus*	Insects and seeds
Red-winged blackbird	*Agelaius phoeniceus*	Insects and seeds
Vesper sparrow	*Pooecetes gramineus*	Grasses, seeds and foliage-dwelling insects
Western meadowlark	*Sturnella neglecta*	Mainly insects with a small amount of seeds and grains

2.7 Focal Species

The focal bird species selected for the refined avian risk assessment commonly occur in and around areas where CPY may be applied and have adequate data to quantify their foraging behavior and diets (Table 1). They have a range of body mass and are insectivorous, granivorous, or both. Each of the focal species is described in more detail in SI Appendix 1.

2.8 Exposure Scenarios

Each focal species was included in a number of exposure scenarios (Tables 2 and 3). The refined assessment focused on application of Lorsban Advanced (the flowable formulation) to the following crops: alfalfa, almond, apple, broccoli, cherry, corn, grape, grapefruit, orange, pecan, soybean, and sweet corn, and application of Lorsban 15G (the granular formulation) to broccoli, corn, onion, peanut, sugar beet, sunflower, sweet corn, and tobacco. These use patterns are inclusive of the range of application rates, methods, regions, and timing on the Lorsban Advanced and 15G labels (Solomon et al. 2014). In all of the modeling simulations, we assumed that the maximum application rates and minimum re-treatment intervals permitted on the Lorsban Advanced and 15G labels were used (Tables 2 and 3). The most important use patterns in terms of mass of CPY applied are included in the refined avian

Table 2 Exposure scenarios for Lorsban® Advanced

Crop	Use pattern	Rate of application		Focal bird species[a]
		kg ha^{-1}	lb A^{-1}	
Alfalfa	Southern Plains—applied broadcast with 10-d interval	1.05	0.94	Dickcissel, Mourning dove, Red-winged blackbird, Vesper sparrow, Western meadowlark
Almond	California—applied airblast post-plant 2× (May, hull-split) with 10-d interval	2.11	1.88	Abert's towhee, American crow, Blue grosbeak, Mourning dove, Red-winged blackbird
Apple	Northwest—applied airblast delayed dormant to petal fall	2.11	1.88	American crow, Blue grosbeak, Mourning dove, Red-winged blackbird
Broccoli	California—applied band at-plant and post-plant 3× with 10-d interval	2.36	2.11	American crow, Horned lark, Mourning dove, Red-winged blackbird
Cherry	Michigan—applied broadcast post-plant	2.11	1.88	American crow, Blue grosbeak, Mourning dove, Red-winged blackbird
Corn	Midwest—applied broadcast at-plant and band post-plant 3× with 10-d interval	1.12	1	American robin, Horned lark, Killdeer, Mourning dove, Northern bobwhite, Red-winged blackbird, Vesper sparrow
Grape	California—applied airblast prior to bud break	2.11	1.88	American crow, Blue grosbeak, Mourning dove, Northern bobwhite, Red-winged blackbird
Grapefruit	Florida—applied airblast post-plant (Apr–Jun)	2.80	2.5	American crow, Blue grosbeak, Mourning dove, Northern bobwhite, Red-winged blackbird
Orange	California—applied airblast post-plant (May–Aug)	6.28	5.6	Abert's towhee, American crow, Blue grosbeak, Mourning dove, Red-winged blackbird
Pecan	Georgia—applied airblast 3× with 10-d interval	1.05	0.94	American crow, Blue grosbeak, Mourning dove, Northern bobwhite, Red-winged blackbird
Soybean	Louisiana—applied broadcast post-plant 3× with 14-d interval (May–Aug)	1.05	0.94	Blue grosbeak, Dickcissel, Horned lark, Indigo bunting, Mourning dove, Red-winged blackbird
Sweet corn	Florida—applied broadcast at-plant and band post-plant 3× with 10-d interval	1.12	1	American crow, Mourning dove, Northern bobwhite, Red-winged blackbird

[a]See Table 1 for scientific names

Table 3 Exposure scenarios for Lorsban® 15G

Crop	Use pattern	Rate of application		Focal bird species[a]
		kg ha^{-1}	lb A^{-1}	
Broccoli	California—applied T-band at-plant	2.52	2.25	Horned lark, Red-winged blackbird, Mourning dove
Corn and sweet corn	Midwest—applied T-band or in-furrow at-plant; applied broadcast or in-furrow postplant	1.46 1.12	1.3 1	Horned lark, Red-winged blackbird, Mourning dove, Northern bobwhite, Common pheasant
Onion	Pacific northwest—applied in-furrow at-plant	1.12	1	Horned lark, Red-winged blackbird, Mourning dove
Peanut	Southeast—applied band postplant	4.48	4	Horned lark, Red-winged blackbird, Mourning dove, Northern bobwhite
Sugarbeet	Midwest—applied broadcast at-plant	1.12	1	Horned lark, Red-winged blackbird, Mourning dove, Northern bobwhite, Common pheasant
Sunflower	Midwest—applied T-band at-plant	1.45	1.3	Horned lark, Red-winged blackbird, Mourning dove, Northern bobwhite, Common pheasant
Tobacco	Southeast—applied broadcast preplant with incorporation	2.24	2	Horned lark, Red-winged blackbird, Mourning dove, Northern bobwhite

[a]See Table 1 for scientific names

assessments for the two formulations. Applications of Lorsban Advanced can be made up to a rate of 6.3 kg ai ha^{-1} (5.6 lb ai A^{-1}) (oranges in California) with a maximum of four applications (alfalfa) per season. Only one application of Lorsban 15G can be made per season with a maximum application rate of 4.5 kg ai ha^{-1} (4 lb ai A^{-1}) (peanut). Lorsban Advanced is generally applied by broadcast, airblast, or banded methods, and Lorsban 15G is applied by T-band, in-furrow, or broadcast methods. For each use-pattern, a "high use" region of North America was determined from regional sales and use data provided in Gomez (2009). Focal species that would likely be in the treated areas for each use and region were selected by reviewing their ranges, preferred habitats, and patterns of seasonal migrations (see SI Appendix 1). The resulting exposure scenarios are shown in Tables 2 and 3.

2.9 Analysis Plan

The refined assessments of risks of Lorsban Advanced and Lorsban 15G to birds were conducted in three phases: assessment of exposure, assessment of effects, and characterization of risks. The following sections outline the analysis plan for Lorsban Advanced (the flowable formulation) and Lorsban 15G (the granular formulation).

Exposure assessment—Lorsban Advanced. The model used in the assessment of exposure for birds was a refinement of EPA's Terrestrial Investigation Model (TIM) (USEPA 2005, 2008) and is known as the Liquid Pesticide Avian Risk Assessment Model (LiquidPARAM). Version 1 of TIM (TIM v1) estimates the fate of each of 20 birds on each of 1,000 fields following an acute exposure (USEPA 2005). TIM v1 is a species-specific model that estimates risks over a defined exposure window of 7-d. The time-step in the model is 12 h. The spatial scale is the treated field where the field and surrounding area are assumed to meet the habitat requirements of a defined cohort of individuals for each focal species. Pesticide contamination of edge or adjacent habitat from drift is assumed to be zero. Version 2 of TIM (USEPA 2008) is similar to TIM v1 except that it has a 1-h time step and includes a more refined puddle exposure algorithm as well as screening-level algorithms for dermal and inhalation exposure.

Major components included in TIM are: (1) Food preferences of selected focal species; (2) Daily ingestion rates of food and water which are randomly assigned from species-specific body mass distributions; (3) Frequency of feeding and drinking on the treated field; (4) Water sources, including dew and puddles; (5) Distributions of residues on food items and in on-field water sources as a function of application rate; (6) Degradation rates of food and water residues over time; and (7) Interspecies distribution-based estimates of dose-response acute toxicity curves for focal species when laboratory-derived toxicity estimates are not available, or the dose-response curve derived from laboratory toxicity tests for focal species (see SI Appendix 2 for additional details on model structure).

For each simulated bird, values are randomly selected for the input parameters in TIM required to estimate exposure. The estimated risk of lethality for each individual bird is calculated from the dose-response curve. Once the fate of an individual on a particular field is determined (i.e., dead or alive), a new individual is carried through the same process. This process is repeated for a total of 20 individuals on the field. The model then moves to the next field. This outer loop continues for a sample size of 1,000 fields, which results in a risk estimate for a total of 20,000 birds on treated fields.

LiquidPARAM shares some of the similar basic structure of TIM (e.g., each model estimates the fate of each of 20 birds on each of 1,000 fields). However, several important refinements have been made and are briefly described below. A more detailed description is given in SI Appendix 3, Sect. 1.

In TIM, concentrations in dietary items within a field are randomly selected from distributions at each time step. Often this leads to situations where concentrations increase several-fold, 4 or more days after application. This situation seems unlikely in normal use given the fairly short half-life of CPY in the field. LiquidPARAM assumes that factors causing variation in concentrations of CPY on dietary items are relatively small within a field at a particular time step relative to those factors that cause variation between fields. Factors affecting relationships between rate of application and concentrations of pesticides on dietary items include: ambient temperature, wind speed, field slope, soil type, rainfall patterns, applicator experience, and type of equipment used to apply the spray. These factors vary only slightly within a field, but can be quite variable among fields within a broad region of the United States. Thus, for each field, LiquidPARAM randomly chooses an initial concentration for each dietary item and then these concentrations decline in the field over time according to the degradation rate for that dietary item. When the model proceeds to the next field, new initial concentrations are randomly selected. The process is repeated for 1,000 fields.

TIM assumes that bands occupy 17% of each field and furrows 5% of each field. However, these factors vary among crops. LiquidPARAM has been customized to have crop-specific row widths and spacing.

TIM uses older allometric equations provided by Nagy (1987) to estimate free metabolic rate. LiquidPARAM uses the more up-to-date allometric equations from Nagy et al. (1999). LiquidPARAM also accounts for uncertainty in estimates of free metabolic rate arising from error due to lack of model fit, while TIM does not.

TIM also does not account for the avoidance behavior that has been observed by birds following initial exposure to CPY (Bennett 1989; Wildlife International 1978). LiquidPARAM accounts for this behavior. Further, TIM only simulates acute exposure following a single pesticide application. LiquidPARAM can simulate both acute and chronic exposures following multiple pesticide applications.

Exposure assessment—Lorsban 15G. Previously, a simulation model was developed that estimated exposure and risk for various bird species that are potentially exposed to the granular formulation of aldicarb (Moore et al. 2010b, c). That model, referred to as the Granular Pesticide Avian Risk Assessment Model (GranPARAM), includes input variables such as: proportion time in the field, rates of ingestion of grit, attractiveness of pesticide granules compared to natural grit, and proportion of soil particles in the grit size range preferred by birds. For input variables that are uncertain, variable, or both, frequency distributions are used rather than point estimates. Monte Carlo analysis is then performed to propagate input variable uncertainties through the exposure model. Similar to LiquidPARAM, GranPARAM determines the fate of 20 randomly chosen birds on each of 1,000 randomly selected fields for the use pattern and region of interest. GranPARAM was revised to be specific to CPY for this refined avian risk assessment (see Sect. 4 and SI Appendix 3, Sect. 2).

Effects assessment. Effects data can be characterized and summarized in a variety of ways, ranging from benchmarks designed to be protective of most or all species to dose-response curves for the focal species of interest. When toxicity data are lacking for a focal bird species, a species sensitivity distribution (SSD) can be used to give an indication of the risk range by varying the dose-response curve from that of a sensitive species to that of a tolerant species. This approach was used by EPA (USEPA 2005) in their avian risk assessment for carbofuran. The SSD approach was also used in this assessment, except when toxicity data were available for the focal species of interest (see Sect. 5).

Effects associated with survival of juveniles or adults were the preferred measure of acute effect because this endpoint was judged to be the most appropriate based on the mode of toxic action of CPY. Gavage studies were used in preference to dietary studies because of problems in estimating dose when avoidance of treated food is a factor. Because flowable CPY may be applied up to four times per growing season, chronic risk was also estimated for that formulation. For the chronic assessment, preferred metrics included the most sensitive of the population-relevant endpoints, viz., survival, growth and reproduction.

The following decision criteria were used in deriving effects metrics for each focal species: (1) If a toxicity study with five or more treatments was available for the focal species or a reasonable surrogate, then a dose-response curve was derived for that species; (2) If multiple toxicity studies that followed a similar protocol and together had five or more treatments were available for the focal species or a reasonable surrogate, then a dose-response curve was derived for that species; (3) For untested focal species, an SSD was derived.

Without toxicity data for a focal species, there is uncertainty regarding the sensitivity of that species to CPY. To deal with this uncertainty, the SSD was used to bound the risk estimates (i.e., assume 5th and 95th centile sensitivity on the SSD) and to estimate median risk (i.e., assume 50th centile sensitivity on the SSD) (USEPA 2005). Dose-response curves were then derived for low (95th centile), median (50th centile) and high (5th centile) sensitivity species by using a distribution of the available LD_{50} data and measured dose-response curve slopes. Because insufficient bird species have been tested for chronic exposure, the most sensitive effects metrics were assumed for all focal species.

Each toxicity study was evaluated, and acceptable studies met the following criteria: (1) Single contaminant exposure only; (2) Gavage (acute) or dietary (chronic) route of exposure; (3) Ecologically-significant endpoint (e.g., survival, reproduction, growth); (4) Adequate statistical design (e.g., five or more treatments including controls, responses spanning most of the range of 0–100% effect including at least one treatment with a partial response) to estimate toxic effect doses; and (5) Study employed acceptable laboratory practices or was previously accepted by EPA (USEPA 2009). Studies that did not meet the above criteria were not used to derive effects metrics.

Risk characterization. Three lines of evidence were used to characterize risks of CPY to birds: (1) Modeling of exposure and effects; (2) Information available from field studies; and (3) Information available from incident reports.

Risk curves were derived for each exposure scenario and focal bird species by determining the percentages of fields that had ≥5% mortality (≥1/20 dead birds per field), ≥10% mortality (≥2/20 dead birds per field), ≥15% mortality (≥3/20 dead birds per field), … , 100% mortality (20/20 dead birds per field). The result was a plot of probability of exceedence versus magnitude of effect. Similar approaches have been used in ecological risk assessments performed for the EPA at the Calcasieu Estuary, Louisiana, the Housatonic River, Massachusetts (USEPA 2002, 2004a) and by others assessing the ecological risk of pesticides (Giddings et al. 2005; Moore et al. 2010a, b, c; Solomon et al. 2001). In this assessment, area under the risk curve (AUC) was estimated for each combination of focal species and exposure scenario. AUC is the area under the curve divided by the sum of the AUC and the area above the curve, with the result multiplied by 100. The AUC was used to

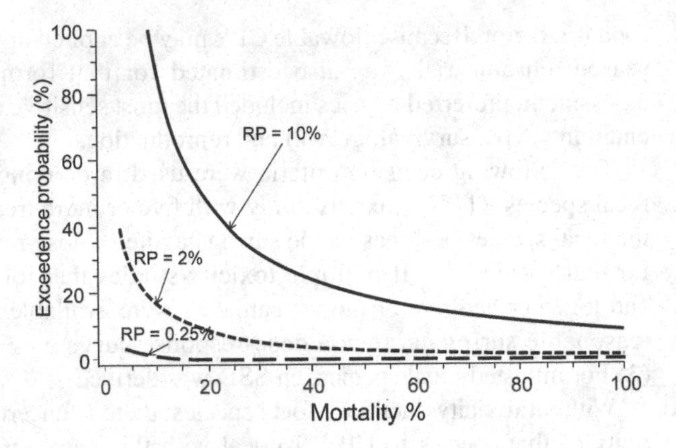

Fig. 3 Risk curves defined by risk products (RP) of 0.25, 2 and 10%

categorize risk as follows: (1) If the area under the risk curve was less than the AUC associated with the curve produced by risk products (risk product = exceedence probability × magnitude of effect) of 0.25% (e.g., 5% exceedence probability of 5% or greater effect = 0.25%), then the risk was categorized as *de minimis*. The AUC for risk products of 0.25% is 1.75%; (2) If the AUC was equal to or greater than 1.75%, but less than 9.82% (i.e., the AUC for risk products of 2%), then the risk was categorized as low; (3) If the AUC was equal to or greater than 9.82%, but less than 33% (i.e., the AUC for risk products of 10%), then the risk was categorized as intermediate; and (4) If the AUC was equal to or greater than 33%, then the risk was categorized as high. The risk curves defined by risk products of 0.25, 2 and 10% are shown graphically in Fig. 3.

Categories of risk were based on a rationale described previously (Moore et al. 2010a, b) and included several considerations: (1) Losses of small numbers of individuals from a local population should not adversely affect the population (Giddings et al. 2005; Moore 1998). One of the foundations of hierarchy theory (Allen and Starr 1982) is that effects at lower levels of ecological organization (e.g., organism level) are not necessarily transmitted to higher levels of ecological organization (e.g., population level); (2) Although there are exceptions, an adverse effect level of 10% is unlikely to be ecologically significant to a local population. Such an effect generally cannot be reliably confirmed by field studies (Moore 1998; Suter et al. 2000); (3) Based on an analysis of EPA regulatory practice, Suter et al. (2000) concluded that decreases in an ecological assessment endpoint of less than 20% are generally acceptable; (4) The curve corresponding to a risk product of 2% passes through the points corresponding to a very low probability (i.e., 10%) of 20% or greater effect, and a low probability (i.e., 20%) of 10% or greater effect. Thus, based on the considerations described above, if risk products are generally less than the 2% boundary for an exposure scenario, then it can almost certainly be considered a low risk scenario; (5) The curve corresponding to a risk product of 10% passes through the points corresponding to a median probability (i.e., 50%) of 20% or greater effect and a 20% probability of 50% or greater effect. In this assessment, exposure scenarios with risk products generally above the 10% boundary were considered to be high risk scenarios because there was a low to median probability of detectable and possibly major impacts on local bird populations. Scenarios with risk curves generally between the low and high boundaries were judged to be intermediate risk scenarios; (6) When there was a very low likelihood of a scenario affecting a focal bird species, risk was categorized as *de minimis*. A 5% probability of exceeding 5% adverse effect lies on the curve defined by the risk product equal to 0.25%. The AUC associated with the risk curve defined by risk products of 0.25% (AUC = 1.75%) was thus used as the upper boundary for the category of *de minimis* risk. The percent protection level was also calculated for each exposure scenario and focal species. Protection level (%) is equal to the number of surviving birds divided by the number of birds included in the model run (20 birds per field × 1,000 fields = 20,000 birds) times 100.

The risk characterization also incorporated available field study results and incident reports into the assessment of the avian risks associated with legal labeled uses of CPY.

3 Exposure Assessment for Flowable Chlorpyrifos

The development of a probabilistic assessment model for risks of flowable pesticides to birds began with the formation of the Ecological Committee on FIFRA Risk Assessment Methods (ECOFRAM). ECOFRAM was tasked with identifying and developing probabilistic tools and methods for ecological risk assessments under the FIFRA regulatory framework. Conclusions and recommendations of the ECOFRAM workgroup on avian exposure models were summarized in the Draft Terrestrial Workgroup Report (ECOFRAM 1999). Subsequently, EPA formed an internal committee to develop tools (e.g., Terrestrial Investigation Model) and to develop an approach for incorporating the ECOFRAM workgroup conclusions and recommendations. That approach was evaluated and endorsed by a United States Environmental Protection Agency (USEPA) Scientific Advisory Panel in 2000 (SAP 2000).

The pilot version of the Terrestrial Investigation Model, version 1 (TIM v1) was developed to evaluate a model pesticide called ChemX. Based on recommendations and comments of another SAP (SAP 2001), the EPA began to refine TIM v1 and produced a draft version of TIM version 2 (TIM v2). TIM v2 was evaluated by the SAP (SAP 2004) and subsequently refined. TIM v1 was the model used by EPA (USEPA 2005) to estimate the risks of a flowable carbofuran formulation to avian species that forage in treated fields and is summarized in SI Appendix 2.

3.1 *Rationale for Developing LiquidPARAM*

Much of the basic structure of the Liquid Pesticide Avian Risk Assessment Model (LiquidPARAM) was based upon TIM v1 and v2. Since the release of TIM, studies have taken place that can be used to refine components of the model. These studies address areas not included in TIM, such as avoidance behavior of birds exposed to CPY (Bennett 1989; Wildlife International 1978) and measured concentrations of CPY on dietary items. Thus, one reason for developing LiquidPARAM was to expand the model structure of TIM to accommodate new information. Changes to TIM were also required to address the recommendations of the SAPs (SAP 2001, 2004) that reviewed the TIM model. Where possible, LiquidPARAM incorporated recommendations of the SAP (SAP 2001, 2004) including, for example:

- Addition of many new focal species and use patterns to ensure better representation of the bird community that forages in agroecosystems.
- Use of a 1-h time step in LiquidPARAM, instead of the 12-h time step used in TIM v1. TIM v2 also moved to a 1-h time step. This refinement was considered necessary to account for the changes in avian foraging behavior, avoidance behavior, and clearance of the pesticide that occur throughout the day.
- For each time step in TIM v1 and v2, the model randomly determines whether a bird is on or off the treated field. The SAP (SAP 2001) felt that this approach misrepresented how birds forage in agroecosystems. LiquidPARAM allows birds to forage on and off fields in each time step. The model also accounts for between-field differences in foraging behavior of bird populations that have been observed due to factors affecting the relative attractiveness of treated fields to birds (e.g., type of edge habitat, availability of cover, etc.). Relative attractiveness of fields to birds can vary dramatically between treated areas.
- The SAP (SAP 2001) observed that TIM v1 confused inter- and intra-field variation by using dietary residue distributions that included both sources of variation. Residue levels in each field in TIM v1 relied on the same distributions, as is the case in TIM v2. However, one would expect larger differences in mean concentrations of residues between fields than within fields because of differences in soil type and topography, operator skill, type of application machinery, etc. Further, birds spatially and temporally average their dietary exposures within fields because they generally make multiple foraging trips within any given 1-h time step (see Sect. 1.2 in SI Appendix 3). LiquidPARAM incorporates a model structure that accounts for the expected variation between fields in mean concentrations of residues in dietary items.
- LiquidPARAM incorporates an avoidance behavior component that was suggested by the SAP (SAP 2001) as being a potentially important factor in reducing risk (see EFSA 2008).
- The SAP (SAP 2001) noted that acute oral studies do not account for the effect of the dietary matrix on adsorption rate of pesticides by birds. LiquidPARAM can account for the difference in toxicity to birds of flowable pesticide administered in water versus a dietary matrix if such data are available.
- Because flowable CPY may be applied up to four times per season, there is a potential for chronic exposure. As a result, LiquidPARAM has been extended to a 60-d model that can be used to estimate chronic risks potentially arising from multiple applications.

- Finally, the SAP (SAP 2001) stated that field validation of a model, particularly a complex model, is critical. This has yet to be done for TIM v1 or v2. An evaluation of model performance was previously done for LiquidPARAM with flowable carbofuran, the results of which are discussed in SI Appendix 3.

It was not possible to incorporate all of the recommendations of the SAP (US EPA SAP 2001, 2004) because information is still lacking in several areas. For example, the SAP (SAP 2001) expressed concern that TIM equates proportion time spent in treated fields as the proportion of diet obtained from the treated fields. The data required to act on this recommendation are not available for North American bird species that forage in agroecosystems.

The SAP (SAP 2001) also noted that there is considerable uncertainty regarding how birds obtain drinking water from treated fields and other nearby habitats. As in TIM v1, LiquidPARAM includes three drinking water scenarios (dew, dew plus puddles on day of application, and dew plus puddles on the day after application). The Panel (SAP 2001) concluded that this approach was reasonable, but recommended that further research be undertaken on: (1) the linkage between time on the field and amount of water consumed, (2) puddle persistence, (3) concentrations in dew and puddles, and (4) consumption of dew by different bird species. The SAP (SAP 2001) noted that field telemetry studies combined with laboratory bird behavior studies could provide the needed data. Because such studies have not yet been conducted, and because drinking water appears to be a minor source of exposure for flowable CPY (see the results of the sensitivity analysis in SI Appendix 3, Sect. 1.4), LiquidPARAM retains the same drinking water scenarios as exist in TIM v1. A graphical description of the structure of the LiquidPARAM model is illustrated in Fig. 4 and details of the model are provided in SI Appendix 3

3.2 Description of the Structure of the LiquidPARAM Model

For acute exposure, LiquidPARAM estimates the maximum retained dose that occurs over a period of 60-d following initial pesticide application in each of 20 birds on each of 1,000 fields (Fig. 4). The model can accommodate up to three applications at intervals specified by the user. The model has a 1-h time step. For each bird, a standard normal Z score is calculated for the maximum retained dose. This Z score determines how extreme the exposure is relative to the appropriate LD_{50} using a log-probit dose-response relationship. The Z score is then compared to a randomly selected value from a uniform distribution with a range of 0–1. If the Z score for exposure exceeds the randomly drawn value from the uniform distribution the bird dies. Otherwise, it survives (Fig. 4).

For species lacking acceptable acute oral toxicity data (all focal species except the northern bobwhite, *C. virginianus*, and red-winged blackbird, *Agelaius phoeniceus*, for CPY), a species sensitivity distribution (SSD) approach is used to generate the effects metrics. With this approach, a regression analysis is first conducted to

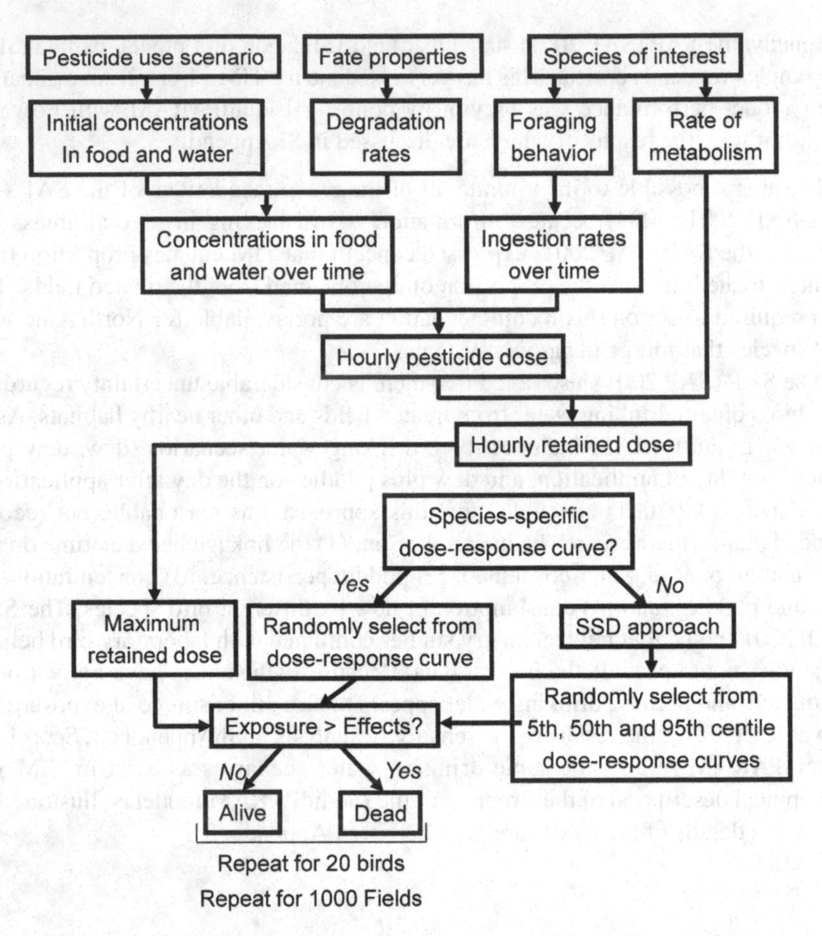

Fig. 4 Components of LiquidPARAM

quantify the relationship between dose and proportion species affected as determined by their LD_{50}s. Hypothetical dose-response curves are then derived for species of high (5th centile species on the SSD), median (50th centile species) and low (95th centile species) sensitivity. The 5th, 50th and 95th centile LD_{50}s are combined with the average slope for tested bird species to parameterize the three hypothetical dose-response curves assuming an underlying log-probit distribution. Section 5 describes the assessment of acute effects in detail.

For chronic exposure, total daily intake (TDI) is estimated for each day in the 60-d model run. TDI is averaged over a period equal to the duration from which the most sensitive effects metric was derived (e.g., gestation period for number of eggs laid). The maximum rolling average from the 60-d model run for each bird is then compared to a randomly drawn TDI from the appropriate chronic dose-response curve, if available, to determine if the bird is adversely affected and, if so, magnitude of effect. In the absence of a chronic dose-response curve, as is the case for CPY,

LiquidPARAM calculates the probability of maximum average TDI across all birds exceeding the chronic NOEL and the corresponding probability for exceeding the LOEL. Section 5 describes the assessment of chronic effects in detail.

4 Exposure Assessment for Granular Chlorpyrifos

The Granular Pesticide Avian Risk Assessment Model (GranPARAM) was used to estimate exposure and fate of birds as a result of consuming pesticide granules in CPY-treated agricultural fields. The model as originally described (Moore et al. 2010b, c) has been updated for this assessment.

GranPARAM simulates the grit ingestion behavior of individual birds and determines how many pesticide granules and the associated dose each bird ingests during the 24-h period immediately following CPY application. Each bird in a GranPARAM simulation is assumed to be actively foraging for grit in and around the agricultural field to which CPY has been applied. The scheme that GranPARAM follows to model granule ingestion behavior is depicted in Fig. 5.

In GranPARAM, each bird is randomly assigned a daily grit intake rate from a large database of grit counts for the species being considered and estimated grit retention rate. This step defines the number of medium- and coarse-sized particles (i.e., particles in the same size range as Lorsban 15G granules) that the individual ingests during the peak day of the simulation. For CPY, the peak day was assumed to be the 24 h immediately following application. The work of Stafford et al. (1996) and Stafford and Best (1997) showed that most pesticide granules are incorporated into soil, and thus unavailable to birds, within 1-d of application. Rainfall accelerates this process (Stafford and Best 1997). GranPARAM relies on estimates of granule counts on the soil surface immediately after application, which clearly represents the maximum possible exposure for birds (Solomon et al. 2001).

Each site of application of the granular formulation is randomly assigned a soil texture (e.g., Silt-Loam) with a probability equal to the occurrence of that texture fraction in the crop-capable acreage in the region of interest. The database in the model was originally for corn, but has been expanded to include other crops and areas to which CPY is applied. Once the soil texture category is assigned, the application site is then randomly assigned a specific soil particle size profile (% of soil mass represented by various particle size categories) from a large soils database of measurements. This step defines the levels of medium- and coarse-sized sand particles available as natural grit.

For each exposure scenario (Table 3), the method of application, rate of application, incorporation efficiency, bird species, region of interest, and other aspects of the analysis included in the simulation were defined (Fig. 5). The rate of application of CPY defines the relative numbers of medium- and coarse-sized granules applied. The method of application (e.g., in-furrow, band, broadcast) determines the spatial placement of these granules and the number available as a source of particles to birds. The choice of bird species determines the number of particles ingested.

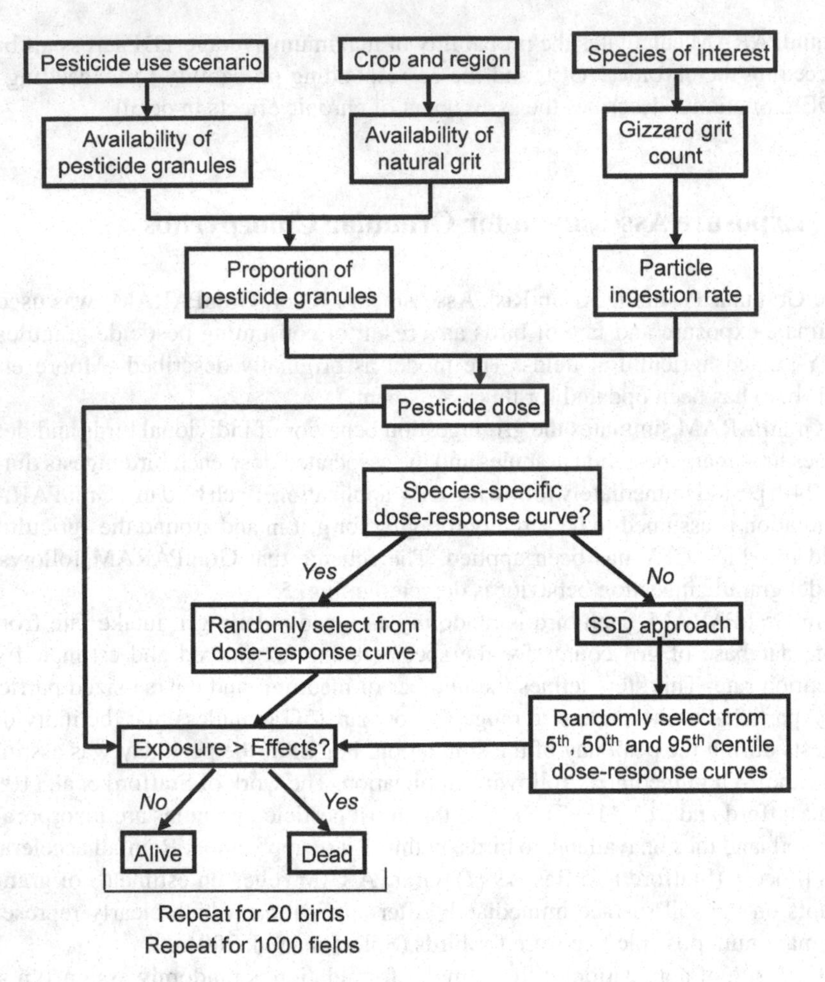

Fig. 5 Components of GranPARAM

Each time a bird feeding in a treated field ingests a particle, the particle is either a granule or a piece of natural grit. The default assumption of the model is that birds forage for particles within each spatial zone randomly, and therefore the probability, *p,* of selecting a pesticide granule is equal to the relative availability of granules in comparison to natural grit particles of the same size. However, birds may select particles non-randomly and show preference for some types of particles over others (Best and Gionfriddo 1994; Best et al. 1996). In GranPARAM, the user has the option to input the relative preference birds have for selecting pesticide granules in comparison to natural grit. If this factor is used, as was the case with CPY, GranPARAM modifies the estimate of *p* accordingly. Birds prefer sand for grit consumption and thus strongly avoid Lorsban 15G granules because of its clay-based formulation (see SI Appendix 3, Sect. 2.1 for additional details). Once *p* is defined, the number of Lorsban 15G granules ingested during the day following application is determined by randomly sampling from a binomial distribution defined by N (number of particles ingested that could be either granules or natural sand) and p. This calculation is made separately for medium- and coarse-sized pesticide granules, and for spatial zones of the field that differ from one another in either the relative availability of granules or relative use by birds (e.g., end rows, field margin, field center). The number of particles the bird obtains from a given zone (N) is estimated from the zone's relative size and use by birds.

The version of GranPARAM described herein estimated exposure to 20 birds on each of 1,000 fields, as previously described for LiquidPARAM. The 1,000 fields were intended to represent the range of soil characteristics for the crop and region of interest. In GranPARAM, characteristics of birds such as grit counts in their gizzard and proportion of time they forage in treated fields are chosen randomly from distributions. Thus, individuals on fields differ from one another and the model is designed to explicitly incorporate the variation observed in nature. Similarly, characteristics of fields such as differences in size distribution and composition of soil are randomly chosen from distributions in GranPARAM. This approach ensures that the variability in field soils observed in nature is reflected in the model.

The outputs from the exposure portion of GranPARAM are estimated acute doses for each of 20 birds on each of 1,000 fields. The effects and risk components of GranPARAM and LiquidPARAM for acute exposure are the same. The risk output from GranPARAM is a bar chart showing the percentages of fields with 0/20 dead birds, 1/20 dead birds, 2/20 dead birds, etc.

The components of the model and input variables for GranPARAM are described in detail in SI Appendix 3. Simulations were run for each of the exposure scenarios listed in Table 3. All simulations were carried out using Latin Hypercube Sampling in Oracle Crystal Ball (2009), Version 7.3.2 with 1,000 trials (i.e., fields) per simulation and 20 birds per field.

5 Effects Assessment

Upon ingestion, CPY is rapidly absorbed and undergoes oxidative metabolism to the oxon form, which is the metabolite primarily responsible for toxicity (Testai et al. 2010). Chlorpyrifos inhibits acetylcholinesterase activity causing acetylcholine to accumulate at nerve terminals and neuromuscular junctions, which leads to cholinergic overstimulation (Testai et al. 2010). In birds, CPY poisoning can be detected biochemically as reduced cholinesterase activity in plasma and brain tissues (Cairns et al. 1991; Parsons et al. 2000; Timchalk 2010). A gavage study with northern bobwhite (*C. virginianus*) found reduced brain cholinesterase activity at concentrations of 47 mg ai kg^{-1} bwt of CPY and greater (Cairns et al. 1991). Cholinesterase activity remained inhibited for at least 24–48 h thereafter.

The following sections present a review of the available acute and chronic effects studies and the derivation of the acute and chronic effects metrics that were used to characterize risks to birds from the use of CPY.

5.1 Acute Toxicity Studies

Oral gavage and dietary studies have been conducted to determine the acute effects of CPY on birds (Table 4). Data on toxicity of CPY to birds were reviewed previously (Solomon et al. 2001) and in the sections that follow, only studies that have been conducted since that review are discussed.

Hubbard and Beavers (2008) administered CPY by oral gavage at 21.6 mg ai kg^{-1} bwt to 19-wk old northern bobwhite (*C. virginianus*) using corn oil as a vehicle and observed signs of toxicity (Hubbard and Beavers 2008). The effects included a ruffled appearance and lethargy. Signs of toxicity were more prevalent and occurred sooner at higher doses. Reduced consumption of food was observed at doses ≥36 mg ai kg^{-1} bwt. The acute oral 14-d LD_{50} was >60 mg ai kg^{-1} bwt, as only 30% mortality occurred at this dose, the largest tested. The no-mortality dose was 36 mg ai kg^{-1} bwt and the NOEL was 13 mg ai kg^{-1} bwt.

The acute oral toxicity of Lorsban 50W was determined by exposing northern bobwhite (*C. virginianus*) (Kaczor and Miller 2000). Twenty-two-wk old birds were dosed once with Lorsban 50W (50.5% purity) and observed for 14-d. Food consumption, body weight, signs of toxicity, and lethality were monitored throughout the observation period. In the group fed the greatest dose (121 mg ai kg^{-1} bwt), all birds died within 24 h of dosing. During the first day or two following dosing, consumption of food by birds exposed to CPY was less that of the controls. However, rates of food consumption quickly returned to those of control birds and, as a result, there was no significant decrease in body weight over the study period for any of the treatment groups. The most prevalent sign of toxicity was lethality. The only other observed sign of toxicity was lethargy and it usually preceded lethality. Necropsies of dead birds revealed gaseous intestines. The LD_{50} was 35.9 mg ai kg^{-1} bwt, the NOEC was 7.6 mg ai kg^{-1} bwt and the LOEC was 15.2 mg ai kg^{-1} bwt.

In other studies, the toxicity of CPY in a variety of formulations was evaluated. These formulations included Lorsban 2.5P (2.48% ai, Brewer et al. 2000b), Lorsban 10.5 LEE (10.5% ai, Brewer et al. 2000a), GF-1668 (18.7% ai, Gallagher and Beavers 2006), and Lorsban Advanced (41.1% ai, Hubbard and Beavers 2008). The lowest LD_{50} from these studies was 12.6 mg ai kg^{-1} bwt (Brewer et al. 2000a) and the smallest NOEL was <4 mg ai kg^{-1} bwt (Hubbard and Beavers 2008).

To determine the importance of duration of acute exposure on a daily basis, 25-wk old northern bobwhite (*C. virginianus*) were pre-conditioned to feed during either a 1- or 8-h period each day (Gallagher and Beavers 2007). Following the pre-conditioning period, the birds were offered food treated with CPY for either 1 or 8 h for 1-d. The birds were observed for 7-d after treatment. Rate of food consumption and body mass decreased with increasing dietary concentration of CPY in both treatment groups. Greater toxicity was observed in the group feeding for only 1 h each day. This result suggests that birds being exposed over an 8-h period had longer to metabolize and detoxify CPY. LD_{50} values were 75 mg ai kg^{-1} bwt for birds receiving their total dose in 1 h and 116 mg ai kg^{-1} bwt for those receiving the dose over an 8-h period. LC_{50} values were 3,697 and 6,986 mg ai kg^{-1} diet for birds

Table 4 LD_{50}s from acceptable oral gavage studies for chlorpyrifos

Species	Life stage	LD_{50} (mg ai kg^{-1} bwt)	Probit slope	SSD input value	Reference
Common grackle (*Quiscalus quiscula*)	Adult	5.62	–	8.55	Schafer and Brunton (1979)
	Adult	13	–		Schafer and Brunton (1971)
Ring-necked pheasant (*Phasianus colchicus*)	Adult, male	8.41	–	12.2	Hudson et al. (1984)
	Adult, female	17.7	–		
Red-winged blackbird (*Agelaius phoeniceus*)	Adult, male	13.1	–	13.1	Schafer and Brunton (1979)
Japanese quail (*Coturnix japonica*)	Adult	13.3	–	15.6	Schafer and Brunton (1979)
	Adult, male	15.9	–		Hudson et al. (1984)
	Adult, male	17.8	–		
Common pigeon (*Columba livia*)	Adult	10	–	16.4	Schafer and Brunton (1979)
	Adult	26.9	–		Hudson et al. (1984)
Sandhill crane (*Grus canadensis*)	Adult, male	25–50	–	25.0	Hudson et al. (1984)
House sparrow (*Passer domesticus*)	Adult	10	–	29.5	Schafer and Brunton (1979)
	Adult, male	21	–		Hudson et al. (1984)
	Adult	122	2.3		Gallagher et al. (1996)
Leghorn chicken (*Gallus domesticus*)	Chick, male	32	–	33.4	McCollister et al. (1974)
	Adult	34.8	–		Miyazaki and Hodgson (1972)
Canada goose (*Branta canadensis*)	Adult	40–80	–	40.0	Hudson et al. (1984)
Chukar (*Alectoris chukar*)	Adult, female	60.7	–	60.9	Hudson et al. (1984)
	Adult, male	61.1	–		
Northern bobwhite (*Colinus virginianus*)	Juvenile	119	3.88	61.7	Kaczor and Miller (2000)
	Adult	32	4.6		Hill and Camardese (1984)
California quail (*Callipepla californica*)	Adult, female	68.3	–	68.3	Hudson et al. (1984)
European starling (*Sturnus vulgaris*)	Adult	75	–	75.0	Schafer and Brunton (1979)
Mallard (*Anas platyrhynchos*)	Adult, female	75.6	–	92.0	Hudson et al. (1984)
	Duckling	112			

receiving the dose over 1 and 8 h, respectively. Concentrations of CPY associated with no mortality were 1,000 and 3,200 mg ai kg^{-1} diet for the two doses, respectively.

5.2 Chronic Toxicity Studies

No chronic toxicity studies on birds have been conducted for CPY since the review by Solomon et al. (2001).

5.3 Derivation of Effects Metrics

The most realistic route of exposure for acute effects is the dietary exposure pathway. This pathway is preferred over oral gavage exposures because the latter are only relevant to situations where active ingredients are ingested rapidly in a single exposure or gorging situation (ECOFRAM 1999). Most, if not all, bird species found in agroecosystems are much more likely to continuously forage for food during the daylight hours (Best 1977; Fautin 1941; Kessel 1957; Kluijver 1950; Pinkowski 1978). Birds foraging near agricultural areas are likely to ingest a mixture of contaminated and non-contaminated food items throughout a day. Studies in which birds were exposed through the diet were available for only one of the focal species, the northern bobwhite (Beavers et al. 2007). However, when dietary exposures for northern bobwhite were converted to dose ingested, there was little evidence of a dose-response relationship. In that study, birds reduced their food intake rates at higher dietary concentrations. Because of this issue, the results of oral gavage studies were used to derive the acute effects metrics in this assessment. Using results of studies that dosed birds by a single oral dose via oral gavage is highly conservative because:

- Doses are administered as one large dose. In the field, most birds feed continuously throughout the day.
- Chlorpyrifos is rapidly biotransformed by birds to less-toxic metabolites. The half-life for metabolism and elimination of CPY is approximately 1-d (Bauriedel 1986). When feeding throughout the day, birds have the opportunity to detoxify and/or eliminate CPY before it accumulates to internal doses that result in lethality.
- Repeated exposure to CPY in the diet leads to avoidance (Bennett 1989; Fink 1978b; Kenaga et al. 1978; Stafford 2010). In the field, birds can switch to sources of food that are not contaminated with CPY or avoid feeding for short periods of time. There can be no avoidance with large single doses administered by intubation during a gavage study.
- In oral exposures, CPY is generally administered in corn oil or gelatin capsules. Such carriers have been shown to result in greater toxicity with other insecticides than occurred when the insecticides were adsorbed to food items consumed by birds in the field (Stafford 2007a, b). Use of corn oil or gelatin carriers maximizes the potential for a pesticide to be absorbed rapidly, more so than would occur in the field where the pesticide is bound to food items. When pesticides are mixed with food, or when consumed at a time when the gastro-intestinal (GI) tract has other food items present, they are absorbed less efficiently than when dosed as a bolus in pure form into an empty GI tract (Lehman-McKeeman 2008).

In this assessment, the preferred effects metrics were dose-response curves for the focal species of interest. However, acute dose-response curves could only be derived for two focal species, the northern bobwhite (*C. virginianus*) and the red-winged blackbird (*A. phoenicieus*). For other focal species, a Species Sensitivity Distribution (SSD) approach was used. With this approach, the 5th, 50th and 95th centiles from the fitted species sensitivity distribution were selected to represent the range of sensitivities of birds to CPY. A hypothetical dose-response curve was derived for each of these centile species by combining the estimated LD_{50} with a probit slope. Because of the toxicity mitigation problems noted above with dietary exposures, the acute effect metrics in this assessment were based upon the results of acceptable acute oral gavage toxicity studies (Table 4).

The following sections describe how the acute and chronic effects metrics were derived. A NOEL and LOEL were selected as the chronic effects metrics because the available chronic toxicity studies did not have a sufficient number of treatments (i.e., five or more) to enable derivation of dose-response curves. There were an insufficient number of tested species to permit development of a SSD for chronic toxicity data.

Acute dose-response relationships for focal species. The LD_{50}, based on oral gavage, for the red-winged blackbird (*A. phoeniceus*) was 13.1 mg ai kg^{-1} bwt (Schafer and Brunton 1971), however, no probit slope was reported. To generate a dose-response curve for red-winged blackbird in this assessment, a geometric mean probit slope of 3.45 was calculated from the studies listed in Table 4. For northern bobwhite (*C. virginianus*), two LD_{50}s have been reported: 32 and 119 mg ai kg^{-1} bwt (Hill and Camardese 1984; Kaczor and Miller 2000). The corresponding probit slopes from these studies were 4.6 and 3.88, respectively. The resulting geometric mean LD_{50} and probit slope were 61.7 mg ai kg^{-1} bwt and 4.22. These values were used to generate the acute dose-response curve for northern bobwhite in this assessment.

SSD for acute toxicity of CPY to untested focal species. The data used in the derivation of the SSD for avian species are shown (Table 4). Multiple toxicity values were reported for several species. Variation in toxicity for a species could be the result of differences in experimental conditions, species strain, and/or test protocol. Using multiple toxicity results for the same species would disproportionately influence the SSD. In these situations, the geometric means were calculated (Table 4). Each bird species was then ranked according to sensitivity and its centralized position on the SSD determined using the Hazen plotting position equation (1) (Aldenberg et al. 2002):

$$PP = \frac{i - 0.5}{N} \quad (1)$$

Where:
PP is the plotting position;
i is the species rank based on ascending LD_{50}s; and
N is the total number of species included in the SSD derivation.

The SSD was derived using SSD Master v2.0, which includes five models: normal, logistic, Weibull, extreme value (=Gompertz) and Gumbel (=Fisher-Tippett) (CCME 2013). All analyses were conducted in log space, except the Weibull model, which was conducted in arithmetic space because a log-Weibull model is the same as the Gumbel model. The log-normal model had the best fit of the five

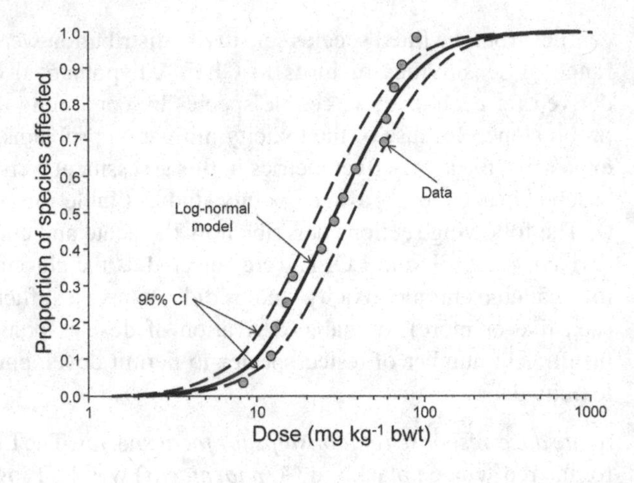

Fig. 6 Species sensitivity distribution for bird species exposed to chlorpyrifos via oral gavage exposure

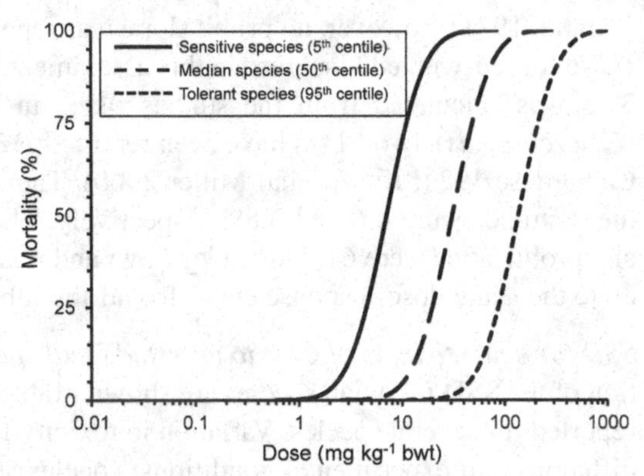

Fig. 7 Hypothetical acute dose-response curves for sensitive, median and tolerant bird species

models tested (Anderson-Darling $A^2 = 0.301$, $p >> 0.1$). The model equation for the two-parameter log-normal cumulative distribution function (CDF) is shown below (2):

$$f(x) = \frac{1}{2}\left(1 + erf\left(\frac{x - \mu}{\sigma\sqrt{2}}\right)\right) \tag{2}$$

Where:

x is the LD_{50} (log mg ai kg^{-1} bwt), and the functional response, $f(x)$, is the proportion of species affected. The location and scale parameters, μ and σ, are the mean and standard deviation of the dataset, respectively, and *erf* is the error function (i.e., the Gauss error function).

Graphical and statistical tests indicated that the homogeneity of variance and normality assumptions of the parametric regression analysis were met. The fitted model parameters were: $\mu = 1.49$ and $\sigma = 0.391$ (Fig. 6). The 5th, 50th and 95th centile LD_{50}s from the best-fit SSD are respectively 7.03, 30.9 and 136 mg ai kg^{-1} bwt. These values were combined with the geometric mean probit slope of 3.45 (Table 4) to generate the hypothetical dose-response curves for avian species of high (5th centile), median (50th centile) and low (95th centile) sensitivity (Fig. 7).

5.4 *Influence of Dietary Matrix on Acute Toxicity*

In a standard acute LD_{50} test, the test chemical is administered via gavage directly into the esophagus or crop of the bird, usually with a carrier such as corn oil, a solvent or water. Use of such carriers maximizes the potential for the chemical to be absorbed rapidly, more so than would occur in the field where the chemical is bound to food items. When pesticides are mixed with food, or consumed at a time when the gastrointestinal (GI) tract has other food items present, they are absorbed less efficiently than when dosed as a bolus in pure form into an empty GI tract (Lehman-McKeeman 2008).

To examine effects of excipient on toxicity, Hubbard and Beavers (2009) administered CPY to 19-wk old northern bobwhite (*C. virginianus*) using either corn oil or a feed slurry as the excipient. Groups of ten birds were randomly assigned to six treatment groups ranging from 0 (control) to 60 mg ai kg^{-1} bwt. The dose was mixed with the chosen excipient and orally intubated into the crop or proventriculus of each bird. For corn oil as excipient, CPY was dispersed in corn oil, and for the feed slurry excipient, treated food was mixed with water at a ratio of 1:2.5. Birds were monitored for 14-d following dosing to evaluate effects on body weight, lethality, consumption of food, appearance, and abnormal behavior. No signs of toxicity were observed in control groups. When corn oil was used as the excipient, signs of toxicity were first observed in the 21.6 mg ai kg^{-1} bwt treatment group, and included a ruffled appearance and lethargy. Body weight of female birds exposed to 21.6 mg ai kg^{-1} bwt decreased as did body weights in both sexes in the greater dose groups. Rate of consumption of food was reduced in males fed 36 mg ai kg^{-1} bwt and in males and females fed 60 mg ai kg^{-1} bwt. The NOEL was 13.0 mg ai kg^{-1} bwt. Lethality only occurred in birds fed 60 mg ai kg^{-1} bwt. Therefore, the LD_{50} was defined as >60 mg ai kg^{-1} bwt when corn oil was used as the excipient. When CPY was diluted with feed slurry as the excipient, signs of toxicity were first observed in individuals exposed to 21.6 mg ai kg^{-1} bwt. The signs of toxicity included a ruffled appearance and 1 lethality. Lethality skewed the observations for both change in body mass and rate of consumption of food. However, a lesser body mass was observed in surviving birds dosed at 21.6 mg ai kg^{-1} bwt or greater. Reduced consumption of food was observed in female birds dosed at 21.6 and 36 mg ai kg^{-1} bwt. The LD_{50} was 29.0 mg ai kg^{-1} bwt and the NOEL was 13.0 mg ai kg^{-1} bwt. In this study, the food-based slurry did not reduce toxicity compared to the corn oil excipient. Therefore, the acute effects metrics derived above were not adjusted to account for the dietary matrix consumed by birds in treated fields.

5.5 *Chronic NOEL and LOEL*

There is an insufficient number of studies to derive a chronic SSD for CPY. Further, there are no chronic studies with a sufficient number of treatments to enable derivation of a dose-response curve. Given the paucity of chronic toxicity studies for birds, we used a conservative approach and derived a NOEL and LOEL from the most sensitive species tested to date, viz., the mallard (*A. platyrhynchos*). In a 1-generation reproduction study (Fink 1978a), reduced reproductive success was observed at a concentration of 125 mg ai kg^{-1} diet but no adverse effects were observed at a concentration of 25 mg ai kg^{-1} diet. The primary response observed when birds were exposed to 125 mg ai kg^{-1} diet was fewer eggs laid per hen. Therefore, rate of intake of food and body mass during the egg-laying phase of the study (i.e., the final 8-wk) were used as the measurement endpoints upon which to base the dietary NOEL and LOEL. Mean rates of intake of food at doses equivalent to the NOEC and LOEC were 0.134 and 0.140 kg diet $bird^{-1}$ d^{-1}, respectively. The corresponding average body masses were 1.12 and 0.934 kg bwt, respectively. The resulting dose-based NOEL and LOEL are 2.99 and 18.7 mg ai kg^{-1} bwt d^{-1}, respectively. The NOEL and LOEL were used as thresholds for chronic effects in this refined assessment of risk.

6 Risk Characterization for Flowable Chlorpyrifos

For each acute exposure scenario, the fate of each bird was determined by converting estimated maximum retained dose to a standard normal Z score from the appropriate dose-response curve and comparing that value to a randomly drawn value from a uniform distribution with a range of 0–1 (see Sect. 3.2). This process was repeated for 20 individuals of each species on each of 1,000 fields. Results were then expressed as a risk curve indicating the percentage of fields that had 5% mortality (1/20 birds died), 10% mortality (2/20 birds died), 15% mortality (3/20 birds died), etc. The Ecological Committee on FIFRA Risk Assessment Methods (ECOFRAM 1999) referred to such plots as "joint probability curves" while others refer to these plots as "risk curves" (e.g., Giddings et al. 2005; Moore et al. 2010a, b). For chronic exposure scenarios, risk was characterized by determining the probabilities that exposure exceeded the NOEL and LOEL for the most sensitive species tested.

The dose-response curve used to estimate acute risk depended on the focal species. If a dose-response curve was available for the focal species of interest, that curve was used (i.e., northern bobwhite (*C. virginianus*), red-winged blackbird (*A. phoeniceus*)). In the absence of species-specific dose-response curves, acute dose-response curves were generated for three hypothetical species representing a range of sensitivities (Fig. 7).

6.1 Modeled Acute Risks from Flowable CPY

The modeling for flowable CPY indicated that, with one exception, all bird species were at low or *de minimis* risk if they had median or lesser sensitivity to flowable CPY applied to alfalfa, almond, apple/cherry, broccoli, corn, grape, grapefruit, orange, pecan, soybean or sweet corn at the maximum application rates and minimum intervals specified on the label (Table 5). Assuming high sensitivity (5th centile on the species sensitivity distribution) to flowable CPY, or using actual dose-response relationships indicated that several species, particularly those that forage extensively in crop fields such as the horned lark (*Eremophila alpestris*), blue grosbeak (*Passerina caerulea*), and red-winged blackbird (*A. phoeniceus*) are at intermediate or high risk in some crops (e.g., grape, grapefruit, orange) if treated at the maximum application rates and minimum treatment intervals.

6.2 Modeled Chronic Risks from Flowable CPY

For all patterns of use, the probability of birds having a total daily intake exceeding the LOEL was <2% (Table 6). For most patterns of use and bird species, the probability of exceeding the NOEL was also small (<5%). However, several species and crop combinations (e.g., vesper sparrow (*Pooecetes gramineus*) in alfalfa, red-winged blackbird (*A. phoeniceus*) in orange) had probabilities of exceeding the NOEL of approximately 20%. The latter scenarios generally involved bird species that forage frequently in treated fields and crops with high maximum application rates (orange) or number of applications (alfalfa). In general, CPY poses little risk to birds from chronic exposure.

6.3 Results of Field Studies for Flowable CPY

Corn. Studies were performed with Lorsban 4E (a flowable formulation) on corn fields in Warren and Madison counties, Iowa (Frey et al. 1994). Lorsban 4E was applied at 3.36 kg ha^{-1} (3 lb ai A^{-1}) during the pre-plant stage (ground broadcast), and at 1.7 kg ha^{-1} (1.5 lb ai A^{-1}) during the emergence (ground broadcast), whorl (aerial broadcast) and tassel (aerial broadcast) stages. Monitoring of fields for birds exhibiting signs of toxicity was done prior to each application and for 13-d following each application, including abundance determinations, carcass search efficiency evaluations, and residue analyses.

Following pre-plant and at-plant applications, collection of moribund birds did not reveal differences among the treated and control fields and invertebrates collected during this period did not have detectable CPY residues. Applications during the emergence test period caused no statistically significant differences in the numbers of dead birds found, but the casualty rate was higher in treated fields (0.14 casualties per search) than in control fields (0.04 casualties per search).

During the whorl test period, bird censuses and mortality rates were similar in control and treated fields. Following tassel stage application, bird censuses did not reveal any differences in mortality among fields. Two robins (*Turdus migratorius*) collected from the fields treated with flowable CPY exhibited signs of toxicity consistent with inhibition of cholinesterase activity. One bird died, while the other bird recovered and was released.

Table 5 Acute risk results for birds exposed to flowable chlorpyrifos

			Birds protected (%)				Risk category			
Crop	Use pattern	Species[a]	5th	50th	95th	Actual	5th	50th	95th	Actual
Alfalfa	Southern plains—1.05 kg ha^{-1} (0.94 lb A^{-1}) applied broadcast 4× with 10-d interval	Dickcissel	79.6	98.8	100	NA	Intermediate	*De minimis*	*De minimis*	NA
		Mourning dove	100	100	100	NA	*De minimis*	*De minimis*	*De minimis*	NA
		Red-winged blackbird	NA	NA	NA	93.0	NA	NA	NA	Low
		Vesper sparrow	64.6	97.0	100	NA	Intermediate	*De minimis*	*De minimis*	NA
		Western meadowlark	95.5	99.8	100	NA	Low	*De minimis*	*De minimis*	NA
Almond	California—2.11 kg ha^{-1} (1.88 lb A^{-1}) applied airblast post-plant 2× (May, hull-split) with 10-d interval	Abert's towhee	100	100	100	NA	*De minimis*	*De minimis*	*De minimis*	NA
		Blue grosbeak	62.3	92.5	99.7	NA	High	Low	*De minimis*	NA
		Common crow	100	100	100	NA	*De minimis*	*De minimis*	*De minimis*	NA
		Mourning dove	100	100	100	NA	*De minimis*	*De minimis*	*De minimis*	NA
		Red-winged blackbird	NA	NA	NA	82.8	NA	NA	NA	Intermediate
Apple/Cherry	Northwest–California—2.11 kg ha^{-1} (1.88 lb A^{-1}) applied airblast during dormant season/ Michigan—2.11 kg ha^{-1} (1.88 lb A^{-1}) applied broadcast post-plant	Blue grosbeak	73.7	95.8	99.9	NA	Intermediate	Low	*De minimis*	NA
		Common crow	100	100	100	NA	*De minimis*	*De minimis*	*De minimis*	NA
		Mourning dove	100	100	100	NA	*De minimis*	*De minimis*	*De minimis*	NA
		Red-winged blackbird	NA	NA	NA	90.5	NA	NA	NA	Intermediate
Broccoli	California–California—2.36 kg ha^{-1} (2.11 lb A^{-1}) applied band at-plant and post-plant 3× with 10-d interval	Common crow	100	100	100	NA	*De minimis*	*De minimis*	*De minimis*	NA
		Horned lark	99.4	100	100	NA	*De minimis*	*De minimis*	*De minimis*	NA
		Mourning dove	100	100	100	NA	*De minimis*	*De minimis*	*De minimis*	NA
		Red-winged blackbird	NA	NA	NA	100	NA	NA	NA	*De minimis*

Corn	Midwest—1.12 kg ha^{-1} (1 lb A^{-1}) applied broadcast at-plant and band post-plant 3× with 10-d interval	American robin	99.6	100	100	NA	*De minimis*	*De minimis*	*De minimis*	NA
		Horned lark	69.5	97.3	100	NA	Intermediate	Low	*De minimis*	NA
		Killdeer	76.0	98.9	100	NA	Intermediate	*De minimis*	*De minimis*	NA
		Mourning dove	99.8	100	100	NA	*De minimis*	*De minimis*	*De minimis*	NA
		Northern bobwhite	NA	NA	NA	100	NA	NA	NA	*De minimis*
		Red-winged blackbird	NA	NA	NA	90.1	NA	NA	NA	Intermediate
		Vesper sparrow	74.5	98.4	100	NA	Intermediate	Low	*De minimis*	NA
Grape	California–California—2.11 kg ha^{-1} (1.88 lb A^{-1}) applied airblast prior to bud break	Blue grosbeak	69.1	95.7	99.9	NA	High	Low	*De minimis*	NA
		Common crow	100	100	100	NA	*De minimis*	*De minimis*	*De minimis*	NA
		Mourning dove	99.6	100	100	NA	*De minimis*	*De minimis*	*De minimis*	NA
		Northern bobwhite	NA	NA	NA	100	NA	NA	NA	*De minimis*
		Red-winged blackbird	NA	NA	NA	84.7	NA	NA	NA	Intermediate
Grapefruit	Florida—2.80 kg ha^{-1} (2.5 lb A^{-1}) applied airblast post-plant (Apr–Jun)	Blue grosbeak	66.1	93.4	99.7	NA	High	Low	*De minimis*	NA
		Common crow	99.9	100	100	NA	*De minimis*	*De minimis*	*De minimis*	NA
		Mourning dove	100	100	100	NA	*De minimis*	*De minimis*	*De minimis*	NA
		Northern bobwhite	NA	NA	NA	100	NA	NA	NA	*De minimis*
		Red-winged blackbird	NA	NA	NA	84.2	NA	NA	NA	Intermediate

(continued)

Table 5 (continued)

Crop	Use pattern	Species[a]	Birds protected (%)				Risk category			
			5th	50th	95th	Actual	5th	50th	95th	Actual
Orange	California—6.28 kg ha^{-1} (5.6 lb A^{-1}) applied airblast post-plant (May–Aug)	Abert's towhee	88.9	99.8	100	NA	Intermediate	*De minimis*	*De minimis*	NA
		Blue grosbeak	45.4	79.7	97.5	NA	High	Intermediate	Low	NA
		Common crow	99.2	100	100	NA	*De minimis*	*De minimis*	*De minimis*	NA
		Mourning dove	100	100	100	NA	*De minimis*	*De minimis*	*De minimis*	NA
		Red-winged blackbird	NA	NA	NA	44.0	NA	NA	NA	High
Pecan	Georgia—1.05 kg ha^{-1} (0.94 lb A^{-1}) applied airblast 3× with 10-d interval	Blue grosbeak	79.4	98.1	100	NA	Intermediate	Low	*De minimis*	NA
		Common crow	100	100	100	NA	*De minimis*	*De minimis*	*De minimis*	NA
		Mourning dove	100	100	100	NA	*De minimis*	*De minimis*	*De minimis*	NA
		Northern bobwhite	NA	NA	NA	100	NA	NA	NA	*De minimis*
		Red-winged blackbird	NA	NA	NA	94.4	NA	NA	NA	Low
Soybean	Louisiana—1.05 kg ha^{-1} (0.94 lb A^{-1}) applied broadcast post-plant 3× with 14-d interval (May–Aug)	Blue grosbeak	78.2	98.7	100	NA	Intermediate	Low	*De minimis*	NA
		Dickcissel	99.4	100	100	NA	*De minimis*	*De minimis*	*De minimis*	NA
		Horned lark	76.6	98.3	100	NA	Intermediate	Low	*De minimis*	NA
		Indigo bunting	99.7	100	100	NA	*De minimis*	*De minimis*	*De minimis*	NA
		Mourning dove	99.9	100	100	NA	*De minimis*	*De minimis*	*De minimis*	NA
		Red-winged blackbird	NA	NA	NA	93.42	NA	NA	NA	Low
Sweet corn	Florida—1.12 kg ha^{-1} (1 lb A^{-1}) applied broadcast at-plant and band post-plant 3× with 10-d interval	Common crow	100	100	100	NA	*De minimis*	*De minimis*	*De minimis*	NA
		Mourning dove	99.9	100	100	NA	*De minimis*	*De minimis*	*De minimis*	NA
		Northern bobwhite	NA	NA	NA	100	NA	NA	NA	*De minimis*
		Red-winged blackbird	NA	NA	NA	89.8	NA	NA	NA	Intermediate

For untested species, risk estimates are shown assuming high (5th centile), median (50th centile) and low sensitivity (95th centile). For tested species, results are based on actual dose-response curves for those species

NA = not applicable

[a]See Table 1 for scientific names

Table 6 Chronic risk results for birds exposed to flowable chlorpyrifos

Crop	Use pattern	Species[a]	Exceedence probability (%)	
			NOEL	LOEL
Alfalfa	Southern plains—1.05 kg ha^{-1} (0.94 lb A^{-1}) applied broadcast 4× with 10-d interval	Dickcissel	0.29	0
		Mourning dove	0	0
		Red-winged blackbird	0.61	0
		Vesper sparrow	18.6	0
		Western meadowlark	0.35	0
Almond	California—2.11 kg ha^{-1} (1.88 lb A^{-1}) applied airblast post-plant 2× (May, hull-split) with 10-d interval	Abert's towhee	0.01	0
		Blue grosbeak	18.7	0.09
		Common crow	0	0
		Mourning dove	0	0
		Red-winged blackbird	9.71	0.01
Apple/ Cherry	Northwest—2.11 kg ha^{-1} (1.88 lb A^{-1}) applied airblast during dormant season/ Michigan—2.07 kg ha^{-1} (1.88 lb A^{-1}) applied broadcast post-plant	Blue grosbeak	7.94	0.01
		Common crow	0	0
		Mourning dove	0	0
		Red-winged blackbird	1.99	0
Broccoli	California—2.36 kg ha^{-1} (2.11 lb A^{-1}) applied band at-plant and post-plant 3× with 10-d interval	Common crow	2.15	0
		Horned lark	1.27	0
		Mourning dove	0.02	0
		Red-winged blackbird	0.32	0
Corn	Midwest—1.12 kg ha^{-1} (1 lb A^{-1}) applied broadcast at-plant and band post-plant 3× with 10-d interval	American robin	1.63	0
		Horned lark	1.89	0
		Killdeer	0.12	0
		Mourning dove	0	0
		Northern bobwhite	0	0
		Red-winged blackbird	0.25	0
		Vesper sparrow	0.43	0
Grape	California—2.11 kg ha^{-1} (1.88 lb A^{-1}) applied airblast prior to bud break	Blue grosbeak	4.86	0
		Common crow	0.39	0
		Mourning dove	0	0
		Northern bobwhite	0	0
		Red-winged blackbird	3.53	0
Grapefruit	Florida—2.80 kg ha^{-1} (2.5 lb A^{-1}) applied airblast post-plant (Apr–Jun)	Blue grosbeak	10.3	0.11
		Common crow	0	0
		Mourning dove	0	0
		Northern bobwhite	0	0
		Red-winged blackbird	5.49	0
Orange	California—6.27 kg ha^{-1} (5.6 lb A^{-1}) applied airblast post-plant (May–Aug)	Abert's towhee	0.08	0
		Blue grosbeak	28.1	1.12
		Common crow	0	0
		Mourning dove	0	0
		Red-winged blackbird	21.0	0.8
Pecan	Georgia—1.05 kg ha^{-1} (0.94 lb A^{-1}) applied airblast 3× with 10-d interval	Blue grosbeak	11.8	0.02
		Common crow	0	0
		Mourning dove	0	0
		Northern bobwhite	0	0
		Red-winged blackbird	4.58	0

(continued)

Table 6 (continued)

Crop	Use pattern	Species[a]	Exceedence probability (%) NOEL	LOEL
Soybean	Louisiana—1.05 kg ha^{-1} (0.94 lb A^{-1}) applied broadcast post-plant 3× with 14-d interval (May–Aug)	Blue grosbeak	0.07	0
		Dickcissel	0.07	0
		Horned lark	4.77	0
		Indigo bunting	5.74	0
		Mourning dove	0	0
		Red-winged blackbird	0.22	0
Sweet corn	Florida—1.12 kg ha^{-1} (1 lb A^{-1}) applied broadcast at-plant and band post-plant 3× with 10-d interval	Common crow	0.37	0
		Mourning dove	0	0
		Northern bobwhite	0	0
		Red-winged blackbird	0.66	0

NA = not applicable
[a]See Table 1 for scientific names

Overall, flowable CPY had minimal effects on birds in treated corn fields (1994). This result occurred in spite of far greater rates of application being used in the field study (3.36 kg ha^{-1} (3 lb ai A^{-1}) during the pre-plant stage, 1.7 kg ha^{-1} (1.5 lb ai A^{-1}) applied broadcast postplant 3×) than currently allowed on the Lorsban Advanced label for corn (1.12 kg ha^{-1} (1 lb ai A^{-1}) applied broadcast at-plant and band post-plant 3× with 10-d interval) The results of the field study suggest that the LiquidPARAM modeling exercise overestimated risks to birds, particularly for the horned lark (*E. alpestris*) and killdeer (*Charadrius vociferous*) (assuming high sensitivity) and the red-winged blackbird (*A. phoeniceus*) (Table 5).

Brassica. Three cabbage fields in central Poland, were chosen to study the effects of CPY application on associated bird communities (Moosmayer and Wilkens 2008). Dursban® 480 EC, a flowable formulation of CPY, was applied twice at a rate of 0.95 kg ha^{-1} (0.85 lb A^{-1}) with an application interval of 14-d. Visual searches for carcasses, monitoring of nests and radio-tracking were used to estimate adverse effects to wildlife. No signs of toxicity were observed during the visual searches or monitoring of nests and no carcasses were recovered from the treated fields. Fifty-three birds were caught, radio-tagged, and tracked over the treatment period. None of the radio-tagged birds experienced adverse effects related to application of CPY.

Although there were no significant effects to birds in the brassica field study, it was not possible to determine whether predictions of LiquidPARAM of little or no risk to birds (Table 5) were reasonable because the application rates in the field study were less than half the rate used in the modeling exercise.

Citrus. Effects of Lorsban 4E applied in California citrus groves to birds were determined by Gallagher et al. (1994). Two application scenarios were included in the study: (1) 1.65 kg ha^{-1} (1.5 lb ai A^{-1}) Lorsban 4E applied post bloom and 6.62 kg ha^{-1} (6 lb ai A^{-1}) after petal-fall and (2) 3.86 kg ha^{-1} (3.5 lb ai A^{-1}) applied post bloom and 4.4 kg ha^{-1} (4 lb ai A^{-1}) post petal fall. The post-bloom applications were made in early spring and the post petal fall applications were made in late spring. The first and second applications were 30–35-d apart.

Casualties among birds residing on fields following post-bloom application (1.65 kg ha^{-1} (1.5 lb ai A^{-1})) from the first application scenario could not be linked to CPY, because no detectable residues of the parent compound were found in the dead birds. There were no bird casualties from the second post-bloom application scenario.

Following applications after petal fall, one dead northern mockingbird (*Mimus polyglottos*) was found on the fields from the first application scenario (6.62 kg ha^{-1} (6 lb ai A^{-1})). The dead bird had measurable residues of CPY on its feathers (5.39 mg ai kg^{-1} bwt). In the second application scenario (4.4 kg ha^{-1} (4 lb ai A^{-1})), concentrations of 3.67 mg ai kg^{-1} bwt CPY were measured in a dead passerine nestling. It is not known if the deaths were treatment related.

In a similar study, three citrus orchards in Spain were spayed twice at a rate of 2.32 kg ha^{-1} (2.1 lb ai A^{-1}) of Dursban® 75 WG with a 14-d re-treatment interval (Selbach and Wilkens 2008a). Birds were captured before each spray application, radio-tagged and released. Birds were then tracked for 3-d before and 7-d after each application. Monitoring of the activities of radio-tagged birds, monitoring of nests, searches for carcasses, and surveys of masses of arthropod biomass were also used to quantify possible adverse effects to birds from application of Dursban. Of the 38 birds tracked during the study, 6 were continuously tracked through both application periods. The tracked birds spent approximately one third of their time in the treated orchards before and after application. Of the 3,751 sightings of birds made during the observation periods, no birds showed signs of toxicity. Three bird carcasses were found. Chlorpyrifos residues of 14 mg ai kg^{-1} bwt were found in the skin and feather matrix of a blackbird and 1.2 mg ai kg^{-1} bwt was detected in the core body matrix. A blackbird (unknown species name) wing was found that contained 6.5 mg CPY ai kg^{-1} bwt in the skin and feathers. Lastly, a dead house martin (*Delichon urbicum*) was found that contained 0.33 mg CPY ai kg^{-1} bwt in the skin and feathers. No CPY was detected in the core body matrix of the house martin or wing of the blackbird. No inhibition of AChE activity in the brain of the house martin was observed. The authors determined that none of the casualties resulted from the CPY application.

Another study was conducted in citrus orchards in Spain to determine the effects of CPY on bird communities and reproductive performance (Dittrich and Staedtler 2010). Observations of communities of birds were made at the end of the main breeding season (July 6 to August 31, 2010) on ten citrus orchards that routinely use CPY to control arthropods (application period: April 1 to June 30, 2010). No additional details on use patterns and application rates were provided. A large diversity and number of birds was observed in the study area. No losses of nests could be attributed to CPY. The bird species most frequently observed were serin, green finch (*Carduelis chloris*), and house sparrow (*Passer domesticus*), while the juveniles most frequently observed were barn swallow (*Hirundo rustica*), nightingale (*Luscinia megarhynchos*), and Sardinian warbler (*Sylvia melanocephala*). Sampling of arthropods following application indicated an abundance of avian food items.

Overall, the bird community residing within the treated citrus fields was considered to be highly viable.

The results of the 3 field studies in citrus orchards indicated that flowable CPY applied at rates comparable to the maximum rate allowed by the label for Lorsban Advanced (i.e., 6.17 kg ha^{-1} (5.6 lb ai A^{-1}) for use in oranges) had no significant adverse effects on birds. As with the corn field studies, the citrus field studies indicate that LiquidPARAM may be overestimating avian risks in citrus orchards (Table 5).

Apple. Dursban 75 WG was applied to three apple orchards at a rate of 0.95 kg ha^{-1} (0.86 lb ai A^{-1}) (Wilkens et al. 2008). Three applications were made to the first orchard with the first and second applications being 14-d apart and the second and third being 28-d apart. Two applications were made to the other two orchards, 14-d apart. Telemetric surveys, visual bird observations, carcass searches, and nest observations were used to quantify the effects of CPY. Radio-tagged birds were tracked for 3-d prior to applications and for 7-d following applications. Birds spent approximately half of their time in the study plots. No tracked birds exhibited signs of toxicity. A total of 3,616 bird observations were made during the study period and no birds exhibited any behavioral abnormalities or signs of toxicity. Only one dead bird was found during the study period. The authors concluded that this death resulted from a collision with a power transmission line. However, the applications did reduce populations of foliage-dwelling pest and non-target arthropods by approximately 87%. There were no significant effects to birds in the apple field study, because the application rates in the field study were approximately half that used in the modeling exercise. Therefore, it cannot be determined whether LiquidPARAM overestimated risks to the blue grosbeak (Table 5) or not.

Grape. Dursban 480 EC was applied twice at a rate of 0.36 kg ha^{-1} (0.32 lb ai A^{-1}) with a 15-d interval to a vineyard in Puy du Maupas, near Puymeras, Vaucluse, Southern France (Brown et al. 2007). The vineyard consisted of eight adjacent fields, with grass growing between the planted rows. The property also contained scrub, woodland, garden, and grassy areas. The area was searched for carcasses of birds prior to each application of CPY and 1, 3 and 7-d following each application. Three to 4-d prior to each application, mist nests were placed in the vineyard and along the boundaries. Collected birds were banded, sexed, measured, and radio-tagged. Tagged birds were tracked for several days prior to treatment and for up to 10-d following treatment. The locations of birds were used to estimate the proportion of time spent on the treated fields and to determine if the birds were alive.

Monitoring of the radio-tagged birds indicated that birds spent a maximum of 20% of their time on the treated fields. Only Cirl buntings (*Emberiza cirlus*), black redstarts (*Phoenicurus ochruros*), stonechats (*Saxicola* sp.), and jays (unknown species name) spent more than 5% of their time there. Birds on the treated crop for the greatest proportion of time were alive at the end of the tracking period. Only one radio-tagged bird was found dead during the monitoring period and, because only a leg was found, it is unlikely that mortality was the result of CPY. Untagged birds found dead during the study had residues of CPY on skin and feather residue levels that were consistent with contact with the treated crop (0.27–1.3 mg ai kg^{-1} bwt). Analysis of AChE activity in the brain of a dead robin (*Erithacus rubecula*) showed no decrease in activity, indicating that mortality was not likely the result of exposure to CPY. There were no indications of short-term negative impacts from CPY on birds in the vineyard during the study. There were no significant effects to birds in the grape field study, however the application rate in this field study was well below that used in the modeling exercise. Therefore, it was not possible to determine whether LiquidPARAM overestimated risks to the blue grosbeak (*P. caerulea*) (Table 5).

Telemetry-based field studies. Brassica, pome fruit and citrus crops were treated with CPY to determine potential effects on wild birds (Wolf et al. 2010). Brassica fields were located near Sochaczew, Poland, pome fields near Belfiore, northern Italy, and citrus groves in Valencia, Spain. Four or five sites were used for each crop type and fields averaged 4 ha (9.9 A) in size. Chlorpyrifos was applied to brassicas using a tractor-mounted boom sprayer at a rate of 0.95 kg ha^{-1} (0.86 lb ai A^{-1}). Three brassica sites received two applications of Dursban 480 EC and two other sites received an application of a formulation not relevant to this assessment (Pyrinex® 25 CS, a microencapsulated formulation). Chlorpyrifos was applied to pome and citrus fruit crops using a tractor-mounted broadcast air-assisted sprayer. Three citrus fields received two applications of Dursban 75 WG at a rate of 2.32 kg ha^{-1} (2.1 lb ai A^{-1}), and the remaining field received two applications of Pyrinex 25 CS. One field of pome fruit received applications of Pyrinex 25 CS and one pome fruit field received three applications of 0.95 kg ha^{-1} (0.86 lb ai A^{-1}) Dursban 75 WG, whereas the other two plots received two applications of 0.95 kg ha^{-1} (0.86 lb ai A^{-1}) Dursban 75 WG. All bird species regularly foraging in the crops were monitored during the study.

Birds were trapped and radio-tagged before each application and tracked for 7-d following each application. Those tagged for earlier applications were monitored during subsequent applications if the radio-tags were still functional. Of the 242 radio-tagged birds, 194 were tracked for the full 7-d period following application. No signs of toxicity or lethality were observed. Un-tagged birds were also observed during the study period. No signs of toxicity were detected. Ten bird carcasses were found during the study, six of which had detectable levels of CPY. Detectable concentrations of CPY on skin and feathers ranged from 0.3 to 14.0 mg ai kg^{-1} bwt. CPY was only detected in the bodies of two birds (1.2 and 0.3 mg ai kg^{-1} bwt). Similarly, core body concentrations of CPY were only detected in two birds at levels of 0.1 and 1.2 mg ai kg^{-1} bwt.

Rates of application used in the pome (e.g., apple) and brassica plots were less than the maximum rates listed on the Lorsban Advanced label for those crops. The rate of application used in the citrus plots (2.32 kg ha^{-1} (2.1 lb ai A^{-1})) was similar to the maximum rate of application for grapefruit on the Lorsban Advanced label (i.e., 2.76 kg ha^{-1} (2.5 lb ai A^{-1})). LiquidPARAM predicted approximately 34% mortality to blue grosbeaks (*P. caerulea*) in grapefruit treated at the maximum rate of application, assuming that this species was highly sensitive (Table 5). All other bird species were predicted to experience little to no mortality. The results for citrus groves indicate that LiquidPARAM might be over-estimating risk to blue grosbeaks (Wolf et al. 2010).

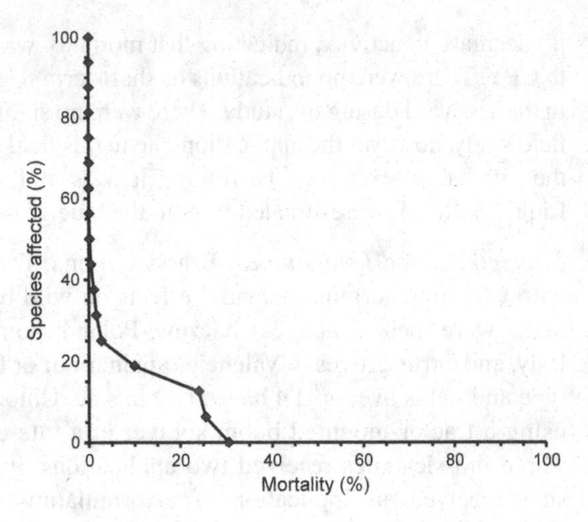

Fig. 8 Percentage of bird species affected versus percent mortality for flowable chlorpyrifos applied broadcast at-plant and band post-plant at a rate of 1.12 kg ha^{-1} (1 lb ai A^{-1}) 3× with a 10-d interval to corn fields

6.4 Discussion of Avian Risks for Flowable CPY

There are a number of bird species that frequent agroecosystems besides those included as focal species in this assessment (see Best and Murray 2003). The USEPA (USEPA 2005) used estimates of mortality for the combination of focal species and bird sensitivities in each modeled exposure scenario to approximate the cumulative distribution of outcomes for the complex of species using treated fields. The approach assumes that the focal species included in the modeling exercise are representative of the birds and their exposures occurring in the fields (USEPA 2005). This assumption is supported by the selection of focal species known to occur on the treated crops by actual survey (e.g., Best and Murray 2003). According to the USEPA (2005), the outcomes of the three modeled sensitivity assumptions (low, median and high sensitivity) "can be viewed as a stratified sample from the population which estimates the limits and mid points of the cumulative risk distribution and therefore provides a reasonable approximation of the distribution."

The resulting cumulative distribution of acute risk for banded application on corn at the maximum application rate of 1.12 kg ha^{-1} (1 lb ai A^{-1}) is shown in Fig. 8. Results of simulations using LiquidPARAM indicate that several species of birds, if highly sensitive, would experience up to approximately 30% mortality. Similar results were predicted for alfalfa, almond, apple/cherry, grape, grapefruit and soybean (Table 5). For orange, somewhat greater risk is expected in the bird community because this crop has the greatest application rate allowed on the Lorsban Advanced label (i.e., 6.27 kg ha^{-1} (5.6 lb ai A^{-1})) (Fig. 9).

Although the results of the LiquidPARAM modeling indicated some acute risk to the most sensitive species for several crops listed on the Lorsban Advanced label (Table 5), the evidence from field studies that used the corresponding application rates (i.e., corn, grapefruit, and orange) indicate that flowable CPY poses little risk to birds (Dittrich and Staedtler 2010; Frey et al. 1994; Gallagher et al. 1994; Selbach and Wilkens 2008b; Wolf et al. 2010). Thus, it would appear that LiquidPARAM

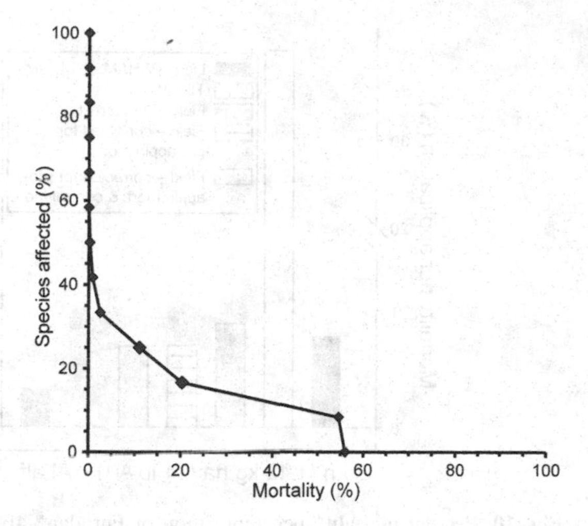

Fig. 9 Percentage of bird species affected versus percent mortality for flowable chlorpyrifos applied airblast to orange orchards at a rate of 6.28 kg ha^{-1} (5.6 lb A^{-1})

may be overestimating the risk CPY poses to birds, particularly in those crops with the highest application rates (i.e., grapefruit and orange). There are several potential reasons why LiquidPARAM may be overestimating acute risks of CPY to birds. For example, acute effects metrics were based on single-dose, oral gavage studies that likely overestimate the toxicity that birds would experience when consuming small amounts over the course of a day, as typically occurs in treated fields (see Sect. 5.3). In addition, the exposure model assumed that proportion time in treated fields equates to proportion diet obtained from treated fields. However, it may be that many bird species obtain a relatively higher proportion of their diet from higher quality edge habitats.

The results of field studies consistently demonstrated that flowable CPY has negligible effects on birds at rates well above the application rate of CPY predicted by Mineau (2002) (i.e., 0.19–0.26 kg ha^{-1} (0.17–0.23 lb ai A^{-1})) to have a 1/10 probability of an avian kill.

6.5 *Strengths of the Refined Risk Assessment for Flowable CPY*

LiquidPARAM explicitly accounts for factors affecting exposure of birds to flowable CPY in the field. These factors include: application rates, number and types of applications, foraging patterns, preferred diets, CPY concentrations on dietary items over time and space, rates of metabolism, and avoidance behavior.

In several instances, LiquidPARAM refined the approach used by EPA's TIM (USEPA 2005). For proportion of time foraging in treated fields, TIM uses data that represent inter-field variability as intra-field variability. In estimating food intake rate, TIM uses distributions for several minor input variables (e.g., gross energy and assimilation efficiency of dietary items), but treats the input variable with the greatest uncertainty (i.e., free metabolic rate, the amount of calories consumed by free-living

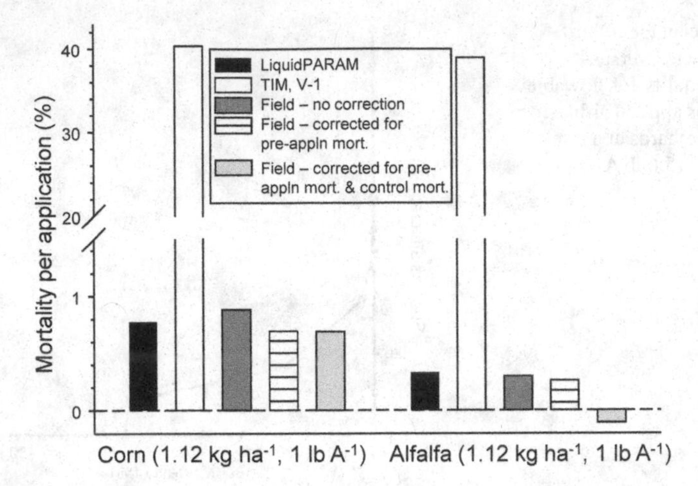

Fig. 10 Percent mortality per application of Furadan® 4F in corn and alfalfa as estimated by LiquidPARAM, TIM (v1), and observed in field studies by Booth et al. (1989) and Jorgensen et al. (1989)

birds per time step) as a point estimate. Free metabolic rate is treated as a distribution in LiquidPARAM.

LiquidPARAM has several capabilities not available in TIM. These include: (1) The ability to model exposure scenarios involving multiple applications of pesticide taking place at a user-specified interval; (2) Addition of many new crops; (3) Addition of ten new focal bird species; and (4) The ability of users to select day length, time of application for first, second, and third applications, and length of time that dew is present on treated fields.

As described in SI Appendix 3, Sect. 1.4, sensitivity analysis and evaluation of model performance has been undertaken for LiquidPARAM. The sensitivity analysis was useful in determining which variables had an important influence on acute and chronic risk for bird species in high exposure scenarios (e.g., choice of drinking water scenario and chronic averaging period in the chronic modeling simulations) and in low exposure scenarios (varying any one factor had little effect on estimated acute or chronic risk to birds regardless of assumed sensitivity).

In addition to the sensitivity analysis, the model for LiquidPARAM has been evaluated. Field studies involving the application of flowable carbofuran, a carbamate pesticide that inhibits brain and plasma acetylcholinesterase activity, and subsequent determination of avian mortality were reviewed to determine those that could be used to evaluate LiquidPARAM performance. Each of the selected studies (Booth et al. 1989; Jorgensen et al. 1989) reported mortality from applications of flowable carbofuran. The exposure scenarios for the selected studies were run in LiquidPARAM to determine how close model predictions were to field observations. For the two field studies selected, LiquidPARAM predictions and field observations of mortality were similar, with LiquidPARAM slightly over-predicting risk. Conversely, EPA's TIM v1 vastly over-estimated risk of carbofuran compared to field observations (Fig. 10). Although LiquidPARAM model predictions and field study results were fairly close for carbofuran, our analyses described herein indicate that LiquidPARAM may be more substantially over-predicting risk of flowable CPY. The small number of incidents (2) involving CPY reported since 2002 suggests that the current labels for CPY are generally protective of birds (SI Appendix 4).

6.6 *Uncertainties of the Refined Risk Assessment for Flowable CPY*

This refined assessment of acute and chronic risks of CPY to birds contains uncertainties. Uncertainties in the problem formulation and assessment of exposure and effects can influence the characterization of risks. It is therefore important to identify the sources of uncertainty in the assessment, and specify the magnitude and direction of their influence.

The following sources of uncertainty were identified in this refined risk assessment for birds (Table 7):

- The refined risk assessment considered exposure of birds to CPY via ingestion of food and water. As discussed in Sect. 2.4, dermal contact, inhalation and preening are unlikely to be important exposure routes for birds in fields treated with flowable CPY. At present, refined models are lacking to quantify these exposure routes in birds.
- The refined risk assessment considered exposure to 15 focal species. Thus, there is a possibility that bird species not considered in this assessment are at risk on or near CPY-treated fields. The focal species were selected because of their affinity for agricultural areas and the crops considered in this assessment. This group of species is more likely to be exposed to flowable CPY than would most other bird species. Furthermore, they span a range of sizes and taxonomic groups, and are representative of bird species found in regions where CPY is used. However, it is conceivable that there are bird species at greater risk to flowable CPY than those included for the 11 crops considered in this assessment.
- When there was uncertainty, these sources were quantified and incorporated in the exposure analyses (e.g., free metabolic rate, initial dietary residue levels following application). Thus, these sources of uncertainty have been explicitly accounted for in the risk estimates described here. Other sources of uncertainty, however, could not be fully accounted for in LiquidPARAM, generally because data were too scarce to reliably parameterize distributions. For example, acute dose-response curves were unavailable for all focal species except the northern bobwhite (*C. virginianus*) and red-winged blackbird (*A. phoeniceus*). The general approach for input variables for which values were uncertain was to use conservative point estimates or rely on surrogate approaches (e.g., the species sensitivity distribution approach to estimate dose-response curves for species of differing sensitivities). The model evaluation exercise indicated that model predictions reasonably replicated patterns of mortality observed in field studies conducted

Table 7 Sources of uncertainty in the avian risk assessment for flowable chlorpyrifos

Area	Source of uncertainty	Action and influence on risk estimates
Exposure scenarios	Not possible to assess all scenarios. Potential to miss high risk scenarios.	High-use crops were assessed assuming maximum application rates and minimum re-treatment intervals. Exposure scenarios likely included upper bound risks posed by CPY to birds.
Routes of exposure	Focus of assessment on ingestion of food and water. Could be other important routes of exposure.	Dermal contact, inhalation and preening unlikely to be important based on results of limited studies. Refined models are lacking for these routes of exposure. Evaluations of model performance against field studies indicated that LiquidPARAM is closely or over-estimating risk to birds. Thus, it is unlikely that routes of exposure other than ingestion of food and water are highly significant.
Risk to non-focal bird species	The refined risk assessment considered exposure to 15 focal species. Thus, there is a possibility that bird species not considered in this assessment are at risk on or near CPY-treated fields.	The focal species were selected because of their affinity for agricultural areas where flowable CPY is used. This group of species is more likely to be exposed to CPY than would most other bird species. Furthermore, they span a range of sizes and taxonomic groups. Thus, there is a low degree of uncertainty associated with overlooking bird species at risk. Additionally, the focal species likely bracket all of the bird species potentially at risk due to CPY application.
Proportion time birds on fields	Information is lacking to fully characterize between-field, between-individual and between-time steps variability in proportion of time birds forage on treated fields. Also assumed that the proportion of time birds spend on treated fields equates to the proportion of their diet that comes from treated fields.	Original data for proportion of time birds spend in treated fields (*PT*) were obtained for corn and alfalfa and analyzed to determine between-field variability for each of the focal species. Without radio-tracking data, however, it was not possible to quantify between-individual and between-time steps variability for any of the focal species. Given the state of knowledge for between-individuals variability, uncertainty was maximized by parameterizing the within-field *PT* distributions to range from 0 to 1, with the best estimate being the randomly chosen most likely *PT* from the between-fields distribution. No variation was assumed between time steps, a reasonable assumption for nesting birds, but not pre-migratory or migratory birds. No data exist to determine whether proportion of time birds spend on treated fields equates to the proportion of their diet that comes from treated fields. The various shortcomings with the *PT* variable could ultimately lead to under- or over-estimation of risk. The results of the evaluation of model performance and the flowable CPY field studies suggest that the risk estimates produced by LiquidPARAM are reasonable or over-estimated.
Quality and quantity of toxicity studies	Toxicity studies for CPY were conducted over several decades using a variety of protocols. Quantity of data is limited for bird species particularly for chronic studies. Gavage method used in acute oral studies does not replicate feeding patterns in the field.	Toxicity studies underwent data quality review. Only data of sufficient quality were used to derive dose-response curves, the species sensitivity distribution and chronic effects metrics. Additional chronic toxicity studies on species other than the mallard (*Anas platyrhynchos*) and northern bobwhite (*Colinus virginianus*) would reduce uncertainty in the chronic risk assessment.

for carbofuran in corn and alfalfa. Overall, it appears that the LiquidPARAM performed well for carbofuran, but may have over-predicted risk for flowable CPY, based on the results of field studies conducted with CPY.

7 Risk Characterization for Granular Chlorpyrifos

For each exposure scenario, acute risk was determined using the same approach as described in Sect. 6 for flowable chlorpyrifos.

As with flowable CPY, the dose-response curve used to estimate acute risk depended on the focal species. If a dose-response curve was available for the focal species of interest, that curve was used (i.e., northern bobwhite (*C. virginianus*), red-winged blackbird (*A. phoeniceus*)). In the absence of species-specific dose-response curves, acute dose-response curves were generated for three hypothetical species representing a range of sensitivities. In this section, the results from the GranPARAM modeling exercise are discussed. In addition, the results of avian field studies are discussed and compared to the results from the modeling exercise. The section concludes with a discussion of sources of uncertainty and strengths of the assessment for granular CPY.

7.1 Modeled Acute Risks from Granular CPY

Simulations conducted for granular CPY indicated that, with two exceptions, all bird species were at *de minimis* risk, even if they had high sensitivity to granular CPY (Table 8). The two exceptions were for horned lark in corn/sweet corn and tobacco, assuming that horned larks (*E. alpestris*) are highly sensitive to CPY. Horned larks forage more in row crops than do any other focal species considered in this assessment (SI Appendix 3, Sect. 1.2). In corn/sweet corn and tobacco, survival of horned larks was predicted to be >95% (Table 8).

7.2 Results of Field Studies for Granular CPY

Avian field studies were performed with Lorsban 15G on corn fields in Iowa (Frey et al. 1994). Lorsban 15G was applied at 2.87 kg ha^{-1} (2.6 lb ai A^{-1}) at-planting (ground banded), and at 1.07 kg ha^{-1} (0.975 lb ai A^{-1}) during the whorl and tassel stages (aerial broadcast). Monitoring of field sites for birds exhibiting signs of toxicity was done prior to each application and for 13-d following each application, including abundance determinations, carcass search efficiency evaluations, and residue analyses.

Table 8 Acute risk results for birds exposed to granular chlorpyrifos

Crop	Use pattern	Species[a]	Birds protected (%)				Risk category			
			5th	50th	95th	Actual	5th	50th	95th	Actual
Broccoli	California—2.52 kg ha^{-1} (2.25 lb A^{-1}) applied T-band at-plant (10 cm (4″) band, 75 cm (30″) row spacing)	Horned lark	99.6	100	100	NA	*De minimis*	*De minimis*	*De minimis*	NA
		Mourning dove	100	100	100	NA	*De minimis*	*De minimis*	*De minimis*	NA
		Red-winged blackbird	NA	NA	NA	100	NA	NA	NA	*De minimis*
Corn and sweet corn	Corn belt—1.46 kg ha^{-1} (1.3 lb A^{-1}) applied T-band (15 cm (6″) band, 91 cm (36″) row spacing)	Horned lark	99.5	100	100	NA	*De minimis*	*De minimis*	*De minimis*	NA
		Common pheasant	100	100	100	NA	*De minimis*	*De minimis*	*De minimis*	NA
		Mourning dove	100	100	100	NA	*De minimis*	*De minimis*	*De minimis*	NA
		Northern bobwhite	NA	NA	NA	100	NA	NA	NA	*De minimis*
		Red-winged blackbird	NA	NA	NA	99.9	NA	NA	NA	*De minimis*
Corn and sweet corn	Corn belt—1.46 kg ha^{-1} (1.3 lb A^{-1}) applied in-furrow at-plant (7.5 cm (3″) furrow, 91 cm (36″) row spacing)	Horned lark	99.6	100	100	NA	*De minimis*	*De minimis*	*De minimis*	NA
		Common pheasant	100	100	100	NA	*De minimis*	*De minimis*	*De minimis*	NA
		Mourning dove	100	100	100	NA	*De minimis*	*De minimis*	*De minimis*	NA
		Northern bobwhite	NA	NA	NA	100	NA	NA	NA	*De minimis*
		Red-winged blackbird	NA	NA	NA	100	NA	NA	NA	*De minimis*
Corn and sweet corn	Corn belt—1.12 kg ha^{-1} (1 lb A^{-1}) applied broadcast post-plant	Horned lark	95.9	98.9	99.9	NA	Low	*De minimis*	*De minimis*	NA
		Common pheasant	99.9	99.9	99.9	NA	*De minimis*	*De minimis*	*De minimis*	NA
		Mourning dove	99.9	99.9	99.9	NA	*De minimis*	*De minimis*	*De minimis*	NA
		Northern bobwhite	NA	NA	NA	100	NA	NA	NA	*De minimis*
		Red-winged blackbird	NA	NA	NA	100	NA	NA	NA	*De minimis*

Onion	Pacific northwest—1.12 kg ha^{-1} (1 lb A^{-1}) applied in-furrow at-plant (7.5 cm (3″) furrow, 45 cm (18″) row spacing)	Horned lark	99.9	100	100	NA	*De minimis*	*De minimis*	*De minimis*	NA
		Mourning dove	100	100	100	NA	*De minimis*	*De minimis*	*De minimis*	NA
		Red-winged blackbird	NA	NA	NA	100	NA	NA	NA	*De minimis*
Peanut	Southeast—4.48 kg ha^{-1} (4 lb A^{-1}) applied band post-plant (15 cm (6″) band, 91 cm (36″) row spacing)	Horned lark	99.0	100	100	NA	*De minimis*	*De minimis*	*De minimis*	NA
		Mourning dove	100	100	100	NA	*De minimis*	*De minimis*	*De minimis*	NA
		Northern bobwhite	NA	NA	NA	100	NA	NA	NA	*De minimis*
		Red-winged blackbird	NA	NA	NA	99.9	NA	NA	NA	*De minimis*
Sugar beet	Midwest—1.12 kg ha^{-1} (1 lb A^{-1}) applied broadcast at-plant	Horned lark	98.7	99.8	100	NA	*De minimis*	*De minimis*	*De minimis*	NA
		Common pheasant	100	100	100	NA	*De minimis*	*De minimis*	*De minimis*	NA
		Mourning dove	99.9	99.9	99.9	NA	*De minimis*	*De minimis*	*De minimis*	NA
		Northern bobwhite	NA	NA	NA	100	NA	NA	NA	*De minimis*
		Red-winged blackbird	NA	NA	NA	100	NA	NA	NA	*De minimis*
Sunflower	Midwest—1.46 kg ha^{-1} (1.3 lb A^{-1}) applied T-band at-plant (15 cm (6″) band, 75 cm (30″) row spacing)	Horned lark	99.7	100	100	NA	*De minimis*	*De minimis*	*De minimis*	NA
		Common pheasant	100	100	100	NA	*De minimis*	*De minimis*	*De minimis*	NA
		Mourning dove	100	100	100	NA	*De minimis*	*De minimis*	*De minimis*	NA
		Northern bobwhite	NA	NA	NA	100	NA	NA	NA	*De minimis*
		Red-winged blackbird	NA	NA	NA	100	NA	NA	NA	*De minimis*
Tobacco	Southeast—2.24 kg ha^{-1} (2 lb A^{-1}) applied broadcast preplant with incorporation	Horned lark	97.9	99.8	100	NA	Low	*De minimis*	*De minimis*	NA
		Mourning dove	99.9	99.9	100	NA	*De minimis*	*De minimis*	*De minimis*	NA
		Northern bobwhite	NA	NA	NA	100	NA	NA	NA	*De minimis*
		Red-winged blackbird	NA	NA	NA	100	NA	NA	NA	*De minimis*

For untested species, risk estimates are shown assuming high (5th centile), median (50th centile) and low sensitivity (95th centile). For tested species, results are based on actual dose-response curves for those species

NA = not applicable

[a]See Table 1 for scientific names

Following the at-plant and post-plant applications, there were no differences in avian mortality between treated and control fields. Remains of birds on granular treated fields were insufficient for residue analysis following application during the tassel stage.

In a similar study (Anderson et al. 1998), Lorsban 15G was applied in a T-band scenario at a rate of 1.34 kg ha^{-1} (1.2 lb ai A^{-1}) during planting of corn crops in Iowa. Nest boxes were erected for starlings (*Sturnus vulgaris*) on both the experimental and control sites. Occupancy of nest boxes was not affected by application of CPY to the experimental field, nor was reproduction. Analysis of 13-d old nestlings showed that there was no difference in AChE activities among birds from nest boxes on the two sites. Wild birds were also caught on the two sites and blood samples taken to measure activity of AChE in plasma. No significant differences between the treated and control sites were detected. Concentrations of CPY measured in food items for nestlings were generally undetectable, but concentrations as great as 10.6 mg ai kg^{-1} wwt were measured in a few samples.

Field studies to determine the potential effects of application of Lorsban 15G to corn corroborate the predictions from GranPARAM of very limited mortality of birds. Rates of application in field studies (1.09–2.91 kg ha^{-1} (0.975–2.6 lb ai A^{-1})) were similar to or exceeded the maximum permitted application rates on the Lorsban 15G label for corn (i.e., 1.46 kg ha^{-1} (1.3 lb ai A^{-1}) applied T-band or in-furrow, 1.12 kg ha^{-1} (1 lb ai A^{-1}) applied broadcast), broccoli (2.52 kg ha^{-1} (2.25 lb ai A^{-1}) applied T-band), onion (1.12 kg ha^{-1} (1 lb ai A^{-1}) applied in-furrow), sugarbeet (1.12 kg ha^{-1} (1 lb ai A^{-1}) applied broadcast), sunflower (1.46 kg ha^{-1} (1.3 lb ai A^{-1}) applied T-band) and tobacco (2.24 kg ha^{-1} (2 lb ai A^{-1}) applied broadcast with incorporation) (Table 3). Only peanuts have a higher maximum application rate (4.48 kg ha^{-1} (4 lb ai A^{-1}) applied in a band) than the highest application rate used in the Frey et al. (1994) field study.

In a study conducted by Worley et al. (1994), Dursban 2.5G granular CPY was applied to plots of turf on golf courses in central Florida to monitor the effects on birds. Granular CPY was applied twice at a rate of 4.48 kg ha^{-1} (4 lb ai A^{-1}) with a 21-d interval. For 13-d following each application, the golf courses were searched for casualties. Two dead birds were found following the granular applications, which was not statistically different from total bird mortality on control sites.

7.3 Strengths of the Refined Assessment for Granular CPY

The refined risk assessment for granular CPY built upon the refined model originally developed by Moore et al. (2010c). The major strengths of this model include:

- Use of an exposure model that explicitly accounted for factors affecting uptake of CPY granules by birds in treated fields. These factors included: availability of natural grit in the size ranges favored by birds, application technique and rate, granule:grit preference factor, spill attraction factor, spill size and concentration, and many others. The method used by EPA (USEPA 2004b) in their screening-level avian assessments for granular pesticides (i.e., estimating LD_{50}s ft^{-2}) does not consider these factors in assessing exposure.

- Derivation of species-specific dose-response curves for northern bobwhite and red-winged blackbird. This approach makes better use of the available toxicity data than does use of a benchmark based on the most sensitive response observed in birds. In the risk analyses, the dose-response curves enabled a determination of the fate of each bird (i.e., dead or alive) in the simulation. The use of the SSD approach permitted exploration of risks for untested bird species by assuming a range of sensitivities to CPY.

7.4 *Uncertainties in the Refined Risk Assessment for Granular CPY*

The refined risk assessment of granular CPY to birds contains uncertainties. In this assessment, conservative point estimates were used when the available data were inadequate to define an input distribution (e.g., daily grit retention). Thus, the assessment erred on the side of conservatism (i.e., over-estimating risk).

The following sources of uncertainty were identified in the refined risk assessment for granular CPY (Table 9):

- The refined risk assessment considered exposure of birds to granular CPY via inadvertent ingestion of grit. Exposure to granular CPY by dermal contact, inhalation or consumption of CPY in water, insects, and plant material were not the focus of the assessment. CPY from granular formulations is not expected to occur at elevated concentrations in the atmosphere, nor is it expected to accumulate or persist in the field environment (Solomon et al. 2001). Given that granular CPY is formulated on clay particles, birds are unlikely to mistake pesticide granules for seeds.
- The refined risk assessment considered exposure to five focal species. Thus, there is a possibility that bird species not considered in this assessment are at risk on or near CPY-treated fields. The focal species were selected because of their affinity for grit and agricultural areas. This group of species is more likely to be exposed to granular CPY than would most other bird species. Furthermore, they span a range of sizes and taxonomic groups, and are representative of species of birds found in regions where granular CPY is used. Thus, there is little uncertainty associated with overlooking bird species at risk.
- GranPARAM has a number of sources of uncertainty. Where possible, these sources were quantified and incorporated in the exposure analyses (e.g., variation in availability of natural grit particles, grit counts in bird gizzards). Thus, these sources of uncertainty have been explicitly accounted for in the risk estimates described here. Other sources of uncertainty, however, could not be accounted for in GranPARAM, generally because data were too scarce. Examples include: granule:grit preference factor, daily grit retention in bird gizzards, and use of the field margin. The general approach for input variables with high uncertainty was to use conservative point estimates. The model evaluation exercise

Table 9 Sources of uncertainty in avian risk assessment for granular chlorpyrifos

Area	Source of uncertainty	Action and influence on risk estimates
Exposure scenarios	Not possible to assess all scenarios. Potential to miss high risk scenarios.	Most highly-used crops were assessed at maximum application rates. Exposure scenarios included upper bound risk posed by granular CPY to birds.
Routes of exposure	Focus of assessment was on inadvertent ingestion of CPY granules for grit by birds. There could be other important routes of exposure.	Clay formulation and short half-life of granules in treated fields suggest that potential exposure routes such as inhalation, dermal exposure and ingestion for food are of minor importance.
Risk to non-focal bird species	The refined risk assessment considered exposure to five focal species. Thus, there is a possibility that bird species not considered in this assessment are at risk on or near CPY-treated fields.	The focal species were selected because of their affinity for grit and agricultural areas. This group of species is more likely to be exposed to granular CPY than would most other bird species. Furthermore, they span a range of sizes and taxonomic groups, and are representative of bird species found in regions where granular CPY is used. Thus, there is a low degree of uncertainty associated with overlooking bird species at risk.
Proportion time birds on fields	See Table 7.	See Table 7.
Granule:Grit preference factor (GGPF)	Only one study quantified GGPF for CPY.	*GGPF* of 0.078 from study on controlled study on house sparrows (*Passer domesticus*). The uncertainty arising from the limited available information on GGPF could lead to under- or over-estimation of risk.
Other variables in GranPARAM	Several variables (e.g., spill concentration factor, size of spills) were difficult to parameterize because of limited data.	Sensitivity analyses involving one-at-a-time manipulations of uncertain variables indicated they had little influence on predicted CPY exposure within parameter ranges that could be reasonably expected to occur in CPY-treated fields. Comparison of model predictions to results of field studies also indicated that GranPARAM performed well, though the database for this comparison was limited.
Quality and quantity of toxicity studies	See Table 7.	See Table 7.

indicated that model predictions reasonably replicated numbers of pesticide granules ingested in by birds in field studies, although this dataset is limited. Overall, it appears that the GranPARAM performs well, despite uncertainties regarding some input variables.

8 Summary

Refined risk assessments for birds exposed to flowable and granular formulations of CPY were conducted for a range of current use patterns in the United States. Overall, the collective evidence from the modeling and field study lines of evidence indicate that flowable and granular CPY do not pose significant risks to the bird communities foraging in agro-ecosystems in the United States. The available information indicates that avian incidents resulting from the legal, registered uses of CPY have been very infrequent since 2002 (see SI Appendix 3). The small number of recent incidents suggests that the current labels for CPY are generally protective of birds. However, incident data are uncertain because of the difficulties associated with finding dead birds in the field and linking any mortality observed to CPY.

Flowable CPY is registered for a variety of crops in the United States including alfalfa, brassica vegetables, citrus, corn, cotton, grape, mint, onion, peanut, pome and stone fruits, soybean, sugar beet, sunflower, sweet potato, tree nuts, and wheat under the trade name Lorsban Advanced. The major routes of exposure for birds to flowable CPY were consumption of treated dietary items and drinking water. The Liquid Pesticide Avian Risk Assessment Model (LiquidPARAM) was used to simulate avian ingestion of CPY by these routes of exposure. For acute exposure, LiquidPARAM estimated the maximum retained dose in each of 20 birds on each of 1,000 fields that were treated with CPY over the 60-d period following initial application. The model used a 1-h time step. For species lacking acceptable acute oral toxicity data (all focal species except northern bobwhite (*C. virginianus*) and red-winged blackbird (*A. phoeniceus*)), a species sensitivity distribution (SSD) approach was used to generate hypothetical dose-response curves assuming high, median and low sensitivity to CPY. For acute risk, risk curves were generated for each use pattern and exposure scenario. The risk curves show the relationship between exceedence probability and percent mortality. The results of the LiquidPARAM modeling exercise indicate that flowable CPY poses an acute risk to some bird species, particularly those species that are highly sensitive and that forage extensively in crops with high maximum application rates (e.g., grapefruit, orange). Overall, most bird species would not experience significant mortality as a result of exposure to flowable CPY. The results of a number of field studies conducted at application rates comparable to those on the Lorsban Advanced label indicate that flowable CPY rarely causes avian mortality. The results of the field studies suggest that LiquidPARAM is likely over-estimating acute risk to birds for flowable CPY.

For chronic exposure, LiquidPARAM estimated the maximum total daily intake (TDI) over a user-specified exposure duration (28-d in the case of CPY). The maximum average TDI was compared to the chronic NOEL and LOEL from the most sensitive species tested for CPY, the mallard. This comparison was done for each of the 20 birds in each of the 1000 fields simulated in LiquidPARAM. The outputs are estimates of the probabilities of exceeding the NOEL and LOEL. LiquidPARAM did not predict significant adverse effects resulting from chronic exposure to flowable CPY. The small number of incidents (2) involving CPY reported since 2002 suggests that the current labels for CPY are generally protective of birds.

Granular CPY is registered for a wide variety of crops including brassica vegetables, corn, onion, peanut, sugar beet, sunflower, and tobacco under the trade name Lorsban 15G. Consumption of grit is required by many birds to aid in digestion of hard dietary items such as seeds and insects. Because CPY granules are in the same size range as natural grit particles consumed by birds, there is a potential for birds to mistakenly ingest granular CPY instead of natural grit. We developed the Granular Pesticide Avian Risk Model (GranPARAM) to simulate grit ingestion behavior by birds. The model accounts for proportion of time that birds forage for grit in treated fields, relative proportions of natural grit versus pesticide granules on the surface of treated fields, rates of ingestion of grit, attractiveness of pesticide granules relative to natural grit and so on. For CPY, each model simulation included 20 birds on each of 1,000 fields to capture variability in rates of ingestion of grit and foraging behavior between birds within a focal species, and variability in soil composition between fields for the selected use pattern. The estimated dose for each bird was compared with randomly chosen doses from relevant dose-response curves for CPY. Our analysis for a wide variety of use patterns on the Lorsban 15G label found that granular CPY poses little risk of causing mortality to bird species that frequent treated fields immediately after application. The predictions of the model have been confirmed in several avian field studies conducted with Lorsban 15G at application rates similar to or exceeding maximum application rates on the Lorsban 15G label.

Acknowledgments The development of LiquidPARAM and this manuscript was guided by the advice of two expert panels, one for carbofuran and the other for CPY. The authors wish to thank the members of the expert panels that included Lou Best, Larry Brewer, Chris Cutler, Jeff Giddings, Don Mackay, John Purdy, and Marty Williams. The authors also thank Don Carlson of FMC Corporation, Dylan Fuge from Latham and Watkins LLP, and Nick Poletika and Mark Douglas from Dow AgroSciences for their helpful contributions to the development and preparation of this manuscript. We thank the anonymous reviewers of this paper for their suggestions and constructive criticism. Prof. Giesy was supported by the Canada Research Chair program, a Visiting Distinguished Professorship in the Department of Biology and Chemistry and State Key Laboratory in Marine Pollution, City University of Hong Kong, the 2012 "High Level Foreign Experts" (#GDW20123200120) program, funded by the State Administration of Foreign Experts Affairs, the P.R. China to Nanjing University and the Einstein Professor Program of the Chinese Academy of Sciences. This study was funded by Dow AgroSciences.

References

Aldenberg T, Jaworska JS, Traas TP (2002) Normal species sensitivity distributions and probabilistic ecological risk assessment. In: Posthuma L, Suter GW, Traas T (eds) Species sensitivity distributions in ecotoxicology. CRC, Boca Raton, FL, pp 49–102

Allen TFH, Starr TB (1982) Hierarchy: perspectives for ecological complexity. The University of Chicago Press, Chicago, IL, 326 pp

Anderson TA, Richards SM, McMurray ST, Hooper MJ (1998) Avian response to chlorpyrifos exposure in corn agroecosystems. Dow AgroSciences, Indianapolis, IN (unpublished report)

Barron MG, Woodburn KB (1995) Ecotoxicology of chlorpyrifos. Rev Environ Contam Toxicol 144:1–93

Baurledel WR (1986) Fate of 14C-chlorpyrifos administered to laying hens. Dow Chemical, USA, Midland, MI (unpublished report)

Beavers JB, Martin KH, Gallagher SP (2007) GF-1668: a dietary LC50 study with the northern bobwhite. Dow Chemical Company, Midland, MI (unpublished report)

Bennett RS (1989) Role of dietary choices in the ability of bobwhite to discriminate between insecticide-treated and untreated food. Environ Toxicol Chem 8:731–738

Best LB (1977) Nestling biology of the field sparrow. Auk 94:308–319

Best LB, Gionfriddo JP (1994) House sparrow preferential consumption of carriers used for pesticide granules. Environ Toxicol Chem 13:919–925

Best LB, Stafford TR, Mihaich EM (1996) House sparrow preferential consumption of pesticide granules with different surface coatings. Environ Toxicol Chem 15:1763–1768

Best LB, Murray LD (2003) Estimating the proportion of diet birds take from treated agricultural fields. CropLife America, Washington, DC

Bidlack HD (1979) Degradation of chlorpyrifos in soil under aerobic, aerobic/anaerobic and anaerobic conditions. Dow Chemical, Midland, MI (unpublished report)

Booth GM, Best LB, Carter MW, Jorgensen CD (1989) Effects of Furadan 4F on birds associated with Kansas and Oklahoma alfalfa fields. FMC Corporation, Philadelphia, PA (unpublished report)

Brewer LW, Kaczor MH, Miller VC (2000a) Dursban 10.5 LEE: 14-day avian acute oral toxicity test with northern bobwhite (*Colinus virginianus*). Dow Chemical Company, Midland, MI (unpublished report)

Brewer LW, Kaczor MH, Miller VC (2000b) Lorsban 2.5P: 14-day avian acute oral toxicity test with northern bobwhite (*Colinus virginianus*). Dow Chemical Company, Midland, MI (unpublished report)

Brown KC, Stamp G, Kitson J (2007) Refinement of the risk to birds following application of chlorpyrifos to vines in Southern France. Dow Chemical Company, Indianapolis, IN (unpublished report)

Cairns MA, Maguire CC, Williams BA, Bennett JK (1991) Brain cholinesterase activity of bobwhite acutely exposed to chlorpyrifos. Environ Toxicol Chem 10:657–664

Campbell S, Hoxter KA, Jaber M (1990) 3,5,6-Trichloro-2-pyridinol: an acute oral toxicity study with the northern bobwhite. Dow Chemical Company, Midland, MI (unpublished report)

CCME (2013) Determination of hazardous concentrations with species sensitivity distributions, SSD master. Canadian Council of Ministers of the Environment, Ottawa, ON, Canada

Chapman RA, Harris CR (1980) Persistence of chlorpyrifos in a mineral and an organic soil. J Environ Sci Health B 15:39–46

Cutler GC, Purdy J, Giesy JP, Solomon KR (2014) Risk to pollinators from the use of chlorpyrifos in the United States. Rev Environ Contam Toxicol 231:219–265

de Vette HQM, Schoonmade JA (2001) A study on the route and rate of aerobic degradation of ^{14}C-chlorpyrifos in four European soils. Dow AgroSciences, Indianapolis, IN (unpublished report)

Dittrich R, Staedtler T (2010) Chlorpyrifos in citrus orchards—field study on the status of bird communities and reproductive performance. Dow AgroSciences, Abingdon, UK (unpublished report)

Dow AgroSciences (2008) Lorsban 15G granular insecticide specimen label. Dow AgroSciences LLC, Indianapolis, IN

Dow AgroSciences (2009) Lorsban advanced insecticide supplemental labeling. Dow AgroSciences LLC, Indianapolis, IN

ECOFRAM (1999) ECOFRAM aquatic final draft reports. United States Environmental Protection Agency, Washington, DC. http://www.epa.gov/oppefed1/ecorisk/aquareport.pdf

EFSA (2008) Scientific opinion of the panel on plant protection products and their residues on a request from the EFSA PRAPeR Unit on risk assessment for birds and mammals. EFSA J 734:1–181

Fautin RW (1941) Development of nestling yellow-headed blackbirds. Auk 58:215–232
Fink R (1977) Eight-week feeding study—mallard duck: chlorpyrifos. Dow Chemical, Midland MI (unpublished report)
Fink R (1978a) One-generation reproduction study—mallard duck: chlorpyrifos. Dow Chemical, Midland, MI (unpublished report)
Fink R (1978b) The effect of chlorpyrifos during a one-generation reproduction study on bobwhite. Dow Chemical, Midland, MI (unpublished report)
Frey LT, Krueger HO, Palmer DA (1994) Lorsban insecticide: an evaluation of its effects upon avian and mammalian species on and around corn fields in Iowa. DowElanco, Indianapolis, IN (unpublished report)
Gallagher SP, Palmer DA, Krueger HO (1994) Lorsban insecticide a pilot year evaluation of its effects upon avian and mammalian species on and around citrus groves in California. DowElanco, Indianapolis, IN (unpublished report)
Gallagher SP, Palmer DA, Krueger HO (1996) Chlorpyrifos technical: an acute oral toxicity study with the house sparrow. Dow Chemical, Midland, MI (unpublished report)
Gallagher SP, Beavers JB (2006) GF-1668: an acute oral toxicity study with the northern bobwhite. Dow Chemical Company, Midland, MI (unpublished report)
Gallagher SP, Beavers JB (2007) Chlorpyrifos technical: a single day dietary exposure with the northern bobwhite. Dow Chemical, Midland, MI (unpublished report)
Giddings JM, Williams WM, Solomon KR, Giesy JP (2014) Risks to aquatic organisms from the use of chlorpyrifos in the United States. Rev Environ Contam Toxicol 231:119–162
Giddings JM, Anderson TA, Hall LW Jr, Kendall RJ, Richards RP, Solomon KR, Williams WM (2005) A probabilistic aquatic ecological risk assessment of atrazine in North American surface waters. SETAC, Pensacola, FL, 432 pp
Gomez LE (2009) Use and benefits of chlorpyrifos in U.S. Agriculture. Dow AgroSciences, Indianapolis, IN (unpublished report)
Hill EF, Camardese MB (1984) Toxicity of acetylcholinesterase insecticides to birds: technical grade versus granular formulations. Ecotoxicol Environ Saf 8:551–563
HSDB (2013) Hazardous substances data bank. Chlorpyrifos. National Library of Medicine. http://toxnet.nlm.nih.gov/cgi-bin/sis/search/f?./temp/~ntQQXy:1. Accessed February 2013
Hubbard PM, Beavers JB (2008) GF-2153: an acute oral toxicity study with the northern bobwhite. Dow Chemical Company, Midland, MI (unpublished report)
Hubbard PM, Beavers JB (2009) Chlorpyrifos: an acute oral toxicity study with the northern bobwhite using two excipients. Dow Chemical Company, Midland, MI (unpublished report)
Hudson RH, Tucker RK, Haegele MA (1984) Handbook of toxicity of pesticides to wildlife, 2nd edn. United States Department of the Interior Fish and Wildlife Service, Washington, DC
Jorgensen CD, Whitmore RC, Booth GM, Carter MW, Smith HD (1989) Effects of Furadan® 4F on birds associated with Nebraska and Texas/New Mexico corn fields. FMC Corporation, Philadelphia, PA (unpublished report)
Kaczor MH, Miller VC (2000) Lorsban 50W: 14-day avian acute oral toxicity test with northern bobwhite (*Colinus virginianus*). Dow Chemical Company, Midland, MI (unpublished report)
Kenaga EE, Fink RJ, Beavers JB (1978) Dietary toxicity tests with mallards simulating residue decline of chlorpyrifos and avoidance of treated food. Dow Chemical Company, Midland, MI (unpublished report)
Kessel B (1957) A study of the breeding biology of the European starling (*Sturnus vulgaris* L.) in North America. Am Midl Nat 58:257–331
Kluijver HN (1950) Daily routines of the great tit, *Parus m major* L 38:99–135
Kunz SE, Radeleff RD (1972) Evaluation of the hazard of chlorpyrifos soil treatments on turkeys. J Econ Entomol 65:1208–1209
Lehman-McKeeman LD (2008) Adsoption, distribution, and excretion of toxicants. In: Klaasen CD (ed) Casaret and Doull's toxicology: the basic science of poisons. McGraw-Hill, New York, NY, pp 131–159
Long RD, Hoxter KA, Jaber M (1990) 3,5,6-Trichloro-2-pyridinol: a dietary LC50 study with the mallard. Dow Chemical Company, Midland, MI (unpublished report)

Long RD, Smith GJ, Beavers JB (1991) XRM 5160 (microencapsulated insecticide): a dietary LC50 study with the mallard. Dow Chemical Company, Midland, MI (unpublished report)
Luttik R, de Snoo GR (2004) Characterization of grit in arable birds to improve pesticide risk assessment. Ecotoxicol Environ Saf 57:319–329
Mackay D, Giesy JP, Solomon KR (2014) Fate in the environment and long-range atmospheric transport of the organophosphorus insecticide, chlorpyrifos and its oxon. Rev Environ Contam Toxicol 231:35–76
McCollister SB, Kociba RJ, Humiston CG, McCollister DD, Gehring PJ (1974) Studies of the acute and long-term oral toxicity of chlorpyrifos (O, O-diethyl-O-(3,5,6-trichloro-2-pyridyl) phosphorothionate). Food Cosmet Toxicol 12:45–61
McGregor WS, Swart RW (1968) Toxicity studies on turkeys confined on soil treated with Dursban. Dow Chemical Company, Lake Jackson, TX (unpublished report)
McGregor WS, Swart RW (1969) Toxicity studies on turkeys with Dursban insecticide (Wettable Powder Formulation TF-137) applied as a spray to range pens. Dow Chemical Company, Lake Jackson, TX (unpublished report)
Mineau P (2002) Estimating the probability of bird mortality from pesticide sprays on the basis of the field study record. Environ Toxicol Chem 21:1497–1506
Miyazaki S, Hodgson GC (1972) Chronic toxicity of Dursban and its metabolite, 3,5,6-trichloro-2-pyridinol in chickens. Toxicol Appl Pharmacol 23:391–398
Moore DRJ (1998) The ecological component of ecological risk assessment: lessons from a field experiment. Human Ecol Risk Assess 4:1103–1123
Moore DRJ, Thompson RP, Rodney SI, Fischer DL, Ramanaryanan T, Hall T (2010a) Refined aquatic risk assessment for aldicarb in the United States. Integr Environ Assess Manag 6:102–118
Moore DRJ, Teed RS, Rodney SI, Thompson RP, Fischer DL (2010b) Refined avian risk assessment for aldicarb in the United States. Integr Environ Assess Manag 6:83–101
Moore DRJ, Fischer DL, Teed RS, Rodney SI (2010c) Probabilistic risk-assessment model for birds exposed to granular pesticides. Integr Environ Assess Manag 6:260–272
Moosmayer P, Wilkens S (2008) Chlorpyrifos (Dursban 480 EC) in brassica crops—field study on exposure and effects on wild birds. Dow AgroSciences, Indianapolis, IN (unpublished report)
Nagy KA (1987) Field metabolic rate and food requirement scaling in mammals and birds. Ecol Monogr 57:11–128
Nagy KA, Girard IA, Brown TK (1999) Energetics of free-ranging mammals, reptiles, and birds. Annu Rev Nutr 19:247–277
Parsons KC, Matz AC, Hooper MJ, Pokras MA (2000) Monitoring wading bird exposure to agricultural chemicals using serum cholinesterase activity. Environ Toxicol Chem 19:1317–1323
Pinkowski BC (1978) Feeding of nestling and fledgling eastern bluebirds. Wilson Bull 90:84–98
PMRA (2007) Probabilistic environmental risk assessment for terrestrial biota exposed to CPY, vol III. Pest Management Regulatory Agency, Environmental Assessment Division, Ottawa, ON
Price MA, Kunz SE, Everett RF (1972) Further evaluation of insecticides to control the chigger *Neoschongastia americana* on turkeys. J Econ Entomol 65:454–455
Racke KD (1993) Environmental fate of chlorpyrifos. Rev Environ Contam Toxicol 131:1–151
Reeves G (2008) Modelling the laboratory soil degradation kinetics of CPY and two metabolites (TCP and TMP) using FOCUS methodology. Dow Agrosciences, Abingdon, UK (unpublished report)
SAP (2000) Implementation plan for probabilistic ecological assessment: a consultation. FIFRA Scientific Advisory Panel Meeting, 5–7 April 2000. United States Environmental Protection Agency, FIFRA Scientific Advisory Panel, Arlington, VA. No. 2000-02
SAP (2001) Probabilistic Models and Methodologies: Advancing the ecological risk assessment process in the EPA Office of pesticide programs. FIFRA Scientific Advisory Panel Meeting, 13–16 March 2001. United States Environmental Protection Agency, FIFRA Scientific Advisory Panel, Arlington, VA. No. 2001-06

SAP (2004) Refined (level II) Terrestrial and aquatic models—probabilistic ecological assessments for pesticides: terrestrial. FIFRA Scientific Advisory Panel Meeting, 30–31 March 2004. United States Environmental Protection Agency, FIFRA Scientific Advisory Panel, Arlington, VA. No. 2004-03

Schafer EW, Brunton RB (1971) Chemicals as bird repellents: two promising agents. J Wildl Manag 35:569–572

Schafer EW, Brunton RB (1979) Indicator bird species for toxicity determinations: is the technique usable in test method development. In: Beck JR (ed) Vertebrate pest control and management materials, vol ASTM STP 680. American Society for Testing and Materials, West Conshohocken, PA, pp 157–168

Selbach A, Wilkens S (2008a) Chlorpyrifos (Dursban 75 WG) in citrus orchards—field study on exposure and effects on wild birds. Dow AgroSciences, Indianapolis, IN (unpublished report)

Selbach A, Wilkens S (2008b) Chlorpyrifos (Dursban 75 WG) in citrus orchards—field study on exposure and effects on wild birds. Dow AgroSciences, Abingdon, UK (unpublished report)

Smith GN, Watson BS, Fischer FS (1967) The metabolism of ^{36}Cl O, O-diethyl O-(3,5,6-trichloro-2-pyridyl) phosphorothioate in rats. J Agric Food Chem 15:132–138

Solomon KR, Giesy JP, Kendall RJ, Best LB, Coats JR, Dixon KR, Hooper MJ, Kenaga EE, McMurry ST (2001) Chlorpyrifos: ecotoxicological risk assessment for birds and mammals in corn agroecosystems. Human Ecol Risk Assess 7:497–632

Solomon KR, Williams M, Mackay D, Purdy J, Giddings JM, Giesy JP (2014) Properties and uses of chlorpyrifos in the United States. Rev Environ Contam Toxicol 231:13–34

Stafford JM (2007a) Assessment of the differential toxicity of carbofuran to northern mallard ducks when dosed as a single aqueous bolus versus the same dose mixed with feed. FMC Corporation, Philadelphia, PA (unpublished report)

Stafford JM (2007b) Assessment of the differential toxicity of carbofuran to northern bobwhite Quail when dosed as a single aqueous bolus versus the same dose mixed with feed. FMC Corporatio, Philadelphia, PA (unpublished report)

Stafford JM (2010) Evaluation of northern bobwhite (*Colinus virginianus*) food consumption relative to the concentration of chlorpyrifos in the diet. Dow AgroSciences, Indianapolis, IN (unpublished report)

Stafford TR, Best LB, Fischer DL (1996) Effects of different formulations of granular pesticides on birds. Environ Toxicol Chem 15:1606–1611

Stafford TR, Best LB (1997) Effects of granular pesticide formulations and soil moisture on avian exposure. Environ Toxicol Chem 16:1687–1693

Suter GW II, Efroymson RA, Sample BE, Jones DS (2000) Ecological risk assessment for contaminated sites. Lewis Publishers, Boca Raton, FL, 460 pp

Testai E, Buratti FM, Consiglio ED (2010) Chlorpyrifos. In: Krieger RI, Doull J, van Hemmen JJ, Hodgson E, Maibach HI, Ritter L, Ross J, Slikker W (eds) Handbook of pesticide toxicology, vol 2. Elsevier, Burlington, MA, pp 1505–1526

Timchalk C (2010) Organophosphorus insecticide pharmacokinetics. In: Krieger RI, Doull J, van Hemmen JJ, Hodgson E, Maibach HI, Ritter L, Ross J, Slikker W (eds) Handbook of pesticide toxicology, vol 2. Elsevier, Burlington, MA, pp 1409–1433

USEPA (1999) Reregistration eligibility science chapter for chlorpyrifos: fate and environmental risk assessment chapter. United States Environmental Protection Agency, Washington, DC.

USEPA (2002) Calcasieu estuary remedial investigation/feasibility study (RI/FS): baseline ecological risk assessment (BERA). United States Environmental Protection Agency, Region 6, Dallas, TX, USA. http://mapping2.orr.noaa.gov/portal/calcasieu/calc_html/pdfs/reports/berassessws.pdf

USEPA (2004a) Ecological risk assessment for General Electric (GE)/Housatonic River Site, Rest of River. United States Environmental Protection Agency, New England Region, Boston, MA. http://www.epa.gov/region1/ge/thesite/restofriver/reports/era_nov04/215498_ERA_FNL_TOC_MasterCD.pdf

USEPA (2004b) Overview of the ecological risk assessment process in the office of pesticide programs: endangered and threatened species effects determinations. United States Environmental

Protection Agency. Office of Prevention, Pesticides, and Toxic Substances, Office of Pesticide Programs, Washington, DC
USEPA (2005) Reregistration eligibility science chapter for carbofuran. environmental fate and effects chapter. environmental risk assessment and human drinking water exposure assessment. United States Environmental Protection Program, Office of Pesticide Programs, Washington, DC
USEPA (2008) Terrestrial investigation model version 2.1. United States Environmental Protection Agency, Office of Pesticide Programs, Washington, DC.
USEPA (2008b) Problem formulation for the environmental fate and ecological risk, endangered species and drinking water assessments in support of the registration review of chlorpyrifos. United States Environmental Protection Agency, Office of Pesticide Programs, Washington, DC
USEPA (2009) Chlorpyrifos final work plan. Registration review. United States Environmental Protection Agency, Office of Pesticide Programs, Washington, DC
USEPA (2010) Screening tool for inhalation risk (STIR) version 1.0 Washington, DC. United States Environmental Protection Agency, Office of Pesticide Programs, Washington, DC.
USEPA (2011) Revised chlorpyrifos preliminary registration review drinking water assessment. United States Environmental Protection Agency, Office of Chemical Safety and Pollution Prevention, Washington, DC, USA. PC Code 059101 http://www.epa.gov/oppsrrd1/registration_review/chlorpyrifos/EPA-HQ-OPP-2008-0850-DRAFT-0025%5B1%5D.pdf
Wildlife International (1978) Eleven-day toxicant 2 x LC50 option with untreated food—mallard duck. Dow Chemical Company, Midland, MI (unpublished report)
Wilkens S, Frese I, Schneider K (2008) Chlorpyrifos (Dursban 480 EC): residues of chlorpyrifos in invertebrates after spray application of Dursban 75 WP in citrus orchards—magnitude and time course of residue decline. Dow AgroSciences, Abingdon, UK (unpublished report)
Williams WM, Giddings JM, Purdy J, Solomon KR, Giesy JP (2014) Exposures of aquatic organisms resulting from the use of chlorpyrifos in the United States. Rev Environ Contam Toxicol 231:77–118
Wolf C, Riffel M, Weyman G, Douglas M, Norman S (2010) Telemetry-based field studies for assessment of acute and short-term risk to birds from spray applications of chlorpyrifos. Environ Toxicol Chem 29:1795–1803
Worley KB, Frey LT, Palmer DA, Krueger HO (1994) Dursban insecticide: an evaluation of its effects upon avian and mammalian species on and around golf courses in fall in Florida. Dow Chemical Company, Midland, MI (unpublished report)

Risk to Pollinators from the Use of Chlorpyrifos in the United States

G. Christopher Cutler, John Purdy, John P. Giesy, and Keith R. Solomon

1 Introduction

Pollinators are crucial species of almost all natural and artificial terrestrial ecosystems (Garibaldi et al. 2013; NAS 2007). While most of the world's food supply, including important crops such as cereals, are mainly wind pollinated, more than three-quarters of angiosperms rely on animals for pollination and approximately 75% of the leading global fruit-, vegetable-, and seed-crops depend at least partially on animal pollination (Klein et al. 2007). Most animal pollination is done by insects, particularly bees. In the United States (U.S.) and Canada, the production of crops that require or benefit from pollination by insects is large. It is estimated that the pollination services of the European honey bee, *Apis mellifera* L. (Apidae), are worth over $15 billion annually to U.S. agriculture, and the value of non-*Apis* pollinators to production of crops is estimated to be over $11 billion (Calderone 2012; Morse and Calderone 2000). In addition to helping ensure a diverse supply of food for humans, pollination plays a critical role in providing the basis for essential ecosystem productivity and services (Kevan et al. 1990; Kevan 1999).

G.C. Cutler (✉)
Department of Environmental Sciences, Faculty of Agriculture,
Dalhousie University, P.O. Box 550, Truro, NS B2N 5E3, Canada
e-mail: chris.cutler@dal.ca

J. Purdy
Abacus Consulting, Campbellsville, ON, Canada

J.P. Giesy
Department of Veterinary Biomedical Sciences and Toxicology Centre,
University of Saskatchewan, 44 Campus Dr., Saskatoon, SK S7N 5B3, Canada

K.R. Solomon
Centres for Toxicology, School of Environmental Sciences, University of Guelph,
Guelph, ON, Canada

J.P. Giesy and K.R. Solomon (eds.), *Ecological Risk Assessment for Chlorpyrifos in Terrestrial and Aquatic Systems in the United States*, Reviews of Environmental Contamination and Toxicology 231, DOI 10.1007/978-3-319-03865-0_7, © The Author(s) 2014

There is concern about potential adverse effects of pesticides on pollinators (EFSA 2012; NAS 2007). Chlorpyrifos (CPY; CAS No. 2921-88-2) is an organophosphorus insecticide and acaricide that is widely used in agriculture and horticulture in the U.S. and other countries to control a wide variety of foliage- and soil-borne insect pests on a variety of food and feed crops (Solomon et al. 2014). Many of the agro-ecosystems where CPY is used contain populations of managed and wild pollinators. In some of these, such as almonds, citrus fruits, and cranberries, pollinators play a critical role in the production of the crop being protected with sprays of CPY. Other crops, such as soybean and corn, which are treated with CPY, do not directly rely on pollinators for production because they are mainly pollinated by wind, but can nonetheless serve as a source of forage for multiple species of pollinators during parts of the season. In addition to food (pollen and/or nectar), pollinators might also obtain nesting materials and occupy nesting sites in habitats exposed to CPY.

In this study, the risk posed by use of CPY to insects that serve as pollinators was assessed. Patterns of use of CPY that are currently registered in the U.S. and Canada were the main focus (Solomon et al. 2014), but tests with formulations used in other countries were considered when relevant data from the U.S. were lacking. Because microencapsulated formulations are not used in the U.S., they were excluded from the assessment. Bees were the focal taxa but other groups of insects were also considered when data were relevant and available, particularly where they are used as surrogate species in regulatory risk assessments. Non-insect pollinators were not considered. Most studies and scenarios explored for the risk assessment were concerned with agricultural systems, but patterns of use of CPY in horticulture and landscaping, such as turf were considered.

2 Problem Formulation

The central question considered in the problem formulation was: Is there sufficient exposure of pollinators to CPY and/or its degradate, chlorpyrifos oxon (CPYO), to present a risk of widespread and repeated mortality or biological impairment to individuals or populations of pollinators? This question forms the basis for the detailed development of the risk assessment in the following sections.

2.1 Use Patterns of Chlorpyrifos: Pollinator Considerations

The uses and properties of CPY are discussed in detail in a companion paper (Solomon et al. 2014). Chlorpyrifos is used to control a wide variety of economically important insect pests in a large number of agricultural and specialty application scenarios throughout the U.S. Several granular and sprayable formulations of CPY are currently marketed in the U.S. and Canada, including Lorsban Advanced® and Lorsban® 15G for agriculture, and Dursban® 50 W for horticultural uses on trees, turf, and ornamental plants. Chlorpyrifos can be applied on foliage, tree bark or soil as a pre- or post-emergent spray in water to control insects or mites. It may also be applied to soil as a spray or as granules to control soil-dwelling insect pests. Maximum single application rates range from 1.12 to 4.5 kg CPY ha^{-1} for granular products and 0.53–6.27 kg CPY ha^{-1} for spray application. Multiple applications of the granular or flowable formulations are allowed on many crops. There are many crops that receive treatment around planting time, post-harvest or during dormancy. For crops such as apple, applications of CPY are delayed until after bloom, and as noted previously, no application is permitted when bees are actively foraging (Solomon et al. 2014).

Chlorpyrifos is widely used on corn, soybeans and wheat in the corn belt that extends from Quebec through the Midwestern U.S. to Manitoba. In the Great Plains regions of North America the main uses are on alfalfa and sunflower. In California, Florida, and Georgia, CPY is used on vegetables, citrus, and tree nuts (Gomez 2009). Other crops treated with CPY include cotton, cranberries, sorghum, strawberries, peanuts and wheat. Some of these crops are highly or partially dependent upon pollinators, or are utilized as forage or nesting material for pollinators (e.g., alfalfa leaf cutting bees). The importance of pollinators in production of tree fruit is well recognized (NAS 2007). For example, it is estimated that over 60% of honey bee colonies in the U.S. are used each year for pollination of almonds (Carman 2011). Cotton and soybeans are not critically dependent on pollinators, but bees will forage readily on the flowers of these crops (Berger et al. 1988; Rhodes 2002) and on extra-floral nectaries of cotton (Willmer 2011).

2.2 *Scope of the Assessment*

The potential for exposure of pollinators to CPY is recognized. Since the primary insect pollinators are bees (superfamily Apoidea), labels for CPY products include warnings not to apply the product or allow it to drift to flowering crops or weeds if bees are visiting the treated area, and advise users to inform local beekeepers prior to application if hives are in or adjacent to fields to be treated. Labels describing restrictions on use and best application practices also include instructions to minimize spray drift to reduce harmful effects on bees in habitats close to the application site (Solomon et al. 2014). For this reason, adverse effects due to negligence or actions contrary to precautions specified on the label were not included in this assessment. It is assumed that applications are made by trained applicators and that all instructions on the label are followed. This assessment focused on incidental exposure during applications to crops listed on the current labels under conditions specified by the labels. A search for documented incidents of harm to commercial beehives from CPY was also conducted through the USEPA and the Health Canada Pest Management Regulatory Agency (PMRA).

Because bees are considered the dominant animal pollinators and are prominent in agricultural landscapes (NAS 2007), toxicity data used in the risk assessment focused mainly on bees. There are more than 17,000 species of bees worldwide

(Michener 2007) and it was not possible to obtain data for all species. Searches of databases incorporated 'chlorpyrifos' with the words such as 'pollinator', 'bee', '*Apis*', '*Bombus*', '*Megachile*', and '*Osmia*'. The later four are major bee genera that exist in the wild and are managed by humans in agricultural settings, and thus are most likely to have associated data on toxicity of CPY to bees. Family names of major bee families, such as 'Apidae', 'Megachilidae', Andrenidae', 'Halictidae', and 'Colletidae' were used. Other potentially important pollinators include the dipteran families Bombyliidae (bee flies) and Syrphidae (hover flies), but no reports examining effects of CPY on these taxa were found. Other insect taxa can pollinate, but generally do so adventitiously, less frequently, and are generally not considered important pollinators of crops in the U.S. to which CPY is applied. For estimates of exposure, data from semi-field or field experiments with leaf-dwelling species was considered a potential source from which to develop point estimates of contact exposure for foliar-applied products (Fischer and Moriarty 2011), since these species may be considered surrogates for bees. Because data on toxicity of CPY to non-*Apis* pollinators were rare, studies were considered that assessed effects of CPY on certain other arthropod taxa that have also been shown to be suitable surrogate species for non-*Apis* bees (Candolfi et al. 2001; Miles and Alix 2012).

Data were collected from sources listing 'chlorpyrifos' or 'chlorpyrifos-ethyl' as the active substance; both are common names of *O*,*O*-diethyl *O*-3,5,6-trichloropyridin-2-yl phosphorothioate. The insecticidal degradate chlorpyrifos-oxon (CPYO) was also considered. Exposure and effects data for 'chlorpyrifos-methyl' (*O*,*O*-dimethyl *O*-3,5,6-trichloro-2-pyridinyl phosphorothioate)—a different compound—were not included in the risk assessment. The assessment of risk was primarily focused on evaluating potential impacts of typical CPY formulations currently registered for crop production in North America on bee pollinators.

2.3 Conceptual Model

Adverse effects on ecosystems result from the interaction of a stressor, in this case CPY and its degradation products, with receptors of concern, such as individual pollinators, hives, nests, or populations. The degradation product of concern for CPY is CPYO, which is also the activated biologically-active product of CPY (Solomon et al. 2014). Other degradates of CPY are of minimal risk to pollinators. A conceptual model can be constructed to illustrate potential routes of exposure during agricultural use, and the taxa and life stages potentially affected (Fig. 1). The conceptual model shows the scope of the risk assessment, guides its development, and illustrates the relationships among the potential exposure pathways. Previous conceptual models for assessments of effects of agricultural chemicals on pollinators have noted the need to quantify exposure within and outside the treated area and to consider the behavioral and biological traits of pollinators (Barmaz et al. 2010).

The conceptual model was developed for foliar spray or granular soil-applied treatments of CPY. For pollinators, exposure is primary if it is to the initial exposed

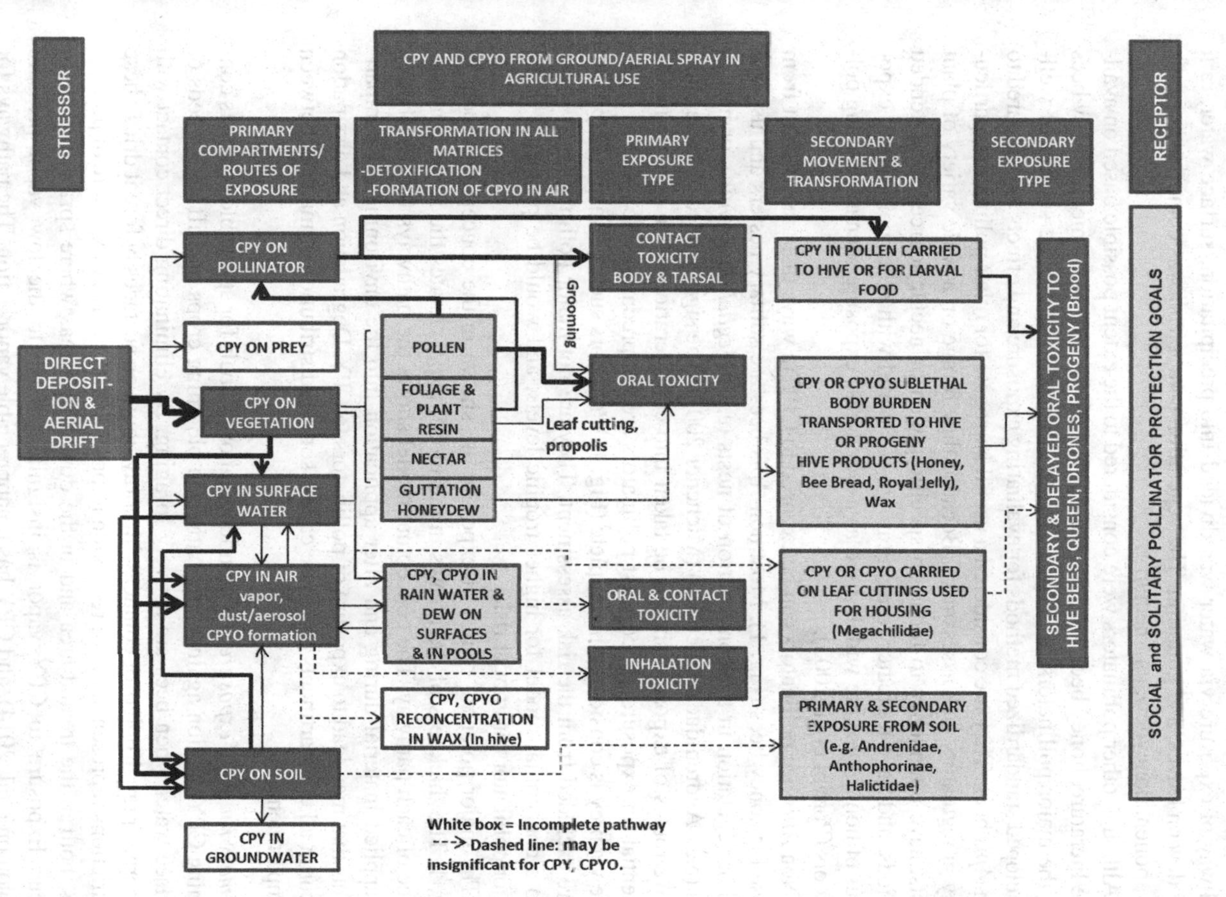

Fig. 1 Conceptual model for exposure of pollinators to chlorpyrifos (CPY) and chlorpyrifos-oxon (CPYO). Thickness of lines indicates relative importance of pathways

individual (e.g., a foraging worker), or secondary if other adults or offspring in a hive or nest are subsequently exposed. Potential exposures through various compartments of air, water, soil, and vegetation are complex and interconnected. The pathway of exposure via water was divided into precipitation, surface water, rain and dew on leaves, and guttation. The vegetation component was comprised of foliage, pollen, honeydew, and nectar.

Although other pollinators were considered to the extent possible based on available literature, honey bees were the main focus of this risk assessment. Honey bees are the major pollinators of crops in North America, and are the subject of well-developed standardized methods for evaluating exposure and effects as compared to non-*Apis* bees. They are considered a useful surrogate for other pollinators, particularly in regulatory risk assessment. Because they forage on a wide variety of plant hosts, have a tendency to focus on specific pollen or nectar sources for extended periods, and have a greater foraging range compared with other pollinators, exposures of honey bees are widely used as a worst-case exposure scenario among pollinators (Porrini et al. 2003).

Non-*Apis* bee pollinators can be exposed to CPY in ways that are different from those for honey bees (Fig. 1). Most non-*Apis* bees are solitary nesters and use soil and/or vegetation in the construction of nests (e.g., *Megachile*, *Osmia*), or nest in soil (e.g., Andrenidae, Halictidae) (Michener 2007). The significance of these alternative routes of exposure should be taken into consideration when comparing the potential for exposure. Since most pollinators are not predators, the route of exposure via prey is considered incomplete (Fig. 1). Predators such as wasps (*Vespa* sp.) were excluded from the risk assessment. They are not major pollinators, fit better into a conceptual model for higher trophic levels, and would be protected if the major pollinators are not at risk.

The major potential routes of exposure are shown in the conceptual model in Fig. 1. The thickness of the arrows in the model approximates the relative importance of each pathway. The conceptual model shows the pathways for distribution of applied material during and after application into the environmental compartments that may lead to exposure of pollinators to CPY. Degradation and dissipation occur in all compartments and there can be some redistribution of material between compartments.

Primary routes of exposure. As mentioned above, labels for sprayable products containing CPY caution against application on blooming crops or drift onto weeds or surface water when bees are actively foraging. By eliminating direct contact with airborne spray droplets or contact with spray liquid on surfaces before it dries, these restrictions represent a major reduction in potential for primary exposure of pollinators, both in the treated area and in the downwind areas where spray drift might occur. Exposure to CPY vapor is insignificant due to the low vapor pressure (Solomon et al. 2014), and CPY has no appreciable vapor action. The pathways for direct exposure of pollinators to airborne spray droplets or vapor are therefore shown as minor pathways in Fig. 1.

For granular CPY products, there should be little or no exposure via drift of dust, or deposition on foliage, pollen, or other surfaces. Chlorpyrifos is non-systemic and is not used as a seed treatment, so there is no contribution from seed dust during planting and there is no translocation and guttation. Volatilization of foliar residues constitutes the most significant source of airborne contamination. However, residues of CPY in air are not persistent, and maximum concentrations found in air monitoring studies were less than 250 ng m^{-3} (Mackay et al. 2014), which, when compared to the toxic dose in bees of about 80 ng bee^{-1} suggests that risks from exposure of this type would be *de minimis*.

Thus, the main route of direct exposure for pollinators is the uptake of CPY from plant surfaces after application. Residues of CPY applied as a spray on vegetation are mostly on foliage, which includes any non-crop flowers open during application. For example, white clover in turf or in groundcover under an orchard or areas adjacent to a treated field is very attractive to pollinators and can be in bloom during foliar application (Barmaz et al. 2010). Pollen in flowers that were open during application remains available for collection by pollinators for some time after treatment, but concentrations of pesticides in pollen and on plant surfaces will decline and become less bioavailable with time, particularly after sprays have dried. Nectar and honeydew were grouped into a sub-compartment of vegetation. Direct oral toxicity due to exposure via nectar and honeydew are incomplete or minimal because CPY is not systemic and is not taken up via roots and translocated upward through the plant (Racke 1993). Relatively smaller amounts of CPY would be expected in honey for the following reasons:

- Nectar is more protected than pollen from exposure to spray droplets by the anatomy of flowers (Willmer 2011).
- Nectar, water, and honeydew are carried internally in the "honey stomach" by bees (Gary 1975; Snodgrass 1975), where residues of pesticides are more likely to be absorbed and metabolized, reducing the amount transferred to the hive. Residues of CPY have been shown to decrease 3-fold when pollen is processed into bee bread (DeGrandi-Hoffman et al. 2013).
- CPY present in nectar would be exposed to water, which would favor hydrolysis and detoxification pathways over oxidation, and formation of CPYO (Solomon et al. 2014).
- Forager bees are initially most exposed to residues in nectar since it is ingested and those individuals could be impaired or killed by greater concentrations before returning to a hive. This potential for toxic effects before returning to a hive would be exacerbated by relatively greater loads of nectar (40–90 mg bee^{-1}) compared to pollen (12–29 mg bee^{-1}) per foraging trip (Gary 1975).
- Residues in honey in the hive are likely the result of transfer of residues from a sublethal body burden of CPY in the adult bees from other sources, such as pollen, nectar and water.

Honey bees actively forage for water to regulate temperature of the hive through evaporative cooling, to prepare larval brood food, and for their own metabolism (Gary 1975; Winston 1987). Exposure of pollinators from large CPY contaminated bodies of water is probably insignificant since bees do not collect water from large areas of open water. The main water sources for bees are wet foliage, dew, and surface water from wet soil and ephemeral pools, which are accessible to bees and can be contaminated with pesticides through rain, runoff, or from soil surfaces. CPY on vegetation and in air can contribute to residues in rainwater and dew on plant surfaces that can be directly toxic, or can be returned to the hive/nest by foragers collecting water as part of the sublethal body burden (Gary 1975). If CPY is present in air, it can appear briefly in rainwater before hydrolysis occurs (Tunink 2010) or the water dries. The release of residues from surfaces of leaves following rain after a spray has dried should be limited due to the high affinity of CPY for nonpolar media (Solomon et al. 2014). No information on collection of water by non-*Apis* pollinators or where they obtain it was found, but scenarios for exposure of honey bees from water should be protective of non-*Apis* taxa, because they provide water for the hive and carry larger amounts.

Foliage and flower parts other than pollen represent a potential source of contact exposure for foraging pollinators. Leaf cutting bees (Megachilidae) may be particularly affected by dried residues of CPY on foliage since they cut and collect leaf discs for construction of their nest cells. Plant resin, e.g. from poplar buds, was included in the foliage compartment since honey bees collect small amounts of this material in making propolis. These materials could contain residues from off target drift, but propolis was considered to represent an insignificant exposure route to honey bee foragers (Fig. 1).

Soil and soil-water represent a potential pathway of exposure to CPY for pollinators that are ground nesters, or use soil in building nest cells, such as mason bees. These exposures can be from sprayable formulations or granular CPY that dissolves into soil-water.

Secondary routes of exposure. With social pollinating insects, such as honey bees and bumble bees, secondary exposure to pesticides can occur in other adults or offspring if the pesticide is brought back to the hive or nest and deposited in food or other materials, or transferred to other individuals (Fig. 1). Solitary bees such as alfalfa leafcutting bees or mason bees would not transfer residues to other adults, but larvae could be exposed orally to residues in food provisions; both larvae and eggs could potentially be exposed by contact with nesting materials that were contaminated in the field. Residues can also be excreted by honey bees in wax, which is produced metabolically and secreted by bees for construction of honeycomb. Residues in wax could originate from the sublethal body burden of CPY in bees as they produce the wax, by partitioning of CPY from contaminated pollen or nectar, and possibly by partitioning of CPY vapor from the air in the hive. As noted, CPY is not persistent in air and the maximum concentration of CPY in air is expected to be less than 250 ng m^{-3}. Thus transfer of CPY from air into wax in the hive is likely an insignificant pathway (Fig. 1). The potential for exposure via transfer of CPY from wax is low because wax is not consumed as food and because CPY is strongly lipophilic, with a Log K_{OW} of 5.0 (Mackay et al. 2014). This predicts that partitioning from wax into eggs, larvae, royal jelly, honey, or stored pollen is unfavorable. Wax is more likely a sink for CPY residues in the hive than a potential pathway for exposure.

The main pathway for secondary exposure is transfer of residues in pollen or nectar into the hive or nest. With CPY, the amount of pesticide in pollen or nectar is limited to what was present on these materials in the treated area of the crop during application, since CPY is non-systemic and is not redistributed within the plant. Some plant species have flowers that provide pollen or nectar for several days after opening and these would present the highest potential for oral exposure. Secondary oral exposure from pollen is not limited to sublethal doses since foragers carry pollen externally and have the potential to bring back to the nest pollen containing lethal pesticide concentrations without being impaired. This is not likely with propolis, nectar, or water, which are carried internally. Direct transfer of residues from propolis to larvae is highly unlikely, but some hive adults can subsequently be exposed when manipulating propolis in the hive.

Potential exposure to contaminated food in the hive depends on the type and amount of food consumed by the various life stages and castes of bees. While pollen likely represents the highest risk of oral exposure, there is a decline in concentration as pollen is processed and used as food in the hive (DeGrandi-Hoffman et al. 2013). Exposure via royal jelly is expected to be minimal because of the large K_{OW} of CPY (Mackay et al. 2014). In field-cage enclosed colonies fed almond pollen, collected from foraging bees in an orchard, the mean concentrations of CPY in bee bread and nurse bees were 32 and 8.3% of that found in the pollen, respectively, and no residues were detected in royal jelly or developing queen bees (DeGrandi-Hoffman et al. 2013). The results in this study show a reduction of at least 1,000-fold between concentrations of CPY in pollen and those in royal jelly and queen larvae. This shows isolation of the queen and larvae from exposure to CPY resulting from the social behavior of the colony, offers significant protection against potential toxicity of CPY.

As mentioned, amounts of CPY in nectar returned to a hive are expected to be less than in pollen since nectaries are less exposed than anthers (Willmer 2011), but this is still a pathway for secondary exposure. Nectar is dehydrated and digested by honey bees to make honey, which is the main source of carbohydrate for the hive. Mature honey in honey comb is capped with wax for later use, alone or mixed with stored pollen to make "bee bread", which is the major protein source for the colony (Winston 1987). Nectar can be consumed directly and is transferred between adult worker bees as food and when communicating forage sources (Butler 1975; Gary 1975). Potential secondary contact exposure of eggs and oral- and contact-exposure of larvae during the first 3 d of development is limited to residues released into royal jelly by nurse bees or transfer of material from beeswax. As noted above, this route is minimal for CPY. Older larvae can receive nectar, but only small amounts of pollen, and no food is offered to pupae. Larval queen bees are fed royal jelly continually and food is left in the capped cell for consumption during the pupal stage (Butler 1975; Dietz 1975). Even after emergence as an adult, the queen depends on nurse bees for food and water. This increases the isolation of the queen from exposure to toxicants in the nectar and pollen and exposure via royal jelly is considered to be less than for other food sources in the hive (USEPA 2012). Overall, in honey bees, greater amounts of pollen are consumed by nurse bees and, to a lesser extent, by larvae. Larger amounts of nectar or honey are consumed by wax-producing bees, brood-attending bees, "winter" bees, and foragers, with foragers consuming relatively large amounts (Rortais et al. 2005 and references therein).

For honeybees, the potential for exposure to CPY can be greater during production of bee bread by worker bees than in other activities in the colony. To make bee bread, workers break newly collected pollen balls deposited by foragers, mix the pollen with saliva and honey, and pack it into cells with their mandibles and tongue (Dietz 1975). It is possible that the appearance of dead bees in front of a hive following accidental overexposure to pesticides could be the result of these bees being exposed to a greater dose than the forager bees (Atkins 1975).

2.4 Endpoints

Assessment endpoints are explicit measures of the actual environmental value or entity to be protected (USEPA 1998). They are important because they provide direction and boundaries in the risk assessment for addressing protection goals and risk management issues of concern. Assessment endpoints were selected *a priori* based on likely pathways of pollinator exposure, patterns of use of CPY, and toxicity, as well as their ecological, economic, and societal value. For honey bees, relevant assessment endpoints are colony strength (population size and demographics) and survival of the colony (persistence), both of which have ecological relevance, are known to be affected by pesticide use, and are directly relevant to the stated management goals (Fischer and Moriarty 2011; USEPA 2012). Productivity of hive products such as honey was also considered as an assessment endpoint and is reflected in hive strength. For wild pollinators, species richness and abundance were considered to be the principle assessment endpoints. In contrast to honey bees, where the loss of a single forager has little impact on a colony as a reproductive unit, the loss of an individual bee of a solitary species represents the loss of a reproductive unit.

Measures of effects are specific parameters that are quantified as indicators of potential effects of stress that are linked to assessment endpoints (USEPA 1998). These measures are obtained from multiple levels of investigations, including laboratory dosing studies, modeling exercises, controlled field application studies, and incidents documented in the field. This approach covered all combinations of toxicity and exposure. In the laboratory, effects of pesticides on bees are mainly measured through survivorship after 24–96 h following acute topical or oral exposure, which is usually expressed as a LD_{50} (dose that kills 50% of the test organisms) or LC_{50} (the exposure concentration that kills 50% of test organisms). Acute exposures are particularly relevant for this risk assessment on pollinators since CPY exerts its toxic effects rapidly and has a relatively short half-life on vegetation (<1 wk) and soil surfaces (≈1 wk) (Mackay et al. 2014; Racke 1993; Solomon et al. 2001).

Chronic and sublethal tests can be conducted in the laboratory but there are no formal guidelines for conducting and interpreting these toxicity tests with pollinators (Desneux et al. 2007; Fischer and Moriarty 2011; USEPA 2012), and consistent linkages to assessment endpoints are lacking (Alix and Lewis 2010; Fischer and Moriarty 2011). Given the limited number of such studies that were found and the high degree of variability in methods among these studies, chronic laboratory studies were not used in this risk assessment. Since the focus of this risk assessment was on endpoints and assessment measures related to survival, development, reproduction, and colony strength, studies that examined effects of CPY on pollinators using endpoints such as oxidative stress (Shafiq ur 2009) and localized cell death (Gregoric and Ellis 2011) were not included.

Available higher-tier semi-field and field-tests provide data on mortality, foraging behavior, brood development, and overall vigor. These should receive greater weight than the results of sublethal testing because the net effect of multiple stressors and modes of action are integrated into these higher-tier tests (Thompson and Maus 2007). Semi-field and field tests were an important line of evidence in this risk assessment.

The analysis conducted here consisted of four parts recommended by the USEPA Risk Assessment Framework (Fischer and Moriarty 2011; USEPA 1998, 2012): (1) characterization of the stressor; (2) characterization of potential exposures by various pathways; (3) characterization of effects in pollinator or surrogate species; and (4) risk characterization.

2.5 *Sources of Information*

Data on exposure and toxicity were mainly obtained from reports in the peer-reviewed literature, the USEPA ECOTOX database (http://cfpub.epa.gov/ecotox/quick_query.htm), and internal reports obtained from Dow AgroSciences. Peer-reviewed articles were searched mainly through the ISI Web of Knowledge and SciVerse Scopus databases. Incident reports for the years 1990 to present were obtained from the Environmental Fate and Effects Division, USEPA Office of Pesticide Programs (USEPA 2013). Additional incident reports were obtained from the Health Canada PMRA.

2.6 *Risk Assessment Approach*

The risk characterization scheme applied was that used by the USEPA Office of Chemical Safety and Pollution Prevention for assessing risks of foliar sprayed pesticides to pollinators (USEPA 2012). The process is iterative, relying on multiple lines of evidence to refine and characterize risk. The scheme incorporates Tier-1 (worst case) screening-level assessments that calculate risk quotients (RQ) based on ratios of estimated exposure by contact exposure and oral uptake of CPY-contaminated nectar and pollen, and effects determined by corresponding toxicity tests. Strictly speaking, a RQ should refer to a value calculated on the basis of probabilities. European terminology favors "hazard quotient" (HQ) to represent this as a deterministic ratio. The OCSPP convention RQ was used in this document.

If the RQ exceeded the level of concern (LOC of 0.4 for acute tests), higher-tier assessments were needed to obtain a more realistic measure of the risk of CPY to pollinators. The Tier-2 process involved more elaborate semi-field or field studies with whole colonies, quantification of residues in pollen and nectar, and modeling. Risks of exposure to CPY through water on wet soil, such as puddles, and wet foliage from rain and dew was assessed by use of simulation models. Tier-3 tests studies were used to resolve important uncertainties identified in Tier-1 and Tier-2 assessments. Incident reports were also considered in the Tier-3 assessment.

Honey bees have long been included in regulatory test requirements as a surrogate for pollinators as well as for terrestrial invertebrates in general (USEPA 1988), and most data on CPY in this risk assessment relate to honey bees. In studies on 21 species of non-*Apis* bees, LD_{50} values for several species are within an order of magnitude of that of the honey bee (Fischer and Moriarty 2011), suggesting *A. mellifera* can be a good surrogate species for other bees (Porrini et al. 2003). Toxicity data for CPY in non-*Apis* pollinators were used when available.

In addition, certain non-target arthropods (NTA) such as *Aphidius* spp. (Hymenoptera: Braconidae), *Typhlodromus* spp. (Mesostigmata: Phytoseiidae), and *Aleochara bilineata* (Coleoptera: Staphylinidae) can be useful in assessing risks to non-*Apis* pollinators (Miles and Alix 2012). Therefore, an attempt was made to find useful toxicity data for CPY with these non-target arthropods and their usefulness as surrogates for non-*Apis* bees was evaluated.

3 Characterization of Exposures

3.1 General Physical and Chemical Properties and Fate

The chemical, physical, and environmental profile of CPY (Giesy et al. 1999; Racke 1993; Solomon et al. 2001, 2014), and its environmental fate on plants, in water and in soil (Mackay et al. 2014; Racke 1993; Solomon et al. 2001), have been well-described by others and is not repeated here.

3.2 General Fate in Insects

The metabolism of CPY in animals consists of transformation and conjugation processes. When not exposed to lethally toxic doses, CPY is readily metabolized and eliminated by most insects (Racke 1993). Activation to CPYO, which is the toxic form of CPY, and deactivation to form trichloropyridinol (TCP) occur simultaneously. Conjugation of the intermediates is a precursor to excretion (Racke 1993). In cockroaches (*Leucophaea maderea*), imported fire ants (*Solenopsis richteri*), and European corn borer (*Ostrinia nubilalis*) larvae, 5.5, 25, and 30.7% of CPY was excreted, respectively (Chambers et al. 1983; Tetreault 1985; Wass and Branson 1970).

3.3 Tier-1 Characterizations of Exposure

Estimates of contact exposure during spray application. CPY is applied as an insecticide and mitigation measures are required to protect pollinators. Such measures are described on product labels. Bee-kill incidents in the U.S. involving direct exposure to CPY are rare (see section on *Incident Reports* below), indicating that the effectiveness and level of compliance with these measures are high. Therefore, the direct contact route of exposure was not considered in the higher tier refinements of the risk assessment.

Estimates of dietary exposure. The USEPA has proposed that doses of pesticide received by bees via food can be calculated from rates of consumption of nectar and pollen estimated for larval and adult worker bees (USEPA 2012). Because toxicity data are expressed as doses (μg CPY bee^{-1}), it is necessary to convert estimated concentrations of CPY in food (mg CPY kg^{-1}) into doses. For honey bee larvae, the proposed total food consumption rate is 120 mg d^{-1}. For adult workers, a median food consumption rate of 292 mg d^{-1} is proposed, based on nectar consumption rates of nectar-foraging bees, which are expected to receive the greatest dietary exposures among different types of worker bees (USEPA 2012). These values are conservative estimates of dietary consumption and are expected to be protective of drones and queens as well. These methods are additionally conservative in that they assume that the pesticides do not degrade in the hive. The USEPA recommends that this Tier-1 exposure assessment covers both honey bees and other non-*Apis* bees (USEPA 2012).

Estimates of pesticide levels in nectar and pollen calculated by the T-REX model have been proposed (USEPA 2012). Based on upper-bound residue values for tall grass, 110 mg CPY kg^{-1} nectar for an application rate of 1.12 kg CPY ha^{-1} is proposed as a conservative (high-end) estimate of dose received by bees consuming nectar. An identical screening value of 110 mg CPY kg^{-1} pollen for an application rate of 1.12 kg CPY ha^{-1} is proposed for pollen. These values assume that concentrations are distributed uniformly in the plant tissues. They are converted to an estimated dietary dose that is based on larval and adult worker bees consuming aforementioned rates of pollen and nectar (120 and 292 mg d^{-1}, respectively). Therefore, the proposed dietary exposure values for larvae and adults are 12 μg CPY bee^{-1} kg CPY ha^{-1}, and 29 μg CPY bee^{-1} kg CPY ha^{-1}, respectively (USEPA 2012). Using these high-end proposed dietary exposure rates with maximum (1.05–6.31 kg CPY ha^{-1}) and minimum (0.26–2.10 kg CPY ha^{-1}) application rates for Lorsban 4E and Lorsban Advanced, gives estimated CPY dietary exposure estimates ranging from 3 to 183 μg CPY bee^{-1}, depending on application rate and life stage (Table 1).

Table 1 Tier-1 estimates of chlorpyrifos (CPY) dietary exposure in honey bees (aggregate nectar and/or honey and pollen consumption) during foliar applications of Lorsban 4E or Lorsban Advanced at minimum and maximum application rates based on T-REX estimates of concentrations in pollen and nectar

Life stage	Minimum dose (μg CPY bee^{-1})	Maximum dose (μg CPY bee^{-1})
Adult	7.5–61	30.5–183
Larvae	3.1–25	12.6–75

Estimates of post-application residual contact exposure. Post-application contact exposure for bees is mainly from exposure to residues on the surface of flowers that were open during application and remain attractive to bees after application, since they are attracted to flowers and do not typically land on leaves or other plant surfaces (Willmer 2011) (Fig. 1). An estimate of the concentration of CPY on the surfaces of plants in units of mass per unit area (e.g., μg cm^2) is required for comparison to the measured endpoint from standard contact toxicity tests. The USEPA guideline does not provide a Tier-1 estimate for this scenario (USEPA 2012). The Kenaga nomogram, as revised by Fletcher et al. (1994), was developed to provide an estimate of exposure on vegetation after application of a pesticide. However, the estimate is in mg AI kg^{-1} fresh weight, making it unsuitable for estimates of post-application contact exposure.

3.4 Tier-2 Characterization of Exposure

Estimates of dietary exposure from field data. Applications of CPY outside the flowering period would not be expected to result in exposure of bees through nectar and pollen, but some flowers, including those on weeds that were open during spray application, may remain available to foraging pollinators after application, leading to both contact and dietary exposure of adult foragers.

No data on concentrations of CPY in pollen and nectar manually collected in the field were found. However, several studies screened pollen or honey collected from honey bee hives for pesticides, including CPY. A broad survey of concentrations of pesticide in samples collected from honey bee hives across 23 states in the U.S., one Canadian province, and several agricultural cropping systems during the 2007–2008 growing season was conducted by Mullin et al. (2010). The survey included both migratory hives moved to multiple crops for pollination and non-migratory hives. Of the 118 pesticides and metabolites surveyed, CPY was the most frequently found insecticide other than those used in the hive as acaricides for mite control, and the third-most detected compound in trapped pollen or beebread samples (153 of 350 samples). The mean concentration was 53.3 ± 10.6 (SEM) μg kg^{-1} in those samples that had positive detects (Table 2). Median and 95th centile CPY pollen concentrations reported by Mullin et al. (2010) were based on calculations that included non-detections.

Table 2 Concentrations of chlorpyrifos (CPY) detected in pollen and honey from honey bee colonies

	Concentration (μg kg^{-1})						
Matrix	Mean[a]	Median[b]	Maximum[b]	95th centile[b]	LOD[c]	% of samples	Reference
Pollen	53.3	4.4	830.0	226.5	0.1	43.7 (153/350)	Mullin et al. (2010)
	35	–	35	–	10.0	0.5 (1/198)	Chauzat et al. (2011)
	955	–	967	–	NA	–	DeGrandi-Hoffman et al. (2013)
	302	–	310	–	NA	–	DeGrandi-Hoffman et al. (2013)
Honey	46	–	80	–	4.0	41.9 (13/31)	Pareja et al. (2011)
	–	–	15	–	0.8	–	Rissato et al. (2007)
	ND[d]	–	ND	–	3.5	0 (0/239)	Chauzat et al. (2011)
	ND	ND	ND	–	5.0	0 (0/51)	Choudhary and Sharma (2008)

[a]Based on positive detections
[b]Based on calculations that included 0 μg kg^{-1} for non-detections
[c]Limit of detection
[d]ND = CPY was included in residue analysis but was not detected

The mean concentration of CPY in almond pollen, collected from pollen traps on honey bee hives in an orchard in California that had been treated 2 wk earlier with Lorsban Advanced at 0.85 kg AI ha^{-1} (0.5 U.S. gal A^{-1}) as a mixture with crop oil, was 955 μg CPY kg^{-1} wet weight (wwt) (DeGrandi-Hoffman et al. 2013). In contrast to the analyses conducted by Mullin et al. (2010), the QuEChERS multiresidue analytical method used by DeGrandi-Hoffman et al. (2013) used external calibration standards, which could not account for matrix effects. This method can give results with large "peak enhancement" errors that may exceed 20% (Kwon et al. 2012). As a result, the concentrations reported can be considered as upper limit values but should be interpreted with caution. This value is 15% greater than the maximum concentration of CPY of 830 μg kg^{-1} and 3.7-fold greater than the 95th centile of 227 μg kg^{-1} (wwt) reported by Mullin et al. (2010). This study also characterized concentrations of CPY in other food components in the hive when the only pollen available to be bees contained residues of CPY. Mean concentrations of CPY in pollen used in the two experiments were 967 and 942 μg kg^{-1} (wwt), and the corresponding concentrations in bee bread were 310 and 293 μg kg^{-1} (wwt), which suggests degradation had occurred (Table 2). No residues were detected in royal jelly or in queen larvae.

In a study examining in-hive concentrations of pesticides in various matrices collected from 24 apiaries in France, 2002–2005, CPY was detected only in one of 198 samples of trapped pollen (Chauzat et al. 2011) (Table 2). In western Uruguay, various honey bee hive matrices were collected from depopulated and healthy honey bee hives (Pareja et al. 2011). Approximately 4,800 samples were obtained from eight depopulated apiaries and approximately 10,000 hive samples were obtained from 29 healthy apiaries. Each set of samples was randomly sub-sampled. CPY was detected

Table 3 Tier-2 estimates of chlorpyrifos (CPY) exposure through daily consumption of pollen and nectar by adult and larval honey bees

Life stage (consumption mg)	Median dose (μg CPY bee^{-1})[a]	95th centile dose (μg CPY bee^{-1})[b]
Adult (292)	1.28×10^{-3}	6.61×10^{-2}
Larva (120)	5.26×10^{-4}	2.72×10^{-2}

[a]Based on median CPY detection in pollen of 4.4 μg kg^{-1} (Mullin et al. 2010)
[b]Based on 95th centile CPY detection in pollen of 226.6 μg kg^{-1} (Mullin et al. 2010)

in honey from just under half of analyzed samples, at a mean concentration similar to that found in pollen in the U.S. by Mullin et al. (2010) (Table 2). CPY was detected in honey in Brazil (Rissato et al. 2007) (Table 2), but the authors did not provide details on the frequency of detection. CPY was not among the many pesticides detected in honey collected from hives in France (Chauzat et al. 2011). Similarly, CPY was not detected in honey samples collected from beekeepers in India, although concentrations of organochlorine, cyclodiene, synthetic pyrethroids, and other organophosphorus insecticides were found (Choudhary and Sharma 2008) (Table 2).

The Tier-1 estimates of exposures of 110 mg CPY kg^{-1} in nectar or pollen (per 1.12 kg^{-1} CPY ha^{-1}) appear to be overly conservative (protective). From the data reported by Mullin et al. (2010), CPY was not detected in most samples and, when detected, it was at concentrations several orders of magnitude below the modeled estimates from T-REX. Based on the data of Mullin et al. (2010) and the aggregate pollen and nectar (honey) consumption rates for larvae (120 mg d^{-1}) and adult worker bees (292 mg d^{-1}), the estimated dose of CPY received by honey bees would be several orders of magnitude below the Tier-1 modeled estimates of 110 mg CPY kg^{-1} from T-REX (Table 3).

The above estimates of oral exposure were based on daily consumption rates of honey bees. There is greater uncertainty regarding rates of nectar and pollen consumption for non-*Apis* bees. However, an analysis of data recently compiled by EFSA (2012) suggests that adult honey bee workers and adult bumble bees have similar consumption rates, while that of adult female European mason bees and alfalfa leaf cutting bees is less. The same trends hold for larvae of these bees. Thus, exposures estimated from consumption of pollen and nectar by adult honey bees should be representative or protective of these non-*Apis* pollinators.

Estimates of post-application residual contact exposure on vegetation. Pollinators can come in contact with residues of CPY on flowers or inflorescences, or in some cases with extrafloral nectaries, following a spray application if flowers that were open during application remain attractive to pollinators after application. This is potentially a major route of exposure (Fig. 1). Residues on flowers are expected to be similar to or less than those found in or on foliage. These will be greatest immediately after spraying of foliage and thereafter dissipate rapidly through volatilization, photolysis, and dilution by growth of the plant. Residual contact exposure will also decline with time as visits of pollinators to older flowers decrease and visits to newer unsprayed flowers increase. This usually occurs within 1–3 d (Willmer 2011).

Table 4 Dissipation and concentration of dislodgeable foliar residues following application of chlorpyrifos (CPY) to different plants

Plant	Application rate (kg ha^{-1})	Half-life (d)	Time (d)	Residue (μg cm^{-2})	Adjusted to 1.12 kg ha^{-1}	Reference
Cotton	1.12	–	0	3.64	3.64	Buck et al. (1980)
			1	0.13	0.13	
			2	0.071	0.071	
			3	0.055	0.055	
			4	0.034	0.034	
Cotton	1.12	<1	0	3.62	3.62	Ware et al. (1980)
			1	0.3	0.3	
			2	0.191	0.191	
			3	0.069	0.069	
			4	0.068	0.068	
Orange	5.6	–	4	0.013	0.003	Iwata et al. (1983)
			10	0.005	0.001	
	11.21	–	4	0.031	0.003	
			10	0.012	0.001	
			17	0.006	0.0006	
	11.21 (ULV)	–	4	0.08	0.008	
			10	0.021	0.002	
			17	0.015	0.001	
			31	0.008	0.0008	
Grape fruit	5.6	2.4	3	0.035	0.007	
	11.21	3.4	3	0.061	0.006	
Cranberry	2.0	3.8	0 (2 h)	52.5	28.9	Putnam et al. (2003)
			3	23.95	13.2	
			15	6.14	3.4	
Kentucky bluegrass	2.2	0.1–0.3	0	0.14	0.07	Goh et al. (1986)
			1	0.04	0.02	
			2	0.03	0.015	
			3	0.018	0.009	
			4	0.013	0.007	
Kentucky bluegrass	2	<1		0.456	0.251	Sears et al. (1987)

An estimate of the upper-bound concentration of CPY likely to be on flowers can be obtained from the results of dislodgeable foliar residue studies (USEPA 2012). These studies show that CPY does not persist on plant surfaces. In some studies, dissipation was too rapid to produce meaningful dissipation curves (Iwata et al. 1983), but the average half-life was 1.5 d (Racke 1993; Solomon et al. 2001). The upper 90% confidence limit on the mean foliar half-life was 3.28 d (Williams et al. 2014). CPY that drifts onto non-target plants should dissipate at a similar rate, but initial concentrations would be less. Initial concentrations recorded for most crops were <4 μg cm^2, but were considerably larger for cranberry (Table 4). In a field study on cotton, the concentration of CPY declined to 3.6% of its initial value at 24 h and probably would not be efficacious for pest control (Buck et al. 1980). This suggests that exposure to pollinators would also be below toxic doses within a day after application. From reported values in Table 4, geometric means for concentrations of dislodgeable CPY at 0 and 3–4 d after spraying were 1.46 and 0.019 μg cm^{-2}, respectively (adjusted for a 1.12 kg ha^{-1} application rate).

Reported effects of irrigation on the concentration of dislodgeable residues on foliar surfaces have been variable (Table 4). Whereas immediate post-application irrigation did not affect the concentration of dislodgeable CPY from leaves of turf grass (Hurto and Prinster 1993), significant reductions in concentrations of CPY on grass foliage were found (4-fold difference after 6 h) following post-spray irrigating with water, as recommended by the product label instructions for most turf insect control situations (Goh et al. 1986).

At least two studies have also examined dislodgeable concentrations of CPYO. After applications at 11.2 kg CPY ha^{-1}, no CYPO was detected with a detection limit of 0.01 μg cm^{-2}. The results also showed very rapid disappearance of the parent insecticide (Iwata et al. 1983). On grapefruit leaves, trace amounts (0.013–0.028 μg cm^{-2}) of CPYO were detected in samples collected 3 d after application. When applied to cranberry at 2 kg ha^{-1}, small amounts of CPYO were initially detected (<7 μg kg^{-1}), but did not accumulate (Putnam et al. 2003). These results indicate that any CYPO formed on foliage is rapidly dissipated and does not accumulate. Given the demonstrated lack of potential exposure, higher Tier-refinement of the potential exposure to CPYO is not required.

3.5 Other Potential Routes of Exposure

Exposure via beeswax. Although wax is not consumed as food, there is direct contact between wax cell surfaces and food or individuals. Residues initially present in the wax could come from sublethal concentrations of CPY inside the body of bees that secrete the wax. After it is secreted, it may accumulate from contact with bees, pollen, nectar or other materials. The transfer of residues into or from wax is reversible and given the nonpolar nature of CPY it is likely that the partition of CPY between wax and bees or food substances tends toward equilibrium with higher concentrations in the wax. The net effect of absorption into wax is to reduce the potential for exposure of bees to CPY.

Several of the North American and European studies mentioned above examined concentrations of pesticides in beeswax collected from honey bee hives. Mullin et al. (2010) found CPY more often in foundation wax than in comb, but at similar concentrations (Table 5). Excluding pesticides that are used by beekeepers within hives to control *Varroa* mite parasites (fluvalinate, coumaphos, and its degradate coumaphos oxon), CPY was the most frequently detected pesticide in beeswax of the 118 pesticides and metabolites analyzed (Mullin et al. 2010). CPY was detected less often in beeswax collected from hives in France (Chauzat et al. 2011) and Spain

Table 5 Concentrations of chlorpyrifos (CPY) detected in comb and foundation beeswax from honey bee colonies

	Concentration (μg kg^{-1})						
Matrix	Mean[a]	Median[b]	Maximum[b]	95th centile[b]	LOD[c]	% of samples	Reference
Comb beeswax	24.5	4.3	890.0	55.7	0.1	63.2 (163/258)	Mullin et al. (2010)
	14.9	–	19.0	–	–	3.5 (3/87)	Chauzat et al. (2011)
	172	–	–	–	6.0	5.6 (1/18)	Serra-Bonvehí and Orantes-Bermejo (2010)
	8	–	15	–	1.0	62 (8/13)	Wu et al. (2011)
	ND[d]	ND	ND	–	1.0	0 (0/31)	Cutler (2013)
Foundation beeswax	22.2	10.0	110.0	76.4	0.1	80.9 (17/21)	Mullin et al. (2010)

[a]Based on positive detections
[b]Based on calculations that included 0 μg kg for non-detections
[c]Limit of detection
[d]ND = CPY was included in residue analysis but was not detected

(Serra-Bonvehí and Orantes-Bermejo 2010) (Table 5). In 31 pooled samples of beeswax (samples from hives from a single site were pooled) collected from the Canadian provinces of Nova Scotia, Prince Edward Island, and New Brunswick, CPY was not detected (Cutler 2013).

In an assessment of effects on bees when exposed to pesticide-contaminated wax, samples of brood comb were taken from hives that were suspected to have died from Colony Collapse Disorder (Wu et al. 2011). Residue analyses were performed on brood comb samples. Of 13 frames of brood comb that contained large concentrations of pesticides, CPY was detected in approximately two-thirds of samples (Table 5).

Concentrations of CPY in beeswax reported by Mullin et al. (2010) were similar to those that were found in pollen and greater than those reported in bees or honey (Johnson et al. 2010; Mullin et al. 2010). Since wax is produced and exuded by bees in the hive, and concentrations are similar to those in pollen, it can be concluded that the concentrations in wax enter the hive mainly on pollen or as the sublethal body burden on forager bees. If wax is indeed a sink for CPY, the presence of CPY in beeswax may not result in exposure (see discussion in Sect. 2.3).

Exposures via soil. Many bees live in or utilize soil for construction of nests. For example, mason bees (*Osmia* spp.: Megachilidae) make compartments of mud in their nests, while mining bees (Andrenidae), digger bees (Anthophorinae), and sweat bees (Halictidae) are solitary underground nesters (Michener 2007). Pollinators that live in or use soil subject to application of pesticides can be exposed to CPY after application of either sprayable or granular formulations (Fig. 1).

Although much has been written on communities of pollinators in agricultural landscapes and factors that influence diversity and abundance of bee populations in these habitats (Williams et al. 2010; Winfree et al. 2009), there is limited information on nesting habits of ground nesting bees within cropping systems (Julier and Roulston 2009; Kim et al. 2006; Williams et al. 2010; Wuellner 1999), and on exposure of ground nesting bees to pesticides. The potential use of data from surrogate species was therefore considered.

The exposure of arthropods to CPY in soil following application of spray or granular formulations on the soil surface has been studied, mainly for characterizing exposure in birds that consume insects (Moore et al. 2014; Solomon et al. 2001). Fewer studies have examined exposure of arthropods to CPY in soil, and these were done to evaluate the efficacy of CPY against pest insects (Clements and Bale 1988; Tashiro and Kuhr 1978; Tashiro 1987). None of these studies included pollinators. Exposures are different for ground-burrowing insect pests that ingest contaminated vegetation or soil, making these data unsuitable for estimating exposures of ground-nesting pollinators. Thus, exposures of soil-dwelling pollinators via this route were not estimated and it remains an area of uncertainty.

Exposure via drinking-water. Water is potentially a significant route of exposure (Fig. 1). In obtaining water for a large number of individuals in a colony, honey bees collect much more water than other bees, and therefore serve as a conservative representative species for this route of exposure. Typical sources include wet foliage, puddles, soil saturated with water, or other sources where they can get access to water without drowning (Gary 1975; Winston 1987). Because CPY is not systemic, exposure to CPY through guttation water is not significant (Fig. 1).

Only a small proportion of the honey bees in a hive are dedicated to foraging for water and recruiting other bees to forage for water (Winston 1987). Water containing CPY brought back to the hive is limited to sublethal levels low enough that the ability of the forager to return to the hive is not affected. When demand for water is large, foraging can continue through the day. Individual loads of water average approximately 25 mg although some loads can be larger, and each load can take approximately 10–12 min to obtain and deliver into the hive. If foraging continues for 10 h, the forager would carry 50 loads or 1,250 mg of water to the hive (Gary 1975) from a source such as a puddle. Honey bees do not forage during rain and the overlap of foraging time with the time when soil is wet enough for bees to collect soil pore water is short. Exposure from puddles is recommended to represent the worst case for collection of water from the soil surface (USEPA 2012).

The time when water can be obtained from wet foliage is also short. It takes approximately 1 h for wet foliage to dry. Foraging after the dew point is reached in the evening is unlikely, but more than one rain event is possible. If water is collected from wet foliage for 2 h each day the forager can carry as much as 250 mg of water from that source. Temperature is lower and humidity is greater when the foliage is wet, and this reduces demand for water in the hive, making this an upper-limit estimate.

The amount of water a honey bee will actually drink is unknown (USEPA 2012) and likely variable. An estimated rate of intake of 47 $\mu L\ d^{-1}$ based on direct measurements of water flux rates of the brown paper wasp (a similar species) is considered reliable by the USEPA for regulatory purposes (USEPA 2012). To simplify the risk assessment, it was assumed that bees collect their full daily requirement from the source with the highest concentration of CPY.

To estimate potential exposure of pollinators to CPY in water, it is also necessary to know or estimate the concentration of CPY in different water sources, which are expected to vary, and the amount of water pollinators derive from each potential source. The main potential sources are evaluated below:

Puddles and soil pore water. Several estimates of the concentration in puddles in the field are available. The USEPA recommends that concentrations of pesticides in puddles located on pesticide-treated fields be estimated using a modified version of the Tier-1 rice model (v. 1.0) (USEPA 2012). The model uses equilibrium partitioning to provide conservative estimates of environmental concentrations and assumes that puddles can be directly sprayed with pesticide and the pesticide will instantaneously partition between a water phase and a sediment phase, independent of the size of the puddle. With this model, the concentration is determined by partition equilibrium and does not increase as the puddle dries out; residues are deposited on the soil during drying to maintain the equilibrium. A sensitivity analysis was done to identify parameters that would give high-end estimates of exposure, and peak estimated concentrations are based on an application rate of 1.12 kg ha^{-1}. The model is represented in equation 1 (USEPA 2012).

$$C_w = \frac{m_{CPY}}{d_w + d_{sed}\left(\theta_{sed} + \rho_b k_d\right)} \quad (1)$$

Where: C_w is the concentration in water (µg L^{-1}), m_{CPY} is mass applied per unit area (kg ha^{-1}), k_d is water-soil partition coefficient (L kg^{-1}) (equivalent to K_{OC} *0.01), d_{sed} is sediment depth, dw is water depth, ρ_b is bulk density (kg m^{-3}), and θ_{sed} is porosity.

Using the mean K_{OC} value of 8216 for CPY (Solomon et al. 2014), the estimated concentration in puddle-water in a field following a spray application of 1.12 kg CPY ha^{-1} is 0.0051 µg CPY L^{-1}. Assuming that the intake rate of water is 47 µL d^{-1} and that bees obtain 100% of their drinking water from such puddles on treated fields, the Tier-1 estimate for CPY dose was 2.40×10^{-7} µg CPY bee^{-1}. For honey bees collecting 1,250 µL d^{-1} the estimated dose was 6.38×10^{-6} µg CPY bee^{-1} d^{-1}.

The maximum concentration of CPY in puddles after application of both the granular formulations and the sprayable formulations was also modeled using PRZM/EXAMS, which provides both puddle and soil pore water concentrations (Williams et al. 2014). The maximum 95th centile of the peak pore water concentrations from the PRZM/EXAMS model among registered uses in the U.S. of the granular and the spray formulations were 571 and 566 µg L^{-1}, respectively, based on a 1.12 kg CPY ha^{-1} application rate and the maximum number of applications per year. These values were obtained from the North Carolina tobacco and California broccoli standard use scenarios for PRZM/EXAMS, respectively (Williams et al. 2014). The greatest peak concentrations predicted for puddle water were 285 µg L^{-1} for granular and 529 μg L^{-1} for spray applications. Given the similarity of the pore water and puddle water values, only the highest value, 571 μg L^{-1} was selected for use in the risk assessment. With a daily intake of 47 μL^{-1}, the predicted 95th centile of the maximum daily dose was 0.027 μg bee^{-1} d^{-1}. The corresponding value for honey bees collecting 1,250 μL of water d^{-1} for the hive is 0.71 μg bee^{-1} d^{-1}. These estimates include peak values after storm events and are much greater than the equilibrium-based values in the Tier-1 Rice model. Exposure to these values is possible but depends on a combination of probabilities, limited to only a few use scenarios. In many use scenarios that were run in the PRZM/EXAMS model, the median predicted puddle concentrations were zero due to the large time interval between application and isolated heavy storm events during the 30-yr simulation interval.

Dew and wet foliage. The USEPA recommends a conservative (protective) equilibrium partition model based on pesticide K_{OC} and plant carbon content to estimate pesticide concentrations in dew (USEPA 2012) (Equation 2).

$$C_{dew(t)} = \frac{C_{plant(t)}}{K_{oc} \times f_{oc}} \tag{2}$$

Where: $C_{dew(t)}$ is the concentration of dissolved pesticide in dew (mg L^{-1}); $C_{plant(t)}$ is the concentration of pesticide on and in plant leaves (mg kg^{-1} (fresh weight)) at time t and was set at 240 mg CPY kg^{-1} foliage, corresponding to T-REX concentrations on short grass; and f_{OC} is the fraction of organic carbon in leaves, set at 0.04 (4% of fresh wt) based on estimates of carbon in plants (Donahue et al. 1983) and water content (Raven et al. 1992). As with the puddle model, this is an equilibrium equation and the concentration does not increase as the water dries on the surface. Partition into rainwater that remains on foliage after a rainfall is expected to be similar without runoff, or less if runoff occurs and reduces the amount of residue left on the leaf surface. Using the mean K_{OC} of 8,216 (Solomon et al. 2014), $C_{dew(0)}$ for a spray application at 1.12 kg CPY ha^{-1} is 730 μg L^{-1}. If a bee consumed 100% of its daily drinking water from contaminated dew and has an intake of 47 μL d^{-1}, this model predicts a point estimate dose of 0.034 μg CPY bee^{-1} d^{-1}. If the intake was 250 μL d^{-1}, the dose would be 0.18 μg CPY bee^{-1} d^{-1}.

A second estimate of exposure for dew and/or wet foliage was obtained using the LiquiPARAM model, which gives both a mean and an estimate of variability (Moore et al. 2014). Using the same data as USEPA (2005), with the K_{OC} for CPY and the f_{OC} value of 0.40 derived for alfalfa, clover, bluegrass, corn stalk, and small grain straw this model predicts a worst-case mean CPY dew concentration (at 09:00 h, immediately after application) of 102 μg CPY L^{-1} and a 95th centile concentration of 210 μg CPY L^{-1}. If a bee consumed 100% of its daily drinking water from contaminated dew and has an intake 47 μL d^{-1}, this model predicts mean and 95th centile daily doses of 0.0048 and 0.0099 μg CPY bee^{-1}, respectively. The corresponding values for collection of 1,250 μL d^{-1} are 0.03 and 0.05 μg CPY bee^{-1}.

A variety of uses of CPY involve application by mixing the product in irrigation water or chemigation. These applications result in wet foliage with water that contains a high concentration of CPY and is available for extended intervals of time during the day. For example at the rate for peppermint, a 1 cm ha^{-1} irrigation with an application at 2.12 kg ha^{-1} (1.88 lb A^{-1}) would give a dose of over 600 ng bee^{-1} for a typical uptake of 30 µL of water by a foraging bee in a single trip. This amount would be lethal, and foraging honey bees would not make it back to the hive to recruit more foragers to the wet foliage as a water source.

Other routes of exposure. Davis and Williams (1990) extended the typical approach of calculating intrinsic toxicity levels and field application rates to consider buffer zones downwind of sprayed areas and provide an estimates of the distance at which bees would encounter an LD_{50} dose from spray drift. These distances were determined using published data on spray depositions under various weather conditions for ground and aerial sprays of crops in Britain. They concluded that ground spraying of CPY at typical application rates would result in exposures of honey bees at the LD_{50} within 36–46 m of the application site at a wind speed of 4 m sec^{-1} (14.4 km h^{-1}). Labels for products containing CPY state that sprays are not to be applied when wind speed exceeds 16 km h^{-1}.

Only one reference on potential toxicity of CPY vapor to pollinators was found, indicating that vapor of Lorsban WP (50% CPY) applied at 0.56 kg product ha^{-1} should not have effects on honey bees (Clinch 1972). There are no fumigant products based on CPY. While the lipophilicity of CPY (log K_{OW}=5.0) (Solomon et al. 2014) is high enough to make accumulation of CPY in honey bee wax from air plausible, the concentrations in air are very small and ephemeral. A maximum of 250 ng m^{-3} has been reported (Mackay et al. 2014). There is little evidence to support the possibility of accumulation of concentrations in wax in the hive from trace concentrations in the air.

4 Toxicity of CPY to Pollinators

4.1 Tier-1 Tests of Effects

Acute toxicity to A. mellifera. Acute toxicity of CPY to *A. mellifera* has been determined, and acute topical LD_{50} values ranged from 0.024 µg bee^{-1} to 0.55 µg bee^{-1} (Table 6). *A. mellifera* appears to be slightly less sensitive to CPY by the dietary route, with oral LD_{50} values ranging from 0.114 µg bee^{-1} with technical product, to 2.15 µg bee^{-1} of formulated product (18.7% CPY) (Table 7).

One study was found that reported the acute toxicity of CPY to honey bee larvae. Atkins and Kellum (1986) carried out studies to determine the potential hazard to honey bee brood of pesticide contaminated food in the hive. Pesticides were added to individual brood cells followed by monitoring of effects throughout the brood cycle and into the adult stage. This resulted in a combined oral and cuticular exposure. For CPY (Lorsban 4E), 5–6 day-old larvae were the most susceptible age-group, whereas 1–2 day-old larvae were the least susceptible. The recorded LD_{50} values for

Table 6 Acute topical toxicity (48 h unless indicated otherwise) of technical and formulated chlorpyrifos (CPY) to the honey bee

Formulation	% Purity	Topical LD_{50} (μg CPY bee^{-1})	Reference
Technical	≥ 95	0.059[a]	Stevenson (1978)
Technical	99%	0.115[b]	Mansour and Al-Jalili (1985)
Technical product geometric mean		0.082	
Lorsban 48E	48	0.024	Carrasco-Letelier et al. (2012)
Dursban F	97.4	0.070	Chen (1994)
Lorsban Advanced	41.1	0.14[c]	Schmitzer (2008)
Dursban 480	48	0.22	Bell (1993)
Dursban WG	75	0.54	Bell (1996)
Formulated product geometric mean		0.123	

[a]Test duration not reported
[b]25 h test duration
[c]Reported as 0.35 μg product bee^{-1}, formulation code GF-2153

Table 7 Acute oral toxicity (48 h unless indicated otherwise) of technical and formulated chlorpyrifos (CPY) to the honey bee

Formulation	% Purity	Oral LD_{50} (μg CPY bee^{-1})	Reference
Technical	≥95%	0.114[a]	Stevenson (1978)
Lorsban 4E	–	0.11[b]	Atkins and Kellum (1986)
Dursban 4	48	0.29	Anonymous (1986)
Dursban 480	48	0.33	Bell (1993)
Dursban F	97.4	0.36	Chen (1994)
Lorsban Advanced	41.1	0.39[c]	Schmitzer (2008)
Lorsban 50 WP	50	0.4[b]	Clinch (1972)
Lorsban 50 W	50.2	0.46[d]	Hahne (2000)
Dursban WG	75	1.1	Bell (1996)
Formulated product geometric mean		0.36	

[a]Test duration not reported
[b]24 h test duration
[c]Reported as 0.94 μg product bee^{-1}, formulation code GF-2153
[d]Reported as 0.91 μg product bee^{-1}

1–2, 3–4, and 5–6 day-old larvae and adults were 0.209, 0.302, 0.066, and 0.11 μg bee^{-1}, respectively (Atkins and Kellum 1986). Based on these data, the LD_{50} geometric mean for all larval stages was 0.146 μg CPY $larvae^{-1}$ (the authors report a mean of 0.051 μg $larvae^{-1}$, although it is unclear how this value was derived), which is approximately twice the topical LD_{50} for adult bees. Therefore, LD_{50} values for adult bees are protective of the larval life stages.

Effects on non-target arthropods (NTA) as a surrogate for non-Apis bees. It has been recommended that Tier-1data for NTA be generated by exposing *Aphidius rhopalosiphi* and *Aphidius pyri* to fresh dried residues of product applied on glass plates to generate LR_{50} values (rate of application of the pesticide causing 50% mortality of the test organisms) (Candolfi et al. 2001). This test is meant to represent a case worse than that experienced on a natural substrate such as a leaf. No reports of toxicity data for CPY to *A. rhopalosiphi* and *T. pyri* using the glass plate technique were found, but data for *Bracon hebetor* (Braconidae) and *A. ervi*, which are related species of wasps were found. The LR_{50} value for *B. hebetor* was 62 g CPY ha^{-1} (Ahmed and Ahmad 2006) and that for adult female *A. ervi* was 0.047 g CPY ha^{-1} (Desneux et al. 2004). These results were obtained with the active ingredient coated on the inside of a 2.3 cm diameter by 9.3 cm glass vial in which the wasp was contained, leading to a greater potential uptake of the dose from the surface than would occur under field conditions. Thus, the results may be useful for comparison of toxicity but are not an indication of toxicity in the field. At present, too few data obtained using this method are available to permit comparisons to be made among species. In addition, the small size (and large surface area to volume ratio) of these wasps suggests that they may experience greater exposures via contact with treated surfaces than the larger pollinators and thus would be poor surrogates.

4.2 Tier-2 Tests of Effects

Semi-field studies. The following semi-field (tunnel tests) studies were conducted in Europe and with formulations not currently registered in the U.S. Nonetheless, these studies were conducted using standard methods with formulations containing amounts of active ingredient similar to that in current US formulations and therefore provide data that are relevant in the assessment of risk of CPY to pollinators.

A semi-field experiment with mini-beehives (approximately 2,000 individuals) in field cages large enough to allow foraging behavior to be assessed in a contained colony was conducted to test effects of exposure of honey bees to CPY and other pesticides at a series of times after application (Bakker and Calis 2003). When potted *Phacelia* plants treated with Dursban 75WG (76.3% CPY) at 1 kg CPY ha^{-1} were added to the cages at night, the number of dead bees collected outside hives was significantly greater compared to control hives on the first day of exposure, but not on subsequent days. Foraging activity of bees was also reduced for up to 4 d following the exposure phase (Bakker and Calis 2003).

In another tunnel test, the effects of aged Dursban 75WG foliar residues on behavior and mortality of foragers, and brood development of *A. mellifera* was examined (Bakker 2000). Dursban 75WG was applied at 1 kg CPY ha^{-1} to potted *Phacelia tanacetifolia* under outdoor conditions at 14, 7, 5, and 3 d before exposure, and the evening before exposure. During the aging process, plants were placed under UV-transparent synthetic foil to protect them from rain. Exposure to aged CPY did not result in a statistically significant increase in the number of dead bees. However, reduced foraging activity was observed in all treatments. Exposure to 1 or 3 d-old CPY residues resulted in an immediate reduction of foraging activity that

Table 8 Vitality of bumble bee, *B. impatiens*, colonies following 2-wk exposure to dry chlorpyrifos (CPY) on mixed stands of turf and flowering white clover (adapted from Gels et al. 2002)

Colony measure	Control	CPY
Weight (g)		
Colony (without hive)	193.4 ± 26.3	107.8 ± 7.2[a]
Workers	23.1 ± 4.9	7.5 ± 1.1[a]
Queen	0.78 ± 0.05	0.78 ± 0.08
No. in colony		
Workers	132.8 ± 19.6	56.8 ± 6.5[a]
Honey pots	41.8 ± 12.9	5.5 ± 3.6[a]
Brood chambers	56.0 ± 5.1	3.5 ± 1.3[a]

[a]Indicates statistically significant ($\alpha = 0.05$) based on analysis with four treatments: control, CPY, carbaryl (not shown) and cyfluthrin (not shown)

lasted the duration of the 4-d post-exposure assessment period. Exposure to older residues of Dursban resulted in a delayed reduction in forager activity. Since the effect persisted longer than in any other study, there could have been a repellent effect from a component of the formulation other than the active ingredient, or the memory by bees of CPY on foliage might have been a retained behavioral influence on foraging, possibly involving the level of demand for food in the hives. No effects on brood development were seen in any treatment (Bakker 2000).

In conditions such as those presented in these studies, where bees are confined to experimental plots with a lack of choice of forage, it appears that CPY is toxic for the first 24 h post-application but only has sublethal effects such as repellency after 24 h. Repellency is considered a sublethal effect but it may be beneficial and is only an indirect adverse effect in that it may result in a reduced food supply to the hive. Avoidance of a pollinator to potentially harmful CPY residues is beneficial.

Toxicity studies with non-Apis pollinators. Acute toxicity data for pesticides and non-*Apis* pollinators is far less common than for *A. mellifera*. No reports of contact and oral LD_{50} values for CPY to *Bombus* spp. were found. However, a semi-field study with *B. impatiens* Cresson was conducted (Gels et al. 2002). Colonies of *B. impatiens* confined in field cages were exposed to dried residues of CPY on weedy turf 24 h after application of Dursban 50 W at 1.12 kg CPY ha^{-1}. Effects on colonies were evaluated at 14 d. Adverse effects on vitality of bumble bee colonies were observed, including fewer worker bees, honey pots (stored food), and brood chambers in hives from treated plots relative to control plots (Table 8). Biomass of workers and weights of colonies were also reduced, and two of the four colonies had no live brood or adults. Reduced foraging activity was also recorded when bumble bee colonies were confined to CPY-treated plots, although endemic bumble bees did not avoid foraging on CPY-treated flowering white clover intermixed with turf (Gels et al. 2002).

Some species of non-*Apis* bees can be exposed to residues of CPY on nesting materials such as foliage or soil collected in or near treated crops. Adults can be exposed when collecting these materials and building nest cells, and immature stages developing in these cells may also be exposed. When caged adult alfalfa leafcutting bees, *M. rotundata*, were exposed to alfalfa plants sprayed with CPY (Lorsban 4E, 2.5 kg ha^{-1}), significant mortality was observed. The population of males was reduced by approximately 90% after only 2 d, and the population of females was reduced by 30% relative to the controls. No significant additional mortality of female *M. rotundata* was observed after the first 4–5 d of exposure (Gregory et al. 1992). It was suggested that male *M. rotundata* were more sensitive to CPY due to their reduced metabolic capacity and smaller surface area to volume ratio compared to females.

Studies to evaluate the toxicity of CPY to three species of bees have been conducted (Lunden et al. 1986). Field applications of 1.12 kg ha^{-1}, followed by 24-h continuous exposure to the treated foliage in small cages was lethally toxic to adult honey bees, alfalfa leafcutting bees (*M. rotundata*), and alkali bees (*Nomia melanderia* Cock.) for 5–7 d, whereas a rate of 0.56 kg ha^{-1} was toxic for 4–6 d. In field tests on several crops, mortality was observed with application of CPY and reduced foraging was observed for 1–7 d (Lunden et al. 1986).

Effects on NTA as a surrogate for non-Apis bees. Tier-2 tests with surrogate NTA species on a natural substrate such as foliage are more realistic than Tier-1 tests that utilize glass plates (Candolfi et al. 2001), but are still conservative because the test organisms are constrained on or near the treated surface. As with the Tier-1 assessment, no contact toxicity data were found for the recommended wasp species *A. rhopalosiphi*. There were data for *B. hebetor* (Ahmed and Ahmad 2006) exposed to CPY via leaves of cotton, but unfortunately the method of treatment, dipping the leaves in an aqueous solution, did not allow the deposition on the surface of the leaf to be calculated, making the data unusable in this risk assessment.

4.3 Tier-3 Field Tests

Several field studies have been conducted to examine the effects on honey bees of applications of CYP to agricultural crops. These effects are summarized in Table 9 and are described in more detail below. The applications made in these field studies did not follow current label restrictions, which prohibit application when bees are actively foraging. The results from application during bloom in a number of crops under various exposure scenarios suggests that CPY remains lethal to honey bees for 1–2 d after application on open flowers and may reduce foraging for several days thereafter.

Dursban was sprayed by helicopter on unreplicated 16-A (approx. 6.5 ha) blooming alfalfa fields that contained 1–3 honey bee colonies A^{-1} (Atkins et al. 1973). When applied at night at the highest rate of 1.12 kg ha^{-1}, Dursban killed an average of 365 bees per colony for 1 d and depressed bee visitation for approximately 3 d. Because honeybee colonies typically contain 30–60 thousand bees in midsummer when

Table 9 Effects of chlorpyrifos (CPY) treatment on honey bees in field studies

Crop	Formulation	Rate (g CPY ha^{-1})	Method	Observations	Reference
Alfalfa	Dursban	280–1,120	Field, aerial	High kill on d-1 and reduced foraging for 2–3 d at all rates; no clear advantage of night applications	Atkins et al. (1973)
	Lorsban 4E	1,120	Aerial	Significant mortality 1–2 d after application; reduced flower visitation of honey bees and alfalfa leaf cutting bees	Lunden et al. (1986)
Carrot	Lorsban 50 W	1,120	Field	Mortality and reduced foraging for 1–2 d; "moderate" bee kill	Lunden et al. (1986)
Citrus	Lorsban 4E	1,700–2,200	Field, aerial	Increased mortality compared to control; classified as "moderately hazardous" in most trials; no clear advantage of night applications	Atkins and Kellum (1993)
Corn	Lorsban 4E	560–1,120	Chemigation	Mortality and reduced foraging	Lunden et al. (1986)
Dandelion	Lorsban 4E	1,120	Orchard	No effects	Lunden et al. (1986)
Raspberry	Lorsban 50 WP	1,680	Field plots	Erratic behavior, reduced floral visitation	Lunden et al. (1986)

alfalfa is in bloom, the loss of this many bees is not likely to be significant for the health of the colony. It is unclear whether reduced visitation was due to mortality of bees, a repellent effect of the treated foliage, or, if the lack of bees returning to the hive to communicate the location of the food resource in the treated area lead to a shift in foraging to areas from which bees did return. Laboratory bioassays with this field-treated foliage killed 100% of exposed bees for 3 d. Unfortunately the concentration of CPY per unit of surface area was not determined. Results were similar but less severe at lesser concentrations of Dursban. No dead bees were found at the colonies when Dursban was applied at 0.28 kg ha^{-1} in the morning, although visitations on the field were moderately depressed for 2 d. Bioassays of foliage aged for 12 h from this treatment killed 100% of bees, and showed 0–31% kill on foliage aged 48–96 h (Atkins et al. 1973).

A number of field trials examining mortality and visitation to flowers in alfalfa, raspberry, dandelion, carrot, and corn were reported by Lunden et al. (1986). In field tests on alfalfa in Washington State, Lorsban 4E was applied by aircraft at 1.12 kg CPY ha^{-1} to several 0.4–0.8 ha plots in the evening (19:00–21:00 h). Each location contained two honey bee colonies and three nesting boards with alfalfa leafcutting bees adjacent to the crop. Mortality of honey bees in the treated plots was five-to eight-fold higher than in controls. A 56–67% reduction in nesting along with reductions in visits to flowers of up to 100% was reported for alfalfa leafcutting bees. The authors concluded that "low-range" honey bee kills occurred (100–200 dead bees per day on an apron in front of a hive), but colonies did not die. Leafcutting bees, which do not have multiple generations per year, would suffer more. It was also suggested that application of CPY to blooming alfalfa would seriously reduce seed set in crops grown for seed. Lorsban 4E applied to single 0.004 ha plots of dandelions in pear and apple orchards at 1.12 kg CPY ha^{-1} caused no reductions in the number of honey bees foraging and no effects on behavior (Lunden et al. 1986).

When Lorsban 50 WP (1.68 kg CPY ha^{-1}) was applied by ground-sprayer to raspberry plots in the evening (Lunden et al. 1986), bees behaved erratically after foraging on blossoms 1 d after treatment, in that they would "mill around, land on leaves and walk in a wobbly fashion". Visitation to flowers was 40% of that observed in the control on d-1 and remained reduced for 7 d. Bioassays conducted in cages with 3-d-old foliage resulted in 70% mortality of honey bees after 24 h. Chlorpyrifos is no longer registered for use on raspberries.

Lorsban 50 W applied to a single, blooming carrot field (8.1 ha) at 1.12 kg CPY ha^{-1} containing adjacent honey bee hives resulted in over 12-fold more dead bees and reduced foraging on the crop the day after application, and threefold more dead bees 2 d after application. The actual number of bees lost was considered to be only a moderate honey bee kill (250–500 dead bees per hive from an Apron type dead bee trap or 500–950 from a Todd type dead bee trap) based on criteria of Mayer and Johansen (1983), and the long-term viability of the hive was not affected (Lunden et al. 1986). In an unreplicated corn field with adjacent honey bee hives, application of CPY resulted in four-fold more dead bees and 95% reduced foraging on corn pollen compared to pre-application counts (Lunden et al. 1986). It is unclear whether there was an overall reduction in foraging, or whether bees simply avoided the CPY-treated plots. Impacts on long-term survival of the hives were not reported.

Several field trials have been done to assess effects of CPY on citrus on honey bees (Atkins and Kellum 1993). Lorsban 4E was sprayed on unreplicated blooming 1.2 A citrus plots at 2.2 kg ha^{-1} in the morning or in the evening. Colonies in plots treated with Lorsban in the morning had "moderate to light" mortality (274 dead bees over 4 d), with 33% fewer dead bees than colonies in plots treated in the evening (395 dead bees over 4 d), which was classified as "moderately hazardous". This rate of bee deaths is not expected to affect the long-term colony survival and the authors concluded that Lorsban could be applied to citrus as an evening or early morning treatment without causing serious honey bee kills.

In a second trial in citrus, honey bee colonies were placed in 10-A plots of blooming citrus and sprayed in the evening with Lorsban 4E at 1.7 kg ha^{-1} (Atkins and Kellum 1993). Treatment suppressed visitation by 64% for approximately 2 d and killed an average of 904 bees colony^{-1} over 1.5 d, suggesting a moderate to high overall hazard, based on the expert opinion of the authors. In another trial, unreplicated 2-A (0.81 ha) plots of citrus in full bloom were aerially treated with Lorsban at 1.7 kg ha^{-1} in the morning or evening using a helicopter. In contrast to what might be predicted, the night treatment of CPY was moderately hazardous to bees, whereas the morning application of CPY was rated as having a low hazard (Atkins and Kellum 1993).

Despite short-term lethal effects on honey bees, colonies should be able to survive such exposure with few long-term effects. The risk is reduced or eliminated if application is not made when flowers are open, since CPY is not systemic and is not translocated to newly opened flowers. Non-*Apis* pollinators with females that annually establish nests that are much smaller than that of the honey bee, are likely to be more sensitive to CPY exposure.

4.4 Other Studies on the Effects of CPY

Toxicity from exposures via beeswax. The potential effect of exposure to CPY contaminated beeswax on honey bees has not been studied extensively. One study reported concentrations of 39 pesticides found in frames of brood comb from hives from the Pacific Northwest, and from colonies provided by the USDA-ARS honey bee laboratory that were suspected to have died from Colony Collapse Disorder (Wu et al. 2011). Worker bees were reared in brood comb containing concentrations of known pesticides that were considered to be high, and in relatively uncontaminated brood comb used as the control. CPY was detected in some comb samples, but no effects were reported where CPY was present (Wu et al. 2011).

Effects on virus titres. In a study with bee colonies placed in field cages large enough to allow foraging in a controlled area but still contain the bees, the frequency of occurrence and titer of viruses in nurse bees, royal jelly, and various life stages of queen bees reared in colonies fed only almond pollen from trees previously treated

Table 10 Concentrations of chlorpyrifos (CPY) detected in honey bees

Concentration (μg kg^{-1})						
Mean[a]	Median[b]	Maximum[b]	95th centile[b]	LOD[c]	% of samples	Reference
3.4	2.2	10.7	9.7	0.1	8.6 (12/140)	Mullin et al. (2010)
ND[d]	–	ND	–	10.0	0 (0/307)	Chauzat et al. (2011)
43	–	57	–	30.0[e]	3.2 (3/92)	Ghini et al. (2004)
77	–	80.6	NA	NA	NA	DeGrandi-Hoffman et al. (2013)

[a]Based on positive detections
[b]Based on calculations that included 0 μg kg^{-1} for non-detections
[c]Limit of detection
[d]ND = CPY was included in residue analysis but was not detected
[e]Limit of quantification

with 955 μg CPY kg^{-1} was compared to that in colonies with free access to flora of the southwestern US desert (DeGrandi-Hoffman et al. 2013). The experiment was repeated using the same pollen to which Pristine® (boscalid + pyraclostrobin) fungicide was added. The authors reported that deformed wing virus (DWV) was not detected in emerged queens grafted from or reared in the reference colonies but was found in all emerged queens grafted from or reared in colonies where pollen contained CPY (DeGrandi-Hoffman et al. 2013). Titres of DWV in queen larvae and emerged queens were less than two-fold those in nurse bees in some treatments.

This study had a number of weaknesses. Treatments were not matched to appropriate controls. Bees exposed to CPY were restricted to almond pollen alone with no reserves of other pollen in the hive. Pollen from almond trees might not have been nutritionally sufficient (Somerville 2005) and contains the natural toxin amygdalin at concentrations that are sublethal for honey bees (Somerville 2005). In contrast, control bees were free to forage on plants of the southwestern desert, which are known to provide the complete nutritional requirement for bees (Ayers and Harman 1992); the correct control should have been almond pollen without CPY. Thus, potential nutritional effects from different pollen sources were confounded with the potential effects of CPY and the applied fungicides. It was not clear whether the uncaged, reference colonies had reserves of pollen or bee bread. Also, there was no evidence of exposure of the queens. CPY was not detected in royal jelly or in queens so exposures, if any, were less the limit of detection (0.1 μg kg^{-1} wwt) and much less than a toxic dose (Table 7). CPY was detected in nurse bees but no symptoms of toxicity were described and they had lower titres of virus, which is the opposite of what would be expected if there were a relationship between CPY exposure and titer of virus. Another weakness of the study was exposure to only one concentration of CPY, which precluded the characterization of a concentration-response, a key factor in the determination of causality.

Concentrations of CPY found in honey bees. Pollinators can be exposed to CPY by contact with spray droplets or residues on surfaces such as pollen, foliage or blossoms. The extent of transfer of these residues to pollinators can be estimated from published residue data for CPY in bees (Table 10). CPY was detected in a small

Table 11 Reported chlorpyrifos (CPY) incidents with honey bees in the U.S., 1990-present (USEPA 2013)

Crop	Reported CPY Incidents with Honey Bees			
	1990–1999	2000–2009	2010-present	No date Specified
Agricultural area (not specified)	4	0	0	2
Alfalfa	1	1	0	
Apple	1	0	0	
Bean	1	0	0	
Carrot	2	0	0	
Carrot seed	1	0	0	
Cherry	0	0	0	2
Corn	0	1	0	
Cotton	0	0	1	
Orchard	4	0	0	
Orchard (unspecified)	5	0	0	
Soybean	0	1	0	
Not reported	7	1	0	
TOTALS	26	4	1	4

portion of bee samples collected from hives throughout the U.S., at concentrations up to 10.7 μg CPY kg^{-1}. Samples consisted of live adult nurse bees removed from brood nests (Mullin et al. 2010). Live worker bees were collected several times during the season from hives in France, but CPY was not detected in any samples (Chauzat et al. 2011). In Italy, CPY in bees was detected in only a small portion of samples (Ghini et al. 2004) (Table 10). Thus, detection of CPY in honey bees collected in the field was infrequent. Assuming an adult worker honey bee weighs 93 mg (Winston 1987), a honey bee worker is estimated to contain up to 9×10^{-4} μg CPY, based on the 95th centile estimate of 9.7 μg CPY kg^{-1} reported by Mullin et al. (2010). The median concentration of 2.2 μg CPY kg^{-1} reported by Mullin et al. (2010) provides an estimate of 2×10^{-4} μg CPY bee^{-1}.

In an experiment where nucleus colonies (five frames with 3000 adults, a queen, larvae) of honey bees were held in cages and fed almond pollen from trees previously sprayed with CPY, or CPY and boscalid + pyraclostrobin, concentrations in bodies of nurse bees were 80.6 and 72.7 μg CPY kg^{-1} (wwt), respectively (DeGrandi-Hoffman et al. 2013). Because bees were held in cages and only had access to trays of pollen from almond trees intentionally treated with CPY, it is not surprising that these values are higher than concentrations reported from field monitoring surveys (Table 10).

Incident Reports. Considering the widespread use of CPY in agriculture in the U.S., data obtained from the USEPA Office of Pesticide Programs show that the number of honey bee incidents reported is very low, and has decreased over the past two decades (USEPA 2013). Reported incidents range from those involving a few to hundreds of honey bee colonies, and involve exposure following registered uses and misuse of CPY. As well, the level of certainty as to whether or not CPY caused the reported incidents was variable, ranging from "unlikely" to "highly probable". The reported incidents since 1990 are listed in Table 11.

Incident report data were also obtained from the Pest Management Regulatory Agency (PMRA) of Health Canada. Since 2007, when reporting of honey bee pesticide incidents was officially initiated, there have been only nine reports potentially implicating CPY as the cause of the incident. All reports were from the province of Saskatchewan in 2012. Four of these incidents were classified as "minor" by PMRA (≤10% of bees suffering lethal or sublethal effects), three were classified as "moderate" (10–30% of bee affected), and two incidents were classified as "major" (≤30% affected).

5 Characterization of Risk of CPY to Pollinators

Hazards and risks were calculated using the margin of exposure method to generate RQ values. For CPY, the hazard to honey bees and other pollinators from direct exposure during spray is well known. This has been dealt with through mitigation measures to protect pollinators through restricted use patterns that minimize direct exposure to spray or spray drift during application (see discussion of Tier-1 exposure above). The following paragraphs cover the calculation of RQ values and assessment of the potential risk to pollinators from post-application exposure.

5.1 Estimates of Risk to Honey Bees

Estimated risk to honey bees through dietary exposure. The geometric mean of LD_{50} values from dietary tests using technical CPY was less than that derived from tests with formulated product (Table 7). The geometric mean of the oral LD_{50} for technical CPY (0.114 μg CPY bee^{-1}) was used as a worst case in calculating RQs.

The upper limit dietary intake of CPY per day can be estimated for different life stages of honey bees based on surrogate T-REX screening values for pollen and nectar (USEPA 2012), or empirical data collected from pollen and honey and/or nectar in the field. The proposed Tier-1 scheme includes an acute oral LOC of 0.4 for adult and larval honey bees that is compared to estimates of RQs for exposure and effects. Using the maximum screening values suggested in the USEPA's proposed risk assessment scheme for pollinators (Table 1), RQ values for oral exposure of adult and larval honey bees following sprays of CPY exceeded the LOC by over three orders of magnitude.

Tier-1 estimates of oral exposures based on T-REX are intended to be conservative, when compared to data for concentrations of CPY collected in the field. Tier-2 estimates are based on measured values of CPY in honey bee food and reflect actual use conditions. Using the monitoring data for concentrations in pollen collected from commercial beehives in the U.S. by Mullin et al. (2010) and food consumption rates established by the USEPA (USEPA 2012), the upper 95th centile dietary exposures were 0.066 μg CPY d^{-1} for adult bees and 0.027 μg CPY d^{-1} for larvae

Table 12 Tier-2 risk quotients (RQs) for oral exposure of honey bees to chlorpyrifos (CPY) via pollen

	Larvae (120 mg)		Adults (292 mg)	
Variable	Median	95th centile	Median	95th centile
Dose (μg CPY bee^{-1})[a]	5.26×10^{-4}	2.72×10^{-2}	1.28×10^{-3}	6.61×10^{-2}
LD_{50}[b]	0.146	0.146	0.114	0.114
RQ[c]	0.004	0.19	0.011	0.580

[a]Reported in Table 3
[b]Technical CPY reported in Table 7 and larvae LD_{50} derived from Atkins and Kellum (1986)
[c]Risk Quotient = Dose/LD_{50} where LD_{50} = 0.146 or 0.114 μg bee^{-1}

(Table 3). The estimates of oral exposure and the oral LD_{50} values of 0.114 μg $adult^{-1}$ bee and 0.146 μg $larva^{-1}$ (Atkins and Kellum 1986) provided RQ values below the LOC of 0.4 for median acute exposures. The 95th centile exposure and larval LD_{50} give an RQ below the LOC, but a RQ slightly above the LOC when the adult LD_{50} was used (Table 12). Based on analysis of CPY in nurse bees (Table 10) fed exclusively a diet of pollen containing CPY at a concentration of 955 μg kg^{-1} (DeGrandi-Hoffman et al. 2013), doses were 7.5×10^{-3} and 6.8×10^{-3} μg bee^{-1} (93 mg) for the two parallel experiments, which would be equivalent of approximately 7% of the LD_{50} for technical grade CPY (Table 7). The authors did not describe symptoms of toxicity in nurse bees so they were apparently unaffected.

The RQ for adults, based on the 90th centile concentration of 140.4 μg kg^{-1} reported by Mullin et al. (2010), is 0.36. Therefore, dietary exposure of adult honey bees to CPY is expected to be below the LOC >90% of the time, while exposures for larvae should be below the LOC >95% of the time. Considering that most pollen samples (56%) collected by Mullin et al. (2010) did not contain CPY, oral exposure to CPY should be of low risk to honey bees, particularly in terms of the protection goals of overall fitness of the colony.

Mullin et al. (2010) did not measure concentrations of pesticides in honey and no studies were found which examined concentrations of CPY in nectar. Only one study was from Uruguay reported detection of CPY in honey. The mean and maximum concentrations of CPY in honey samples that were positive for CPY (42%) were 46 and 80 μg kg^{-1}, respectively (Pareja et al. 2011). Using the recommended consumption rates of 120 mg honey d^{-1} for larvae and 292 mg honey d^{-1} for adults, honey bee larvae exposed to a concentration of 80 μg CPY kg^{-1} honey would consume 0.0096 μg CPY d^{-1}, whereas adults would be expected to consume 0.023 μg d^{-1}. At an LD_{50} of 0.114 μg bee^{-1}, corresponding RQ values for larvae and adults would be 0.08 for larvae and 0.2 for adults. These values are below the LOC of 0.4 and suggest little risk to honey bees from acute exposure to CPY via honey.

Exposure estimates through consumption of nectar and pollen are assumed to be conservative representations of potential exposures through honey and bee bread, respectively (USEPA 2012). The estimates assume that pesticides do not degrade while honey and bee bread are stored in the hive. They also assume that rates of consumption of pollen and nectar and resulting exposures are protective of exposures through consumption of royal jelly and brood food, since concentrations of pesticides in food consumed by nurse bees are 2–4 orders of magnitude greater than concentrations measured in royal jelly (Davis and Shuel 1988).

Estimated risk through consumption of water. Pollinators can be exposed to CPY in drinking water from small ponds, puddles, or on foliage wet from rain or dew. Exposure from wet foliage occurs with sprayable formulations only, while exposure from puddles or small ponds occurs with both sprayable and granular formulations of CPY. Using the modified rice paddy model recommended by USEPA to provide estimates of pesticide exposure to bees through puddles in treated fields, estimated worst-case daily doses of CPY in puddles were $<2.4 \times 10^{-7}$ μg CPY bee^{-1}. With the oral LD_{50} value of 0.114 μg CPY bee^{-1}, this provides a RQ of 2.1×10^{-6} (Table 13), which is well below the LOC of 0.4, indicating a *de minimis* risk to pollinators relative to other potential exposure routes.

Using the PRZM/EXAMS puddle 95th centile concentrations, which apply to both sprayable and granular applications of CPY, the predicted peak concentrations were much higher than the values obtained using the Tier-1 rice paddy model, since they include storm runoff events within hours after application. It is unlikely that bees would be exposed to these concentrations as storm events are rare, and such events within hours after application are even less common. Bees are not likely to go into fields to collect water, given the high humidity and availability of water around a hive after such a storm.

The model recommended by the USEPA to predict concentrations of pesticide in dew estimated a worst-case dose of 0.034 μg CPY bee^{-1} (USEPA 2012). This model and dose also provides a RQ less than the LOC (Table 13) and suggests low risk to pollinators through consumption of contaminated dew. This scenario also applies to wet foliage from rain or irrigation as well as from dew. Residues of CPY in dew come from the leaf surface, and the concentration is determined by partition between the leaf surface and the water. The maximum potential concentration occurs when there is no runoff of rain or dew to carry material away from the leaf. As water dries, the residues partition back onto the leaf surface. Summation of exposures via food and water from Tables 12 and 13 also suggests that the risks from the combined sources would be small for most bees.

As described in the exposure section, honey bees may collect water for direct consumption, to prepare food, or to control temperature in the colony. Assuming transport of 1,250 mg d^{-1} of water to the colony from a source such as a puddle or 250 mg d^{-1} from a more temporary source such as dew, with 100% uptake of CPY from the water being carried, the modified USEPA puddle model gave an RQ well below the LOC (Table 13), but the dew model gave an RQ of 1.6. The RQ values calculated from puddle concentrations obtained using the PRZM/EXAMS model also exceeded the LOC (Table 13). These values are based on conservative approximations, and in the case of the PRZM/EXAMS predictions, have low probability of occurrence. Using the refinements in the Liqui-PARAM model, the 95th percentile RQ for collection of water from wet foliage was reduced to a level essentially the same as the LOC. The median RQ for both the OCSPP model and the Liqui-PARAM model estimates of exposure were well below the LOC (Table 13).

Table 13 Risk quotients for water consumption of chlorpyrifos (CPY) (LOC = 0.4) based on concentrations determined by different models

	Median		95th centile or maximum			
Variable	OCSPP 2012 dew/wet foliage[a]	Dew/wet foliage Liqui-PARAM[b]	Tier-1 rice model[a]	PRZM/EXAMS puddle/pore water[c]	OCSPP 2012 dew/wet foliage	Dew/wet foliage Liqui-PARAM
Predicted concentration at 1.12 kg CPY ha^{-1} ($\mu g\ L^{-1}$)	140	102	5.10×10^{-3}	571	730	210
Dose from drinking[d] water ($\mu g\ bee^{-1}$)	0.007	0.005	2.40×10^{-7}	0.027	0.034	0.010
Tier 1 RQ[e] drinking water	0.06	0.04	2.10×10^{-6}	0.24	0.30	0.09
Dose based on water[d] collected ($\mu g\ bee^{-1}$)	0.035	0.03	6.38×10^{-6}	0.71	0.18	0.05
RQ[e] for water collection by honey bees	0.31	0.22	5.59×10^{-5}	6.26	1.60	0.46

[a] Described in OCSPP (2012)

[b] Moore et al. (2014)

[c] Williams et al. (Williams et al. 2014). Based on a NC tobacco scenario run for 30 yr from 1961 to 1990

[d] Dose = concentration × volume, where volume = 47 μL for drinking water, 1,250 μL for puddle, and 250 μL for dew/wet foliage (see discussion above)

[e] Risk Quotient = Dose/LD_{50} where LD_{50} = 0.114 $\mu g\ bee^{-1}$

There are other reasons to expect contaminated dew or wet foliage exposure will be insignificant to honey bees. All the models used are conservative, since they are based on the assumption that honey bees will consume 100% of the water they need from a given water source, whereas in reality they can obtain 7–100% of their required water from food (USEPA 2012). Given the short foliar half-life of CPY (Williams et al. 2014), the peak concentrations associated with the higher RQ values are present for a short time.

Assessment of aggregate risks to honey bees through semi-field and field tests. Several studies have investigated the concentrations of CPY on foliage of treated plants and this material was used in laboratory bioassays with bees (Atkins et al. 1973; Lunden et al. 1986). Their data generally showed that foliage treated with CPY at the label rate can remain lethal to honey bees, alfalfa leaf cutting bees, and alkali bees for several days after application. Residual toxicity was determined by calculating a RT25 value, which is the residue-degradation time required to bring bee mortality down to 25% or less (Lunden et al. 1986). For CPY, the RT25 was longer than 72 h. For comparison, an RT25 of 8 h that was suggested as indicative of a product that poses little risk to bees (Lunden et al. 1986). Although this result from caged bioassays is expected to overestimate uptake of material from the surface and the duration of effects, it corresponds with results of multiple semi-field and field studies that indicate residual CPY on plant foliage poses a risk to honey bee survival 1–2 d after application.

With one exception, there are no guidelines for pollinator-safe post-spray periods for CPY. The exception is citrus crops in California, for which CPY must be applied from 1 h after sunset until 2 h before sunrise (see Atkins and Kellum 1993), giving a 2-h minimum post-spray interval. Assessments of effects after field applications indicate that some mortality may occur 1–2 d after application on flowering crops, and reduced foraging may persist for up to a week, but residues remaining after 7 d have no impact. Rapid, normal turnover of foragers in honey bees colonies and availability of alternate foraging sites should buffer out these short-term effects. There were no reports of adverse effects of CPY on honey bee brood development.

The field tests of Lunden et al. (1986) indicate that concentrations of CPY on alfalfa during flowering remain lethal to alfalfa leafcutting bees for at least 1 d after application, with reduced nesting observed. Because leafcutting bees are univoltine, Lunden et al. (1986) suggested applications to blooming alfalfa could have a "substantial" effect on this pollinator. Current label precautions preclude application when bees are foraging and are intended to mitigate this risk.

Assessment of risk through exposure to contaminated beeswax. A few studies have reported contamination of beeswax with CPY. Concentrations in beeswax reported by Mullin et al. (2010) were similar to amounts found in pollen. Although there can be a risk of sublethal effects through this route of exposure for some pesticides, one study found this was not the case for CPY (Wu et al. 2011). No data were available on the uptake of pesticides into larvae from contaminated wax and this is an area of uncertainty.

5.2 Estimates of Risk to Non-Apis Pollinators

Studies with non-Apis pollinators. No acute direct contact or oral toxicity data (LD_{50}) based on laboratory tests were reported for non-*Apis* pollinators. In a greenhouse experiment, Gregory et al. (1992) exposed alfalfa leafcutter bees (*M. rotundata*) to CPY-treated alfalfa and found that mortality only occurred during the first 3 d after treatment. This is in agreement with semi-field and field studies with honey bees (see discussion above). Because females can only construct nests if doses of CPY on foliage are sublethal, and because the bioavailability of CPY on foliage drops rapidly, it is expected that few eggs and larvae will be exposed to hazardous amounts of CPY from nesting materials.

The only semi-field or field study with a non-*Apis* pollinator was performed by Gels et al. (2002). Detrimental effects were seen in bumble bee (*B. impatiens*) colonies exposed for 2 wk to CPY-treated clover. However, bumble bees were confined within tunnels for the duration of the entire experiment (a worst-case scenario). In an open system, effects would likely be less severe. Following label precautions to avoid application when bees are present, mowing flower heads before treatment, and weed management with herbicides are useful tactics to alleviate such hazards from applications of CPY (Gels et al. 2002).

No studies on the exposure of bee flies, Bombyliidae, and hover flies, Syrphidae, to CPY were found. Their potential for exposure is dominated by foraging at flowers since they are nonsocial insects that do not build nests, and feed only themselves. Honey bee foragers must visit more flowers, and bee behavior while on the flower leads to a much higher potential for transfer of material. Therefore, the honey bee may be considered a conservative surrogate for these taxa. Measures that are taken to protect honey bees are expected to be protective of these pollinators.

Estimated risk to non-Apis pollinators using NTAs as surrogates. Since toxicity data for CPY in NTAs is limited and the suitability of these small wasps as surrogates for wild pollinators is questionable (see above), this risk assessment was not conducted. However, Addison and Barker (2006) found that although *Microctonus hyperodae*, another parasitic wasp, was initially (1 h post-treatment) highly susceptible to foliage treated with CPY, no 24-h mortality was observed with 2-d old foliage at rates up to 100 g CPY ha^{-1}. Bioassays such as these, where insects are confined to cages are conservative, because most flying insect pollinators do not spend much time on foliage. The risk posed by CPY is still an area of uncertainty because of the lack of data for non-*Apis* pollinators.

Estimated risks through exposure to contaminated soil. Although honey bees can be a good surrogate for many flying insect pollinators, ground nesting bees and mason bees can experience exposure via soil (Fig. 1), which is not encountered by honey bees. Mason bees collect soil and use it for the construction of nests, and ground-nesting bees nest below the soil surface. They might dig their own burrows or they may use existing cavities built by other animals such as mice (Michener 2007). CPY is toxic to soil dwelling insects and is used in the management of soil dwelling pests. European chafer grubs (*Amphimallon majalis*) and leatherjackets (*Tipula* spp.) are known to have suffered significant mortality at field-relevant CPY soil concentrations

(Clements and Bale 1988; Tashiro and Kuhr 1978). However, the potential for use of these data to provide a worst-case estimate of CPY exposure to ground nesting bees through soil is limited, because it is unclear how much of this mortality was due to contact with soil, versus ingestion of CPY on plant roots. Different behavioral and physiological differences between soil pests and ground nesting pollinators might influence uptake of CPY from soil.

Several factors would likely reduce the risk of CPY to ground nesting bees. The tendency of CPY to adsorb to soil surfaces reduces bioavailability of CPY in most soil environments (Racke 1993), but the time for toxicity to drop below levels that cause mortality or sublethal effects on ground nesting pollinators is unknown. The dearth of studies on nesting of bees within agriculture fields (Julier and Roulston 2009; Kim et al. 2006; Williams et al. 2010; Wuellner 1999) suggests that most species nest outside cultivated fields where the risk of exposure is low. Many of these are solitary bees that forage specifically on non-agricultural plants and nest in non-agricultural soil (Willmer 2011). Nonetheless there are some important pollinators including squash bees (*Peponapis* sp.) that nest within crops (Julier and Roulston 2009), where the potential for exposure to CPY is much higher. Although CPY is not registered for use on squash or cucurbit crops pollinated by squash bees, there may be other bees that do nest in crops treated with CPY.

Exposure might also vary depending on the architecture of cells within the nest. Although cells of some nests are unlined excavations into the soil, those of other species are lined with a cellulose- or wax-like material (Michener 2007) that possibly provides a barrier from direct contact with soil. Even without a lining, the potential for transfer of residues from soil to the insect during entry and exit from a nest is much lower than when the insect is digging through the soil. Immature life stages are expected to be less exposed than adults. For example, bumble bees usually nest underground, but larvae develop within their own cocoons. Eggs and larvae cannot actually contact the soil at all until late development and adult emergence (Michener 2007), and their dietary exposure is limited to the levels that can be successfully brought to the hive by adult foragers. Contact with freshly contaminated soil is not likely to be a major contribution to aggregate CPY exposure for ground-nesting bees, but it has not been characterized or quantified and remains an area of uncertainty.

No studies were found, in which soil was collected from mason bee (*Osmia* spp.) nests from within areas exposed to CPY. Exposure via soil in this group is limited by the rapid degradation and limited bioavailability of CYP on soil. As with other nesting bees, CPY exposure of immature life stages is limited to concentrations that are collected by adults, and because larvae do not emerge until long after the nest is built, CPY residues on this soil would be negligible.

5.3 *Strengths and Uncertainties*

The current assessment delineated potential exposure pathways of CPY to pollinators in detail. Sufficient exposure and effects data relevant to honey bees were available and permitted a satisfactory characterization of the risk of CPY to them.

Data on non-*Apis* bees was scant, and data available for NTA as surrogates for non-*Apis* pollinators was not usable in the risk assessment. However, in many cases the honey bee is a suitable surrogate for exposure and effects in other pollinating insects. Since all insect pollinators have in common certain aspects of their behavior, biology, and ecology, worst-case exposures for honey bees should generally be protective of non-*Apis* bees. However, there are a number of biological and ecological characteristics of these taxa that can influence risk. Some of these have been described previously and relate to: the role of the queen in founding nests in the spring; increased susceptibility due to smaller colony nest size (i.e., less redundancy); the smaller size of some of non-*Apis* bees that leads to greater potential for exposure (i.e., greater surface area:volume ratio); the smaller foraging range; and the location and construction of nests (EFSA 2012; Thompson and Hunt 1999).

We identified several data gaps and areas of uncertainty in our assessment of CPY on pollinators. Below, we summarize key research topics that deserve more research attention, many of which are relevant to other insecticides:

- Given the increasing recognition of the significant role that wild bees have in agricultural and natural ecosystems (Garibaldi et al. 2013), more data are needed on non-*Apis* species to accurately evaluate the risk of CPY to these taxa as part of higher tier testing.
- More information on sublethal effects of CPY on pollinators is needed, in view of the recent increased focus on behavioral effects such as navigation to and from the hive. However, accepted guidelines for sublethal tests are also required.
- The stability and rate of degradation of CPY residues in nectar, pollen, and bees wax should be determined. Area-wide concentrations have been reported in monitoring studies, but the concentrations of CPY in nectar and pollen over time, following a defined field application, have not been quantified. Concentrations are expected to be lower on pollen and nectar than foliage for non-systemic insecticides like CPY, but the Tier-1 assessment models assume the same levels are present in all parts of the plant. In addition, depending on floral phenology, pollen present at the time of application will likely be available or attractive to foraging pollinators for only a few days after application. Quantification of concentrations of CPY in pollen and nectar over time after application would help to refine the risk assessment and facilitate testing in the laboratory with environmentally relevant concentrations and routes of exposure.
- How CPY partitions and transfers between wax and bee brood or the food stored in wax cells is unknown. It is possible that wax represents a sink for CPY in the colony and that the residues are not bio-available when present at the concentrations that have been reported in wax. Although there can be a risk of sublethal effects from residue in wax for some pesticides, this was not the case for CPY (Wu et al. 2011).
- Partitioning of CPY from wax, in the range of concentrations that have been reported, into the airspace of a colony should be quantified. The physical properties of CPY and its strong propensity to partition into nonpolar substances makes

it unlikely that volatility from wax in a honey bee hive is a significant exposure pathway. Partitioning of vapor from air into wax is more likely.

- The transfer of CPY from soil or foliage during nest construction is expected to be minor or insignificant, except in the first few hours after application. However, tests on the transfer of CPY from nesting materials on a variety of species is needed to confirm that this route of exposure is negligible, and it may be that the exposure is so low that differences in species sensitivity are unimportant. As this route of exposure is one of the key differences between the exposure pathways encountered by foraging honey bees and many solitary pollinators, new research results would clarify the usefulness of honey bees as a surrogate for other species.
- The significance of extra-floral nectaries as a food source and potential route of exposure appears to be minor, but has not been quantified.

6 Summary

CPY is an organophosphorus insecticide that is widely used in North American agriculture. It is non-systemic, comes in several sprayable and granular formulations, and is used on a number of high-acreage crops on which pollinators can forage, including tree fruits, alfalfa, corn, sunflower, and almonds. Bees (Apoidea) are the most important pollinators of agricultural crops in North America and were the main pollinators of interest in this risk assessment.

The conceptual model identified a number of potential exposure pathways for pollinators, some more significant than others. CPY is classified as being highly toxic to honey bees by direct contact exposure. However, label precautions and good agricultural practices prohibit application of CPY when bees are flying and/or when flowering crops or weeds are present in the treatment area. Therefore, the risk of CPY to pollinators through direct contact exposure should be small. The main hazards for primary exposure for honey bees are dietary and contact exposure from flowers that were sprayed during application and remain available to bees after application. The main pathways for potential secondary exposure to CPY is through pollen and nectar brought to the hive by forager bees and the sublethal body burden of CPY carried on forager bees. Foraging for other materials, including water or propolis, does not appear to be an important exposure route. Since adult forager honey bees are most exposed, their protection from exposure via pollen, honey, and contact with plant surfaces is expected to be protective of other life stages and castes of honey bees.

Tier-1 approaches to estimate oral exposure to CPY through pollen and nectar/honey, the principle food sources for honey bees, suggested that CPY poses a risk to honey bees through consumption of pollen and nectar. However, a Tier-2 assessment of concentrations reported in pollen and honey from monitoring work in North America indicated there is little risk of acute toxicity from CPY through consumption of these food sources.

Several models were also used to estimate upper-limit exposure of honey bees to CPY through consumption of water from puddles or dew. All models suggest that the risk of CPY is below the LOC for this pathway. Laboratory experiments with field-treated foliage, and semi-field and field tests with honey bees, bumble bees, and alfalfa leafcutting bees indicate that exposure to foliage, pollen and/or nectar is hazardous to bees up to 3 d after application of CPY to a crop. Pollinators exposed to foliage, pollen or nectar after this time should be minimally affected.

Several data gaps and areas of uncertainty were identified, which apply to CPY and other foliar insecticides. These primarily concern the lack of exposure and toxicological data on non-*Apis* pollinators. Overall, the rarity of reported bee kill incidents involving CPY indicates that compliance with the label precautions and good agricultural practice with the product is the norm in North American agriculture. Overall, we concluded that, provided label directions and good agricultural practices are followed, the use of CPY in agriculture in North America does not present an unacceptable risk to honeybees.

Acknowledgements We thank Dow AgroSciences, LLC for their support of this study, in particular Nick Poletika for assistance in gathering technical information, and Mark Miles for discussion on assessments using non-target arthropods. We thank Dwayne R.J. Moore for providing a model used in estimation of CPY in dew, Marty Williams for providing data on concentrations in puddles, and anonymous reviewers for useful comments and suggestions. Prof. Giesy was supported by the Canada Research Chair program, a Visiting Distinguished Professorship in the Department of Biology and Chemistry and State Key Laboratory in Marine Pollution, City University of Hong Kong, the 2012 "High Level Foreign Experts" (#GDW20123200120) program, funded by the State Administration of Foreign Experts Affairs, the P.R. China to Nanjing University and the Einstein Professor Program of the Chinese Academy of Sciences. Funding for this project was provided by Dow AgroSciences.

References

Addison PJ, Barker GM (2006) Effect of various pesticides on the non-target species *Microctonus hyperodae*, a biological control agent of *Listronotus bonariensis*. Entomol Exp Appl 119:71–79

Ahmed S, Ahmad M (2006) Note: toxicity of some insecticides on *Bracon hebetor* under laboratory conditions. Phytoparasitica 34:401–404

Alix A, Lewis G (2010) Guidance for the assessment of risks to bees from the use of plant protection products under the framework of Council Directive 91/414 and Regulation 1107/2009. EPPO Bull 40:196–203

Anonymous (1986) Bee-toxicological test of Dursban 4 (EF-747). Station for Wildlife and Nature Conservancy Facankert. 29 pp

Atkins EL, Greywood EA, Macdonald RL (1973) Effects of pesticides on agriculture. University of California Riverside, Riverside, CA, USA. Report No. 1499

Atkins EL, Kellum D (1986) Comparative morphogenic and toxicity studies on the effect of pesticides on honeybee brood. J Apic Res 25:242–255

Atkins EL, Kellum D (1993) A compilation of data concerning the effects of LORSBAN applications on honey bees. University of California, Riverside, CA, Unpublished Report
Atkins LE (1975) Injury to honey bees by poisoning. In: Veatch E (ed) The hive and the honey bee. Dadant & Sons, Hamilton, IL, pp 663–696
Ayers GS, Harman JR (1992) Bee forage of North America and the potential for planting for bees. In: Graham JM (ed) The hive and the honey bee, revised edition. Dadant & Sons, Hamilton, IL, pp 437–535
Bakker F (2000) Effects of Dursban 75 WG on honeybees, *Apis mellifera* L., when applied to *Phacelia tanacetifolia* 1, 3, 5, 7 and 17 days before exposure, determined in a cage test. Dow AgroSciences, Indianapolis, IN, Unpublished Report
Bakker F, Calis J (2003) A semi-field approach to testing effects of fresh or aged pesticide residues on bees in multiple-rate test designs. Bull Insectol 56:97–102
Barmaz S, Potts SG, Vighi M (2010) A novel method for assessing risks to pollinators from plant protection products using honeybees as a model species. Ecotoxicol 19:1347–1359
Bell G (1993) EF 1042 (Dursban 480) acute toxicity to honey bees (*Apis mellifera*). DowElanco Europe, Letcombe Regis, Unpublished Report
Bell G (1996) EF-1315: acute oral and contact toxicity to honey bees (*Apis mellifera*). DowElanco Europe, Letcombe Regis, p 28, Unpublished Report
Berger LA, Vaissiere BE, Moffett JO, Merritt SJ (1988) *Bombus spp*. (Hymenoptera: Apidae) as pollinators of male-sterile upland cotton on the Texas High Plains. Environ Entomol 17:789–794
Buck NA, Estesen BJ, Ware GW (1980) Dislodgable insecticide residues on cotton foliage: Fenvalarate, permethrin, sulprofos, chlorpyrifos, methyl parathion, EPN, oxamyl, and profenofos. Bull Environ Contam Toxicol 24:283–288
Butler CG (1975) The honey bee colony life history. In: Veatch E (ed) The hive and the honey bee. Dadant & Sons, Hamilton, IL, pp 39–74
Calderone NW (2012) Insect pollinated crops, insect pollinators and US agriculture: Trend analysis of aggregate data for the period 1992–2009. PLoS ONE 7:e37235
Candolfi MP, Barrett KL, Campbell PJ, Forster R, Grandy N, Huet M-C, Lewis G, Oomen PA, Schmuck R, Vogt H (2001) Guidance document on regulatory testing and risk assessment procedures for plant protection products with non-target arthropods. Society of Environmental Toxicology and Chemistry Europe, Brussels
Carman H (2011) The estimated impact of bee colony collapse disorder on almond pollination fees. ARE Update 14:9–11
Carrasco-Letelier L, Mendoza-Spina Y, Branchiccela MB (2012) Acute contact toxicity test of insecticides (Cipermetrina 25, Lorsban 48E, Thionex 35) on honeybees in the southwestern zone of Uruguay. Chemosphere 88:439–444
Chambers JE, Redwood WT, Trevathan CA (1983) Disposition of metabolism of ^{14}C-chlorpyrifos in the black imported fire ant, *Solenopsis richteri* Forel. Pestic Biochem Physiol 19:115–121
Chauzat M, Martel A, Cougoule N, Porta P, Lachaize J, Zeggane S, Aubert M, Carpentier P, Faucon J (2011) An assessment of honeybee colony matrices, *Apis mellifera* (Hymenoptera: Apidae) to monitor pesticide presence in continental France. Environ Toxicol Chem 30:103–111
Chen JL (1994) Volatility control for foliage-applied chlorpyrifos by using controlled release emulsions. J Control Release 29:83–95
Choudhary A, Sharma DC (2008) Pesticide residues in honey samples from Himachal Pradesh (India). Bull Environ Contam Toxicol 80:417–422
Clements RO, Bale JS (1988) The short-term effects on birds and mammals of the use of chlorpyrifos to control leatherjackets in grassland. Ann Appl Biol 112:41–47
Clinch PG (1972) Chlorpyrifos found to be highly toxic to honey bees. Agricultural Department—Wallaceville Animal Research Centre, Wallaceville, New Zealand, 19903
Cutler GC (2013) Pesticides in honey bee hives in the maritime provinces: residue levels and interactions with *Varroa* Mites and *Nosema* in colony stress. Dalhousie University, Truro, NS, Unpublished Report

Davis AR, Shuel RW (1988) Distribution of ^{14}C-labeled carbofuran and dimethoate in royal jelly, queen larvae and nurse honeybees. Apidologie 19:37–50

Davis BNK, Williams CT (1990) Buffer zone widths for honeybees from ground and aerial spraying of insecticides. Environ Pollut 63:247–259

DeGrandi-Hoffman G, Chen Y, Simonds R (2013) The effects of pesticides on queen rearing and virus titers in honey bees (*Apis mellifera* L.). Insects 4:71–89

Desneux N, Rafalimanana H, Kaiser L (2004) Dose–response relationship in lethal and behavioural effects of different insecticides on the parasitic wasp *Aphidius ervi*. Chemosphere 54:619–627

Desneux N, Decourtye A, Delpuech J-M (2007) The sublethal effects of pesticides on beneficial arthropods. Annu Rev Entomol 52:81–106

Dietz A (1975) Nutrition of the adult honey bee. In: Veatch E (ed) The hive and the honey bee. Dadant & Sons, Hamilton, IL, pp 125–126

Donahue RL, Miller RW, Shickluna JC (1983) Soils, an introduction to soils and plant growth. Prentice Hall, Englewood Cliffs, NJ, 768 pp

EFSA (2012) EFSA panel on plant protection products and their residues (PPR); scientific opinion on the science behind the development of a risk assessment of plant protection products on bees (*Apis mellifera*, *Bombus* spp. and solitary bees). EFSA J 10:2668

Fischer D, Moriarty T (2011) Pesticide risk assessment for pollinators: summary of a SETAC Pellston Workshop. Society of Environmental Toxicology and Chemistry, Pensacola, FL

Fletcher JS, Nellessen JE, Pfleeger TG (1994) Literature review and evaluation of the EPA food-chain (Kenaga) nomogram, an instrument for estimating pesticide residues on plants. Environ Toxicol Chem 13:1383–1391

Garibaldi LA, Steffan-Dewenter I, Winfree R, Aizen MA, Bommarco R, Cunningham SA, Kremen C, Carvalheiro LG, Harder LD, Afik O, Bartomeus I, Benjamin F, Boreux V, Cariveau D, Chacoff NP, Dudenhöffer JH, Freitas BM, Ghazoul J, Greenleaf S, Hipólito J, Holzschuh A, Howlett B, Isaacs R, Javorek SK, Kennedy CM, Krewenka KM, Krishnan S, Mandelik Y, Mayfield MM, Motzke I, Munyuli T, Nault BA, Otieno M, Petersen J, Pisanty G, Potts SG, Rader R, Ricketts TH, Rundlöf M, Seymour CL, Schüepp C, Szentgyörgyi H, Taki H, Tscharntke T, Vergara CH, Viana BF, Wanger TC, Westphal C, Williams N, Klein AM (2013) Wild pollinators enhance fruit set of crops regardless of honey bee abundance. Science 339:1608–1611

Gary NE (1975) Activities and behavior of honey bees. In: Veatch E (ed) The hive and the honey bee. Dadant & Sons, Hamilton, IL , pp 185–264

Gels JA, Held DW, Potter DA (2002) Hazards of insecticides to the bumble bees *Bombus impatiens* (Hymenoptera: Apidae) foraging on flowering white clover in turf. J Econ Entomol 95:722–728

Ghini S, Fernández M, Picó Y, Marín R, Fini F, Mañes J, Girotti S (2004) Occurrence and distribution of pesticides in the province of Bologna, Italy, using honeybees as bioindicators. Arch Environ Contam Toxicol 47:479–488

Giesy JP, Solomon KR, Coats JR, Dixon KR, Giddings JM, Kenaga EE (1999) Chlorpyrifos: ecological risk assessment in North American aquatic environments. Rev Environ Contam Toxicol 160:1–129

Goh KS, Edmiston S, Maddy KT, Meinders DD (1986) Dissipation of dislodgeable residues of chlorpyrifos and dichlorvos on turf. Bull Environ Contam Toxicol 37:27–32

Gomez LE (2009) Use and benefits of chlorpyrifos in US agriculture. Dow AgroSciences, Indianapolis, IN, Unpublished Report

Gregoric A, Ellis JD (2011) Cell death localization in situ in laboratory reared honey bee (*Apis mellifera* L.) larvae treated with pesticides. Pestic Biochem Physiol 99:200–207

Gregory DA, Johnson DL, Thompson BH, Richards KW (1992) Laboratory evaluation of the effects of carbaryl and chlorpyrifos baits and sprays on alfalfa leafcutting bees (*Megachile rotundata* F.). J Agric Entomol 9:109–115

Hahne R (2000) Lorsban 50 W: acute oral toxicity test with the honeybee, *Apis mellifera*. Dow AgroSciences, Indianapolis, IN, Unpublished Report

Hurto KA, Prinster MG (1993) Dissipation of foliar dislodgeable residues of chlorpyrifos, DCPA, diazinon, isofenphos, and pendimethalin. In: Racke KD, Leslie AR (eds) Fate and significance of pesticides in urban environments, vol ACS, Symposium Series 522. American Chemical Society, Washington, DC, pp 86–99

Iwata T, O'Neal JR, Barkley JH, Dinoff TM, Dusch E (1983) Chlorpyrifos applied to California citrus: residue levels on foliage and on and in fruit. J Agric Food Chem 31:603–610

Johnson RM, Ellis MD, Mullin CA, Frazier M (2010) Pesticides and honey bee toxicity—USA. Apidologie 41:312–331

Julier HE, Roulston TH (2009) Wild bee abundance and pollination service in cultivated pumpkins: farm management, nesting behavior and landscape effects. J Econ Entomol 102:563–573

Kevan PG, Clark EA, Thomas VG (1990) Pollination: a crucial ecological and mutualistic link in agro-forestry and sustainable agriculture. Proc Entomol Soc Ont 121:43–48

Kevan PG (1999) Pollinators as bioindicators of the state of the environment; species, activity and diversity. Agr Ecosyst Environ 74:373–393

Kim J, Williams N, Kremen C (2006) Effects of cultivation and proximity to natural habitat on ground-nesting native bees in California sunflower fields. J Kans Entomol Soc 79:309–320

Klein A-M, Vaissière BE, Cane JH, Steffan-Dewenter I, Cunningham SA, Kremen C, Tscharntke T (2007) Importance of pollinators in changing landscapes for world crops. Proc R Soc B Biol Sci 274:303–313

Kwon H, Lehotay SJ, Geis-Asteggiante L (2012) Variability of matrix effects in liquid and gas chromatography–mass spectrometry analysis of pesticide residues after QuEChERS sample preparation of different food crops. J Chromatogr A 1270:235–245

Lunden JD, Mayer DF, Johansen CA, Shanks CH, Eves JD (1986) Effects of chlorpyrifos insecticide on pollinators. Am Bee J 126:441–444

Mackay D, Giesy JP, Solomon KR (2014) Fate in the environment and long-range atmospheric transport of the organophosphorus insecticide, chlorpyrifos and its oxon. Rev Environ Contam Toxicol 231:35–76

Mansour SA, Al-Jalili MK (1985) Determination of residues of some insecticides in clover flowers: a bioassay method using honeybee adults. J Apic Res 24:195–198

Mayer DF, Johansen CA (1983) Occurrence of honey bee (Hymenoptera: Apidae) poisoning in Eastern Washington. Environ Entomol 12:317–320

Michener CD (2007) The bees of the world. John Hopkins University Press, Baltimore, MD, 913 pp

Miles MJ, Alix A (2012) Assessing the comparative risk of plant protection products to honey bees, non-target arthropods and non-*Apis* bees. In: Oomen P, Thompson H (eds) Hazards of Pesticides to Bees—11th International Symposium of the ICP-Bee Protection Group, 2011, vol 437. Julius-Kühn-Archiv, Wageningen, pp 30–38

Moore DRJ, Teed RS, Greer C, Solomon KR, Giesy JP (2014) Refined avian risk assessment for chlorpyrifos in the United States. Rev Environ Contam Toxicol 231:163–217

Morse RA, Calderone NW (2000) The value of honey bees as pollinators of U.S. crops in 2000. Bee Cult (March 2000):2–15.

Mullin CA, Frazier M, Frazier JL, Ashcraft S, Simonds R, vanEngelsdorp D, Pettis JS (2010) High levels of miticides and agrochemicals in North American apiaries: implications for honey bee health. Plos One 5:e9754

NAS (2007) Status of pollinators in North America. The National Academies Press, Washington, DC, 322 pp

Pareja L, Colazzo M, Perez-Parada A, Niell S, Carrasco-Letelier L, Besil N, Cesio MV, Heinzen H (2011) Detection of pesticides in active and depopulated beehives in Uruguay. Int J Environ Res Pub Hlth 8:3844–3858

Porrini C, Sabatini AG, Girotti S, Fini F, Monaco L, Celli G, Bortolotti L, Ghini S (2003) The death of honey bees and environmental pollution by pesticides: the honey bees as biological indicators. Bull Insectol 56:147–152

Putnam RA, Nelson JO, Clark JM (2003) The persistence and degradation of chlorothalonil and chlorpyrifos in a cranberry bog. J Agric Food Chem 51:170–176
Racke KD (1993) Environmental fate of chlorpyrifos. Rev Environ Contam Toxicol 131:1–154
Raven PH, Evert RF, Eichhorn SE (1992) Biology of plants. Worth Publishers, New York, NY, 800 pp
Rhodes J (2002) Cotton pollination by honey bees. Aust J Exp Agric 42:513–518
Rissato SR, Galhiane MS, de Almeida MV,MG, Apon BM (2007) Multiresidue determination of pesticides in honey samples by gas chromatography–mass spectrometry and application in environmental contamination. Food Chem 101:1719–1726
Rortais A, Arnold G, Halm M-P, Touffet-Briens F (2005) Modes of honeybees exposure to systemic insecticides: estimated amounts of contaminated pollen and nectar consumed by different categories of bees. Apidologie 36:71–83
Schmitzer S (2008) Effects of GF-2153 (acute contact and oral) on honey bees (*Apis mellifera* L.) in the laboratory. Dow AgroSciences, Indianapolis, IN, Unpublished Report
Sears MK, Bowhey C, Braun H, Stephenson GR (1987) Dislodgeable residues and persistence of diazinon, chlorpyrifos and isofenphos following their application to turfgrass. Pestic Sci 20:223–231
Serra-Bonvehí J, Orantes-Bermejo J (2010) Acaricides and their residues in Spanish commercial beeswax. Pest Manag Sci 66:1230–1235
Shafiq ur R (2009) Evaluation of malonaldialdehyde as an index of chlorpyriphos insecticide exposure in *Apis mellifera*: Ameliorating role of melatonin and α -tocopherol against oxidative stress. Toxicol Environ Chem 91:1135–1148
Snodgrass RE (1975) The anatomy of the honey bee. In: Veatch E (ed) The hive and the honey bee. Dadant & Sons, Hamilton, IL, pp 75–124
Solomon KR, Giesy JP, Kendall RJ, Best LB, Coats JR, Dixon KR, Hooper MJ, Kenaga EE, McMurry ST (2001) Chlorpyrifos: ecotoxicological risk assessment for birds and mammals in corn agroecosystems. Human Ecol Risk Assess 7:497–632
Solomon KR, Williams WM, Mackay D, Purdy J, Giddings JM, Giesy JP (2014) Properties and uses of chlorpyrifos in the United States. Rev Environ Contam Toxicol 231:13–34
Somerville D (2005) Fat bees, skinny bees: a manual on honey bee nutrition for beekeepers—A report for the Rural Industries Research and Development Corporation of Australia. NSW Department of Primary Industries, Goulburn, NSW, Australia
Stevenson JH (1978) The acute toxicity of unformulated pesticides to worker honey bees (*Apis mellifera* L.). Plant Pathol 27:38–40
Tashiro H, Kuhr RJ (1978) Some factors influencing the toxicity of soil applications of chlorpyrifos and diazinon to European chafer grubs. J Econ Entomol 71:904–907
Tashiro H (1987) Turfgrass insects of the United States and Canada. Comstock Pub. Associates, Ithaca, NY, 391 pp
Tetreault GE (1985) Metabolism of carbaryl, chlorpyrifos, DDT, and parathion in the European corn borer: Effects of microsporidiosis on toxicity, Department of Entomology. University of Illinois, Urbana-Champaign, IL, 79 pp
Thompson HM, Hunt LV (1999) Extrapolating from honeybees to bumblebees in pesticide risk assessment. Ecotoxicol 8:147–166
Thompson HM, Maus C (2007) The relevance of sublethal effects in honey bee testing for pesticide risk assessment. Pest Manag Sci 63:1058–1061
Tunink A (2010) Chlorpyrifos-oxon: determination of hydrolysis as a function of pH. DowAgroScience, Indianapolis, IN, Unpublished Report
USEPA (1988) Nontarget insect data requirements, 40 CFR 158.590. http://www.gpo.gov/fdsys/pkg/CFR-2004-title40-vol22/pdf/CFR-2004-title40-vol22-sec158-590.pdf. Accessed Dec 2012
USEPA (1998) Guidelines for ecological risk assessment. Fed Reg 63:26846–26924
USEPA (2005) Reregistration eligibility science chapter for carbofuran. Environmental fate and effects chapter. Environmental risk assessment and human drinking water exposure assessment. United States Environmental Protection Agency, Office of Pesticide Programs, Washington, D.C.

USEPA (2012) White paper in support of the proposed risk assessment process for bees. United States Environmental Protection Agency Chemical Safety and Pollution Prevention, Office of Pesticide Programs, Environmental Fate and Effects Division. Washington, DC

USEPA (2013) United States Environmental Protection Agency, Office of Pesticide Programs Ecological Incident Information System (EIIS) Database. Queried on 11 April 2013

Ware GW, Estesen BJ, Buck NA (1980) Dislodgable insecticide residues on cotton foliage: acephate, AC 222,705, EPN, fenvalerate, methomyl, methyl parathion, permethrin, and thiodicarb. Bull Environ Contam Toxicol 25:608–615

Wass MN, Branson DR (1970) Comparative metabolism of insecticides. II. The fate of O, O-diethyl O-(3,5,6-trichloro-2-pyridyl) phosphorothioate in Madeira cockroaches. DowElanco, Indianapolis, IN, Unpublished Report

Williams WM, Giesy JP, Solomon KR (2014) Exposures of aquatic organisms to the organophosphorus insecticide, chlorpyrifos resulting from use in the United States. Rev Environ Contam Toxicol 231:

Williams NM, Crone EE, Roulston TH, Minckley RL, Packer L, Potts SG (2010) Ecological and life-history traits predict bee species responses to environmental disturbances. Biol Conserv 143:2280–2291

Willmer P (2011) Pollination and floral ecology. Princeton University Press, Princeton, NJ, 828 pp

Winfree R, Aguilar R, Vázquez DP, Lebuhn G, Aizen MA (2009) A meta-analysis of bees' responses to anthropogenic disturbance. Ecol 90:2068–2076

Winston ML (1987) The biology of the honey bee. Harvard University Press, Cambridge, MA

Wu JY, Anelli CM, Sheppard WS (2011) Sub-lethal effects of pesticide residues in brood comb on worker honey bee (*Apis mellifera*) development and longevity. PlosONE 6:e14720

Wuellner CT (1999) Nest site preference and success in a gregarious, ground-nesting bee *Dieunomia triangulifera*. Ecol Entomol 24:471–479

Index

J.P. Giesy and K.R. Solomon (eds.), *Ecological Risk Assessment for Chlorpyrifos in Terrestrial and Aquatic Systems in the United States*, Reviews of Environmental Contamination and Toxicology 231, DOI 10.1007/978-3-319-03865-0, © The Author(s) 2014

Printed in the United States
By Bookmasters